Systems and Synthetic Biology

Vikram Singh • Pawan K. Dhar
Editors

Systems and Synthetic Biology

Springer

Editors
Vikram Singh
Centre for Computational Biology
and Bioinformatics
Central University of Himachal Pradesh
Dharamshala
India

Pawan K. Dhar
Department of Life Sciences
School of Natural Sciences
Shiv Nadar University
Gautam Budh Nagar
India

ISBN 978-94-024-0804-1 ISBN 978-94-017-9514-2 (eBook)
DOI 10.1007/978-94-017-9514-2
Springer Dordrecht Heidelberg New York London

Softcover reprint of the hardcover 1st edition 2015

Printed on acid-free paper

Springer is part of Springer Science+Business Media (www.springer.com)

Preface

The emergence of high throughput biological technologies in 1990s encouraged scientists to ask bigger questions. People moved their attention from parts to interactions. Though immense power was offered by modern technologies, they also posed major challenges of data capture, analysis, integration and interpretation.

To be successful in this new kind of science, one required good understanding of biology, mathematics and computation to address issues at the network level. In the mid 1990s a popular opinion was that mathematizing biology would be easy and straightforward. However, it turned out that finding patterns in biomolecular information pathways was non-trivial, due to a combination of various data types presenting themselves in a background of highly contextual and emergent phenomena. This grand challenge to connect parts behavior with network properties and a corresponding phenotypic outcome gave rise to what we know as Systems Biology.

In the early decade of 2000, scientists asked if it was possible to construct organisms the way lego blocks were put together to design toys of various shapes. This idea was fueled by publications that demonstrated the feasibility of running microbial genetic circuits as applets. The first conference at MIT in 2004 formally announced the emergence of Synthetic Biology. Biologists began thinking like engineers in search of rules and standards to compose organisms. The idea of well characterized parts, stitched together to make modules and networks found experimental support. The science of engineering biology was born.

This book is targeted, mainly, at under-graduate and graduate students. However, researchers who are planning to contribute in these emerging areas may also find the information helpful. It would help to have a basic knowledge of molecular biology to enjoy the science discussed in this book. Given the vastness of these areas, it was difficult to do justice to everything that is important in systems and synthetic biology. Readers are advised to consult more specialized journals and books for in depth information on various topics.

We hope that this first version provides a suitable primer for extending thoughts in search of good questions in systems and synthetic biology. We would consider our efforts successful, if good research problems are identified by readers after reading the chapters.

Our sincere gratitude to all the authors, critical reviewers and family members who helped us compile this work.

January 2015

Vikram Singh
Pawan K. Dhar

Contents

Contributors

Jahoor Alam Center for Interdisciplinary Research in Basic Sciences, Jamia Millia Islamia, New Delhi, India

Preethi Badrinarayan Centre for Molecular Modeling, CSIR-Indian Institute of Chemical Technology, Hyderabad, India

Priyanka Baloni Department of Biochemistry, Indian Institute of Science, Bangalore, India

Manju Bansal Molecular Biophysics Unit, Indian Institute of Science, Bengaluru, Karnataka, India

Silpa Bhaskaran Dept. of Computational Biology and Bioinformatics, University of Kerala, Kerala, India

Pooja Bhatia Kusuma School of Biological Sciences, Indian Institute of Technology Delhi, Delhi, India

Nagasuma Chandra Department of Biochemistry, Indian Institute of Science, Bangalore, India

Dharmendra Kumar Chaudhary National Bureau of Fish Genetic Resources, Lucknow, India

Chinmayee Choudhury Centre for Molecular Modeling, CSIR-Indian Institute of Chemical Technology, Hyderabad, India

Archana Chugh Kusuma School of Biological Sciences, Indian Institute of Technology Delhi, Delhi, India

Pawan K. Dhar Department of Life Sciences, School of Natural Sciences, Shiv Nadar University, Dadri, U.P, India

Department of Computational Biology and Bioinformatics, Centre for Systems and Synthetic Biology, University of Kerala, Trivandrum, Kerala, India

Shayoni Dutta Department of Biochemical Engineering and Biotechnology, Indian Institute of Technology (IIT) Delhi, New Delhi, India

Soma Ghosh Department of Biochemistry, Indian Institute of Science, Bangalore, India

Alessandro Giuliani Environment and Health Department, Istituto Superiore di Sanitá, Roma, Italy

Gurpreet Kaur Grewal Genomics and Molecular Medicine Unit, CSIR-Institute of Genomics and Integrative Biology (IGIB), Delhi, Delhi, India

Rajdeep Kaur Grewal Bose Institute, Kolkata, India

Sandeep Grover Genomics and Molecular Medicine Unit, CSIR-Institute of Genomics and Integrative Biology (IGIB), Delhi, Delhi, India

Meenal Gupta Genomics and Molecular Medicine Unit, CSIR-Institute of Genomics and Integrative Biology (IGIB), Delhi, Delhi, India

Naira Hovakimyan University of Illinois at Urbana-Champaign, Champaign, IL, USA

Bin Hu Theoretical Biology and Biophysics Group, Theoretical Division, Los Alamos National Laboratory, Los Alamos, NM, USA

Aastha Jain Kusuma School of Biological Sciences, Indian Institute of Technology Delhi, Delhi, India

Hua Jiang University of Minnesota, Minneapolis, USA

Mukta Joshi Centre of Systems and Synthetic Biology, Department of Computational Biology and Bioinformatics, University of Kerala, Trivandrum, India

Evgeny Kharisov University of Illinois at Urbana-Champaign, Champaign, IL, USA

Remya Krishnan Department of Computational Biology and Bioinformatics, University of Kerala, Trivandrum, Kerala, India

Ritushree Kukreti Genomics and Molecular Medicine Unit, CSIR-Institute of Genomics and Integrative Biology (IGIB), Delhi, Delhi, India

Vishwesh V. Kulkarni University of Minnesota, Minneapolis, USA

Aditya Kumar Molecular Biophysics Unit, Indian Institute of Science, Bengaluru, Karnataka, India

Shankar Kundapura Centre of Systems and Synthetic Biology, Department of Computational Biology and Bioinformatics, University of Kerala, Trivandrum, India

Indra Mani National Bureau of Fish Genetic Resources, Lucknow, India

Department of Biochemistry, Faculty of Science, Banaras Hindu University, Varanasi, India

Achuthsankar S. Nair Dept. of Computational Biology and Bioinformatics, University of Kerala, Kerala, India

Umesh P. Dept. of Computational Biology and Bioinformatics, University of Kerala, Kerala, India

Keshab Parhi University of Minnesota, Minneapolis, USA

Mrugainduta Patil Symbiosis School of Biomedical Sciences, Symbiosis International University, Pune, India

Thirtha Poovaiah Centre of Systems and Synthetic Biology, Department of Computational Biology and Bioinformatics, University of Kerala, Trivandrum, India

Shidhi PR Department of Computational Biology and Bioinformatics, University of Kerala, Trivandrum, Kerala, India

Mark Riedel University of Minnesota, Minneapolis, USA

Soumen Roy Bose Institute, Kolkata, India

Ushasi Roy Bose Institute, Kolkata, India

G. Narahari Sastry Centre for Molecular Modeling, CSIR-Indian Institute of Chemical Technology, Hyderabad, India

Yumnum Silla Genomics and Molecular Medicine Unit, CSIR-Institute of Genomics and Integrative Biology (IGIB), Delhi, Delhi, India

R. K. Brojen Singh School of Computational and Integrative Sciences, Jawaharlal Nehru University, New Delhi, India

Vijai Singh Molecular Diagnostics & Biotechnology Laboratory, Division of Crop Protection, Central Institute for Subtropical Horticulture, Rehmankhera, Lucknow, U.P., India

Vikram Singh Centre for Computational Biology and Bioinformatics, Central University of Himachal Pradesh, Dharamshala, India

Durai Sundar Department of Biochemical Engineering and Biotechnology, Indian Institute of Technology (IIT) Delhi, New Delhi, India

Puneet Talwar Genomics and Molecular Medicine Unit, CSIR-Institute of Genomics and Integrative Biology (IGIB), Delhi, Delhi, India

Vipin Thomas Department of Computational Biology and Bioinformatics, University of Kerala, Trivandrum, Kerala, India

Deepthi Varughese Department of Computational Biology and Bioinformatics, University of Kerala, Trivandrum, Kerala, India

Navya Vinod Department of Computational Biology and Bioinformatics, University of Kerala, Trivandrum, Kerala, India

Venkata Rajesh Yella Molecular Biophysics Unit, Indian Institute of Science, Bengaluru, Karnataka, India

Part I
Systems Biology

Chapter 1
Introduction to Systems Biology

Bin Hu and Pawan K. Dhar

Abstract In the mid 1990s when Leroy Hood reintroduced the term "Systems Biology", the fusion of ideas gave rise to confusion to such an extent that there used to be special talks on 'what is systems biology'? Over the last decade, Systems Biology has undergone directed evolution leading to the emergence of personalized versions of this term. Irrespective of this, strong computational dependency and a significant increase in the scale of investigation often appear as constant features in the systems biology background. In our opinion, Systems Biology is an approach that involves the following (a) experimental and computational studies describing collective behavior of molecules in relation to the pathway and networks, and with the higher-level physiological outcome (b) new experimental and mathematical methods important to study group behavior of interacting components. This chapter describes the origin and evolution of systems biology, as a formal discipline, steps and challenges in building models and their potential applications.

Keywords Modeling in Biology · Simulation · System · Biological complexity · Pathways · Networks

1.1 Introduction

The traditional approach of doing science has mainly centered around the twin strategy of observation and classification i.e., observe some measurable quantity, say flower color, height of plant and so on, collect data from a large number of plants and try to find some non-obvious pattern. At least in biology, the role of analytical techniques has rarely been pursued as a serious scientific discipline. This is due to the fact that in the traditional setting biological data was easily countable and available to human analysis and interpretation. The science of taxonomy was built upon the

P. K Dhar (✉)
Department of Life Sciences, School of Natural Sciences,
Shiv Nadar University, Dadri, U.P, India
e-mail: pawan.dhar@snu.edu.in

B. Hu
Theoretical Biology and Biophysics Group, Theoretical Division,
Los Alamos National Laboratory, Mail Stop K710, Los Alamos, NM, USA
e-mail: hubin.keio@gmail.com

V. Singh, P. K. Dhar (eds.), *Systems and Synthetic Biology*,
DOI 10.1007/978-94-017-9514-2_1

foundation of finding common patterns among a large number of samples and categorizing them hierarchically. The strategy was that a higher-level abstraction should be shared by all the members of the group, which can be further sub-sorted into various bins based on some additional parameter. Thus, you see kingdom, families, genus and species as a top-down flow of information in taxonomy. Charles Darwin stretched the idea of 'finding patterns from external observations' further, and ended up his long and careful study by proposing the theory of Natural selection. Lamarck and other scientists extended the story further and tried to make his story predictive.

However, in all these situations classifying organisms did not explain how they worked. There was a need to adopt a different approach. Mendel made the first bold attempts to look beyond a horizontal (population-based) plane of vision and vertically move down from phenotype to causal elements. He assumed a linear correlation between a causal element and a phenotypic observation. It was a groundbreaking work. In absence of any high-resolution physical device, he could generate accurate rules and predictions of inheritance simply by looking at the external phenotypes.

After Mendelian era, the science of biology got predominately biochemical and microscopic. Technological developments helped scientists move from external phenotype to cell interiors. However, due to technical complexity and cost of data generation, biological data was mostly qualitative, studied at the level of human analysis and did not require special mathematical techniques and computational infrastructure for interpretation.

As the technological tools got more sophisticated, scientists moved from external observations to the study of cells, chromosomes, DNA, protein and so on. Having seen so many parts co-existing in a small cellular space, there was a natural curiosity—how are these parts created, used, retired, recycled. What is the role of these parts in determining higher order behavior?

Two parallel efforts were aggressively pursued: (i) uncover as many parts and modules (collection of parts employed for a single purpose) as possible, and (ii) find the role of each part in determining a given phenotype. We call this strategy as 'reductionist biology' i.e., reduce a system to a set of components and study each component separately. The Human Genome Sequencing Project was started precisely keeping the first aim of reductionist biology in mind i.e., if we know our genetic blueprint, we will figure out everything about ourselves. In parallel to this, a large body of mutations and chromosomal aberrations was collected from diseased tissue to correlate abnormal physiological/morphological conditions with the underlying genetic cause.

However, soon people realized that reductionist approach was unhelpful beyond a point. There were so many incidences where a visible genetic variation/mutation did not lead to a corresponding change in the phenotype. Worse still, in many instances a so-called important gene when knocked-out did not result in the expected outcome. Organisms employed even unrelated genes take over the function of a missing one. Thus, to learn biological decisions there was a need to invent a novel approach.

The trigger for paradigm shift came when microarray technology was invented in the early 1990s. Suddenly huge real-time data was generated. There was no direct way to understand this data, the underlying hidden patterns and correlations. Instead of focusing on one gene, people could now study hundreds of gene expression events together. The impact of even one gene knock-out could be studied in relation to hundreds of unrelated genes. The point of focus moved from sequence level to the expression level. From low throughput human readable data, the scientific community moved to automated, high throughput, machine analyzable data. This was a real **phase shift** in biology. One could ask questions about the whole system and not about just few parts. By mid 1990s, Systems Biology had truly arrived. This is not to suggest that Systems Biology started in the mid 1990s. The original seeds of thoughts were sown much earlier.

In 1944, Norbert Weiner foresaw the need for systems approach. Unfortunately, the time was not ripe for Systems Level analysis due to data scarcity. Even if all the data were available at that time, the lack of sufficient computational resources would have still precluded scientists to make best use of it. The idea of systems analysis slowly moved from theoretical to practical realm. In the mid 1960s and 1970s, metabolic control analysis gained prominence. The hope was to study the flow of metabolites through the network and find steps that exerted maximal control over metabolic flux in the network. This came to be known as Biochemical Systems Theory. A number of key concepts we use today in flux and control analysis can be traced back to the earlier work (on computational analysis of metabolic networks) by Michael Savageau and co-workers.

Probably the situation wouldn't have changed much, but for a new technology invented in the early 1990s. Dr. Stephen Fodor (later Chairman and CEO of Affymetrix) and his colleagues published a ground-breaking work in Science in 1991. Biology suddenly underwent a paradigm shift, from low-throughput to high-throughput science. At the same time, computer technology got more advanced, the microprocessors got faster and the storage got cheaper. Time was ripe to collect large amounts of data and store it in computers for analysis.

In the background of technological developments, Leroy Hood formalized this new integrated biology approach and called it 'systems biology'. For several years people were confused (and probably still are) about: what is systems biology? The community has gone through significant brainstorming on how to define Systems Biology? Though Leroy Hood projected it a specialized field of science, generally people like to view Systems Biology as an "approach" than an independent discipline (Hao et al. 2003). Given the existence of so many flavors of systems biology, probably it is best to describe the properties of Systems Biology than to give it a rigid definition.

1.2 Systems Biology—A Primer

1.2.1 What Is Systems Biology?

First, we need to define the term 'System'. A System is composed of several elements and is defined by the scope of investigation. For example, to study photosynthesis as a systems biology problem, one would need to describe all the genes and molecular networks involved in the process of photosynthesis. It is not necessary for example to model lipid synthesis, if one is investigating photosynthesis as a systems biology problem. Likewise, one can omit photosynthetic pathways if one is modeling lipid metabolism. In other words, the boundary conditions of a system are determined by the components that are directly involved in the process under study. This is not to say that a system is a space constrained by rigid boundary conditions. In reality, a system is a flexible term, described by the availability of data and by the kind of questions.

Systems Biology is a formal approach to understand higher-level behavior as a result of group interaction of the constituent elementary components. As it involves a large variety and scale of data, computational modeling and analysis is frequently employed to store, understand and find meaningful correlations. Systems Biology starts from experiments, goes through computational route and ends at experiments i.e., experimental data → Statistical treatment and modeling → Correlations → Predictions → Experiments. The key difference between systems biology and traditional biology is the focus on group behavior of molecules as against single molecular correlation in the latter.

1.2.2 Why Is Systems Biology Necessary?

In physical sciences, modeling and simulation, in addition to theoretical and experimental studies, is the third indispensable approach because not all hypotheses are amenable for confirmation or rejection by experimental observations. In biology, researchers are facing the same or maybe even worse situation. On one hand experimental study is unable to produce enough data for theoretical interpretation; on the other hand, due to data insufficiency and inaccuracy, theoretical research cannot provide substantial guidance and insights for experimentation. To meet this need, computational modeling takes a more important role in biology.

1.2.3 What Is a Model?

A model is a formal or abstract representation of a system, usually in the form of a set of objects and the relations between them. It is a skeleton of the real system but not a replica, built with key components based on a combination of assumptions and existing knowledge. The key to modeling is the identification of elements that can reflect key global properties with incomplete information. Modeling is an iterative process that repeats until a model reaches its final stage and is validated by experiments. In the process, different prototypes are often developed for validation. A model may be formal, with mathematical representation, or conceptual, with diagrams or even concepts only. It may be mechanistic (cause-effect relationship), or phenomenological i.e., based upon a combination of observed phenomena and expert knowledge. Mathematical models are commonly divided into deterministic (responses to given inputs are predictable) and stochastic (responses are picked up based on probability distribution), quantitative and qualitative, and linear and non-linear.

1.2.4 Is Modeling in Biology New?

Biological modeling is both old and new. Originating from modeling concepts in physical systems, it has a history of several decades. However, due to the distinctive differences between biological systems and physical systems, biological modeling presents itself with additional challenges and calls for new strategies and tools. To model biological systems at various levels i.e., molecular, cell, tissue and organ different strategies and techniques are needed.

Modeling and simulation appeared on the scientific horizon much before the emergence of molecular and cellular biology. Early on the objective of modeling was to explore the features of black boxes e.g., heart, brain, and circulation system, a concept borrowed from physical systems. In such scenario, the main challenge was to understand and predict the behavior of a system without knowing the microscopic details. The strategy was to reproduce observed phenomena at high level with simplified description of internal structures. Though inferring microscopic details was necessarily a major goal, one needed to know how to understand the system as a whole and utilize this understanding in clinical practice. The cases in point are: the inverse modeling of cardioelectrical (Gulrajani et al. 1988) whose simulation results were used to improve diagnosis of heart and brain diseases.

Two interesting methodological features emerged at this stage. First, since biological systems were treated as physical systems or even structure-less systems, many methods and tools were directly borrowed from engineering fields such as FEM (finite element method) and BEM (boundary element method) to compute biological systems (Bradley et al. 2001). Electrical activity of cardiac cells was abstracted to dipoles with different moment and direction. The second feature was high-level abstraction based on inverse approach. Cellular electrical activity was abstracted as an attribute of dipoles [6]. Consequentially, complex numerical techniques for ODE (ordinary differential equation) and PDE (partial differential equation) solution were developed. Both black box assumption and inverse modeling, though suitable for modeling mechanical systems, suffer from major problems when applied to biological systems. The first one is that many inverse problems are mathematically ill-posed. Even if the available data are adequate and precise, unique solution is not always guaranteed and special techniques like regularization are employed (Johnston and Gulrajani 1997). The second assumes that that the internal structure is static, does hold true when a system evolves with time. Thus, this method cannot describe growth process with gene regulation, for the system undergoes state transition while an inverse solution is searched for. Complex internal structure and evolution are key features that differentiate biological systems from mechanical systems. The top down approach doesn't work very well in biological systems due to absence of information at various levels. Even the bottom up approach (from molecular modeling to organs) encounters the same problem. The solution is to start at an information level and expand vertically upwards/downwards.

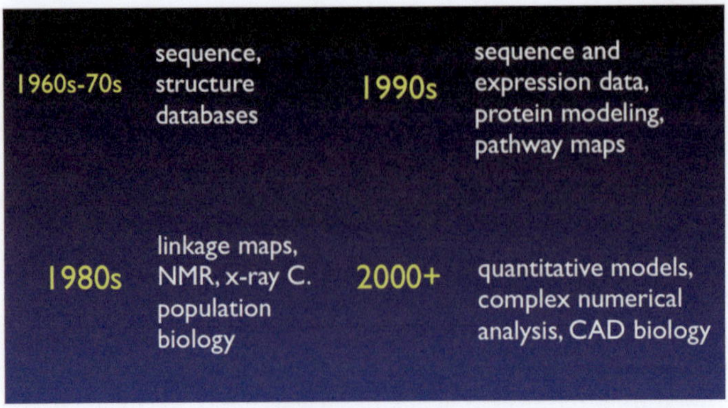

Fig. 1.1 Predominant computational approaches in biology

1.3 Modeling Pathways and Networks

Mendel used simple elementary mathematics of addition and division to obtain laws of inheritance. However, with the arrival of large amount of biochemical and molecular data, mathematical treatments and computer applications got more and more sophisticated (Fig. 1.1). Currently, the predominant phase in biology is process analysis and systems engineering. Process analysis is what we know as Systems Biology and Systems Engineering is commonly referred to as Synthetic Biology.

Table 1.1 describes some of the commonly used resources and tools in computational systems biology. Modeling is one of the activities in systems biology. It is easy to understand why? Modeling helps address "what-if" questions, facilitate rejection of false hypothesis, and predict future system state in response to a perturbation. Good models are experimentally validated, analyzable and open for manipulation and optimization.

1.4 Steps in Model Building

Step One Make a parts list (collect data from literature and experiments). Take into consideration the measurements made, protocols followed, perturbations applied, constraints during experiment and error bar. Was the data independently confirmed? In case of conflicting results, pick up the data from the most reliable group and iterate with the next.

Step Two Draw an interaction map. The pathway representation should be robust and represent events like translocation, transformation and binding. A pathway map typically consists of nodes (molecules) and edges (interactions). In a standard textbook diagram all the interactions drawn on a uniform background canvas, may (a) belong to different cellular compartments and also (b) occur at different time points. Thus, in reality a standard metabolic/signaling map represents spatially and temporally overlapped data.

Table 1.1 Resources and Tools for Computational Systems Biology. (This list is not exhaustive. We recommend readers to consult relevant scientific literature for more information)

Resource
For visualizing/construction
Pathfinder (online graphical representation of cell signaling pathways) http://www.sigmaaldrich.com/life-science/cell-biology/learning-center/pathfinder.html
ArrayXPath (mapping and visualizing microarray gene-expression data) http://www.snubi.org/software/ArrayXPath/
HighChem (a suite of interconnected modules containing tools for constructing, visualizing and analyzing biochemical and metabolic pathways) http://www.highchem.com/leading-edge-technologies/biochemical-pathways.html
Pre-constructed pathway maps
IUBMB-Nicholson minimaps http://www.tcd.ie/Biochemistry/IUBMB-Nicholson/
Kyoto encyclopedia of genes and genomes http://www.genome.ad.jp/kegg/
PUMA2 (High throughput comparative and evolutionary analysis of genomes and metabolic networks with Grid computational backend) *http://compbio.mcs.anl.gov/puma2/*
The seed (An annotation/analysis tool) *http://theseed.uchicago.edu/FIG/index.cgi*
Biopathways consortium http://www.biopathways.org
BioCyc (Collection of 507 Pathway/Genome Databases. Each database in the BioCyc collection describes the genome and metabolic pathways of a single organism) http://www.biocyc.org
BioCarta (Interactive graphic models of molecular and cellular pathways) http://www.biocarta.com
Enzyme databases
BRENDA http://www.brenda-enzymes.info/
ExPASy http://www.expasy.ch/
Tools
170 modeling and simulation tools listed http://sbml.org/SBML_Software_Guide/SBML_Software_Summary

Step Three Converting map into a model. Actually, map itself is a model—a connectivity model. However, to understand dynamic nature of the system a connectivity representation must be converted to a quantitative model. Gene expressions are stochastic and may be modeled with stochastic equations. Metabolic pathways are modeled with Ordinary Differential Equations. Even though Michaelis Menton kinetics is the most accepted way of modeling metabolic events, the MM equation is

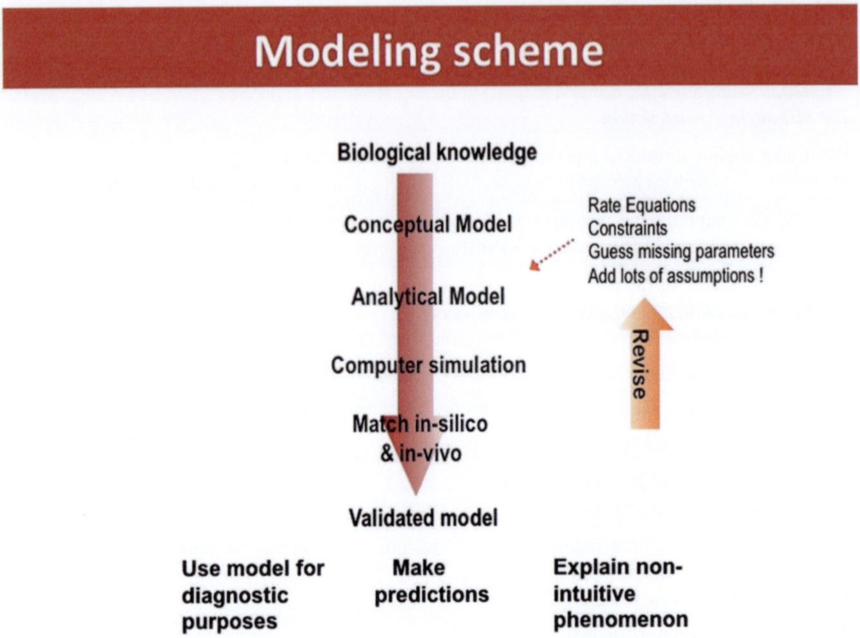

Fig. 1.2 General modeling scheme

itself based on assumptions, some of them are not true e.g., well mixed reaction chamber. Figure 1.2 describes a general modeling scheme. Based on the questions asked and system under investigation, distinct modeling approaches are used (Table 1.2).

Step Four Animate the static model. A large number of tools developed for free are available currently (http://www.sbml.org). Most of them offer exchange of results based on the standard SBML output (SBML—Systems Biology Markup Language).

Some of the desirable features of an effective software tool for Systems Biology from both computational and software viewpoints are presented below.

a. **Algorithmic Support**. Algorithms form the core of any tool. We have seen that there are a number of formalisms and algorithms each with its own strengths and weaknesses. Flexibility to quickly use different algorithms from the same environment would be critical for reducing the cycle time of building large and complex models. We further classify algorithmic support into three divisions:
b. **Modeling and Simulation Support**. Abstractions of different cellular processes require different information about the target systems such as Gene Regulatory Network, Signal Transduction Network or spatial diffusion. These are based on system specific inputs and implementation of the underlying algorithms. Table I lists details of some of the processes. The whole cell modeling tool must eventually provide support for handling and processing this information.

Table 1.2 Commonly used kinetic modeling formalisms

Process	Input	Mathematical formalism
Gene expression	Quantitative time series data	Stochastic equation
Metabolic reaction	Concentrations, rate constants	Ordinary differential equation
Gene regulatory networks	Network topology, stoichiometry, rate constants, number of particles, rules, thresholds	Boolean, rule based, stochastic master equation
Signaling network[a]	Network topology, stoichiometry, rate constants, number of particles, rules, thresholds	Boolean, stochastic (gillespie, stochsim, petrinets),
Metabolic pathway	Network topology, stoichiometry, kinetic rate laws, initial concentrations, algebraic rules	Non linear ordinary differential equations, s-systems
Membrane transport and other spatial processes	Initial spatial concentrations, diffusion constants	Reaction diffusion, deterministic partial differential equations, spatial stochastic master equation

[a]Recently rule based modeling approach has gained prominence. GetBonNie is a good tool for building rule-based models of signaling networks (http://getbonnie.cs.unm.edu/GetBonNie/). This is particularly useful since qualitative data are the most frequent/dependable form of data obtainable from signaling networks. As an extension, I would strongly encourage readers to go through Dr.Eric Davidson's work on modeling embryonic development. (http://www.its.caltech.edu/ mirsky)

c. **Analysis Support**. An important aspect of a typical modeling project in Systems Biology is analysis of the qualitative and quantitative features of the network. Parameter estimation, network optimization, flux balance analysis, bifurcation analysis, extreme pathways and metabolic control analysis are some of the strategies being used currently. Figure 1.3 shows the kind of data used in quantitative model. Parameter estimation algorithms are indispensable for complementing the limited knowledge that can be obtained from experimentation. These algorithms can be used for estimating the unknown rate constants for reproducing an experimentally observed time series. Flux Balance Analysis and Metabolic Control Analysis have a long history of application to metabolic networks. Stoichiometric Network analysis and Extreme Pathways are used to extract qualitative information about a network such as the critical paths.

d. **Visualization**. Powerful visualization tools are necessary for improving the efficiency of the modeling process and understanding the output of the simulation. Some of the desirable features of a visualization tool are:
   - Graphical User Interface for constructing the network and entering various input parameters. A text-based input does not give a good idea of the network topology. Graphical interface becomes particularly desirable for representing spatial features of a model such as compartmentalization and localization. Visualization is required for monitoring the dynamics of a model such as evolution of the network topology through a change in the network layout or the relative concentration of the species through a color code.

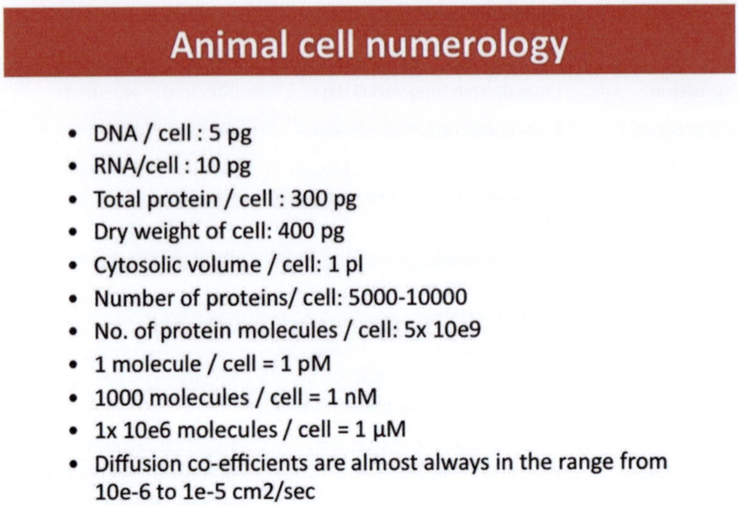

Fig. 1.3 shows typical quantities used in a kinetic model

  – Powerful graph plotters. The outputs of most of the simulation algorithms are some form of time series. As a result in-built support for powerful plotters is very important for analysis of the output.

e. **Software Architecture**. Simulation and analysis of large-scale models are invariably computationally expensive and often need high performance distributed computing.
   Some tasks, amenable to and can benefit from distributed computing, are genetic algorithms based parameter estimation, multiple simulations for parameter sweep and parallel PDE solvers for spatial simulation.
f. **Modeling Language**. Model building is complex activity requiring collaboration between various research groups, both experimentalists and theorists. Thus development of a common language for smoother information exchange is imperative. Some of the ongoing efforts in this direction are BioPAX, SBML and CellML.

1.4.1 Challenges in Building Reliable Models

- Lack of accurate and adequate biological data
- A general lack of quality control with respect to strain, culture conditions and protocols
- A cell is a gel, shows gradients, non-uniform distribution of substances in compartments. Frequently, a model does not consider these variables.
- Parameter values are often inaccurate or taken in special culture/harvesting conditions. To fill in the gap, deterministic and stochastic parameter estimation methods have been developed. However, none of the methods guarantees an accurate answer. Also, given that good data is often less frequently available, the parameter search space is almost always significantly large. The larger the search space, the lesser the possibility of finding an accurate answer.

- Unknown reaction kinetics
- Temporal inactivation/degradation of enzymes in generally left out during modeling process
- Metabolic channeling effects
- Emergent phenomena

1.5 Capturing Biological Complexity

The grand challenge of twenty first century is to understand and model complexity of biological systems. Though complexity has been extensively discussed subjects at different levels (Lynch and Conery 2003; Yang et al. 2003), there is no operational definition of complexity for the biological systems (Adami 2002). Some hallmarks of complexity, e.g., linearity and non-linearity, number of parameters, order of equations and evolution of network, come into existence only when a system is formalized in specific ways. Furthermore, from what has been clear, there are two kinds of complexity in biological systems: functional and structural, or dynamic and static; both encountered by modelers. The identification and measurement of biological complexity is a very big task for experimental biologists.

As Adami pointed out, the popular measure of complexity for dynamical systems, computational complexity (for example, the complexity of a sequence can be inferred from what finite state machine can produce), is unsuitable for biological systems. Even though it characterizes the amount of information necessary to predict the future state of the machine it fails to address their meaning in a complex world. Yet the meaning or semantics of molecular interaction really makes sense in signaling processes. An alternative approach may be to think about the complexity issue at higher level and in much larger scope. Recently, the complexity of networks has attracted interests of researchers with different background (Bhalla and Iyenger 1999; Strogatz 2001; Wagner and Fell 2001). Since the topological structure of molecular network, consisting of active genes and proteins, undergoes significant evolution within cells in biological development, to measure complexity of molecular systems, both static and dynamic, according to such evolution may be a practical way, because it is easier to identify and abstract information from it (Bornholdt 2001).

Features in topological structure are also helpful in identifying modularity of molecular interaction. In a large, multicellular landscape, the speed and scope of parallel network evolution in cells, if measured properly, can effectively reflect the complexity of biological systems. Another widely used index of complexity in both physical and biological systems is non-linearity, including parameter sensitivity and initial value sensitivity (Savageau 1971). In evolvable systems, it often implicates the speed of evolution and the appearance of emergent events. Last but not least, the existence of stochasticity and noise increase the complexity of the system even further by introducing issues of robustness, noise resonance and bi-model behaviour.

1.5.1 Computational Challenges in Building Stochastic Models

Experiments have conclusively proved that molecular activity, including gene regulation, are stochastic (Elowitz 2002). The intrinsic stochasticity of biochemical processes such as transcription and translation generated intrinsic noise; and the fluctuations in the amounts or states of other cellular components lead indirectly to variation in the expression of a particular gene and thus represented extrinsic noise (Swain 2002). There are also opinions that the stochasticity contributes much to system complexity.

To describe the stochasticity, intrinsic and/or extrinsic, two strategies have been developed. The first is to design specific stochastic simulation algorithms that can cut down the computational burden; the second is to use stochastic differential equations, which are modified ODE with stochastic flavor. We first describe these two approaches, then, turn to methods of reducing time consumption of stochastic modeling.

The CME formalism employs an equation for every possible state transition and solves all equations simultaneously. Generating one state transition trajectory is straightforward. However, when the dimensionality of a system increases, the possible trajectories of the state transition, or the state space, explode combinatorially, rendering the system intractable. In view of this serious limitation, Gillespie devised a more efficient algorithm to generate all trajectories (Gillespie 1977). Instead of writing all the master equations explicitly, he generated trajectories by picking up reactions and time intervals according to correct probability distributions so that the probability of generating a given trajectory is exactly the same as the solution of the master equation. For a homogeneous, well-mixed chemical system, Gillespie has proposed two exact Stochastic Simulation Algorithms (SSA), namely the Direct Reaction Method and First Reaction Method to solve the chemical master equations.

Although Gillespie algorithm solves the master equation exactly, it requires substantial amount of computational effort to simulate even a small system. Each of following three factors contributes to a considerable increase of time consumption:

- Increase in the number of reaction channel
- Increase in the number of molecules for the species
- Faster reaction rate of the reaction channels

These factors cause scalability problem, which is similar to the stiffness problem in usual ODE description i.e., whenever reaction rates between different reaction channels vary in magnitude, computation slows down considerably. In the stochastic algorithms, whenever the complexity of a system increases through the augmentation of any of the abovementioned factors, a smaller should be adopted to reflect the true nature of the system, i.e., to maintain the exactness of simulation. The difference in time scale between different reaction channels is a cause for its large computational complexity.

In 1998, Morton-Firth and Bray developed Stochsim algorithm, treating bio- logical components, for examples, enzymes and proteins, as individual interactive objects based on probability distribution derived from experimental data. In this scheme, in each round of computation, a pair of molecules is checked for potential reaction. Due to the probabilistic treatment of interactions between molecules, Stochsim is capable of reproducing realistic stochastic phenomena in biological systems. Though both Gillespie algorithm and Stochsim algorithm are based on the identical, fundamental physical assumptions, an important feature of the latter is the concept of "pseudo-molecules", which serves as a numerical treatment to maintain the accuracy of the algorithm. Furthermore, in this algorithm, the number of pseudo-molecules can be optimized to overcome the stiffness problem.

In contrast to the variable time step in Gillespie algorithm, Stochsim algorithm uses fixed time step that can be optimized to the desired accuracy. However, the convenience of this measure comes with an additional burden of using empty time step i.e., a time step in which zero events occur. Another limitation of the Gillespie algorithm is its computational infeasibility for multi-state molecules. For example, a protein with ten binding sites will have a total of 210 states and it requires the same amount of reaction channels to simulate this multi-state protein. Considering the scaling feature of Gillespie algorithm with the number of reaction channels, it is impossible to perform such a simulation on with available computational power. Stochsim algorithm can be modified to overcome this problem by associating states to molecules without introducing much computational burden.

Several strategies have been adopted to improve the efficiency of stochastic modeling. Gillespie and Gibson (2001) were the first to modify the SSA to improve efficiency of the algorithms.

Gibson proposed the Next Reaction Method as a revised approach to Gillespie's First Reaction Method for simulation efficiency. The algorithm has been applied for simulation of the Bacteriophage Lambda model. In 2001, Gillespie presented the Tau-Leap Method to produce significant gains in the computational speed with acceptable loss in accuracy (Gillespie 2001). In the original version of Gillespie Algorithm, master equations were solved exactly to produce precise temporal behavior of systems by generating the exact timing of the firing of each reaction channel. However, it is sometimes unnecessary to obtain so much detail from simulation. Instead of finding out which reaction happens at which time step, one may like to know, how many of each reaction channels are fired at certain time intervals. If the time interval is large enough for many reactions to happen, one can expect substantial gain in the computational speed.

However, the method still possess the inherit disadvantages of supressing stochasticity in fast reaction and the computational efficiency of Implicit Tau Leap method is still unexamined for a large biological pathway model. Another way of improving efficiency of SSA is to adopt multi-scale integration.

1.5.2 The Rise of Hybrid Modeling

Pure stochastic modeling deals with biological systems as physical systems without biological semantics. Besides the huge burden of time consumption, specific semantic of gene/protein interaction is often buried under low level biochemical reactions.

Hybrid modeling can have multiple meanings. First of all, a model containing metabolic and signaling networks is a hybrid model. Actually these two networks are not independent of each other. For example, in Type II diabetes, the weakened transduction of insulin signal and the changed metabolism activity in cells are closely coupled. In such model, very often, different description methods should be employed to disclose different aspects or parts of a biological system, because, when ODEs are used to describe deterministic events, the basic assumption on continuity and determinism in ODE methods hamper the true representation of noise and stochastic events in cellular environment [64]. Finally, different cellular processes, like gene expression and biochemical reaction and different biochemical reactions, ask for description not only different in methods but also at different time-scales. For a successful simulation, various techniques should be implemented to ensure the feasibility of computation, including the multiple time-scale integration of different equations like ODE, SSA, and SDE [62].

Biological systems in nature undeniably involve multi-scale activities. Algorithms discussed earlier tackle the problem by obtaining solution for the scale of interest while eliminating the other scales in the problem. However, these algorithms produce results of less fidelity in the situation when different scales are heavily coupled together. Furthermore, these algorithms may not be computationally feasible for the scenario as well. One of the methods to reduce simulation time of these algorithms will be to combine different algorithms that handle different scales (Welnan and Engquist **?**). The idea of mixing different algorithms to handle hybrid system is not new and has been first adopted in ODE system of equations. Anders [66] presents multi-adaptive-galerkin methods for solving stiff ODE system. The method showcases the possibilty of applying different time-steps and algorithms for different equations in the system and highlights the potential of hybrid methods. However, the method is derived for solving ODE system only and therefore insufficient in tackling the problem in computational cell biology. Recently, Haseltine and Rowlings (2002) presented a method for performing mixed ODE/SSA calculation to approximate system dynamics. The approach are theoretically based on the the equivalence of stochastic and deterministic assumption at the thermodynamic limits, where N and V become infinite. The methods offer insight into integration of the mesoscopic and macroscopic timescale but fail in providing a robust control mechanism and exact mathematical solutions. In addition to that, the methods adopt switches to partition the system into either stochastic or deterministic regime which resulted in sharp transition of the dynamics. This is unnature and unrealistic as compared to the dynamics in the cells which exhibit smooth transition of states from microscopic scale to macroscopic scale.

Integration of diffusion and biochemical pathway has been attempted recentlym (Stundzia and Lumsden 1996). The method derives the reaction-diffusion master equation and simulate the system with SSA. These approaches produce interesting insight about the dynamics between diffusion and chemical reactions. However, the computational requirement is enormous and not feasible for realistic model. Furthermore, the methods do not consider concentration gradient and therefore are not accurate in simulating diffusion processes.

A recent version of Stochsim algorithms includes a 2- dimensional lattice to model the interaction among neighboring molecules. In this approach, spatial information is added as an attribute of each molecular species. The algorithm has been applied for studying the dynamics of signaling proteins associated with the chemotactic receptors of coliform bacteria. MCell [59] has also introduced another way of simulating stochastic diffusion by directly approximate the Brownian movement of individual molecules. In MCell, random numbers are used to determine the motion and direction of molecules during simulation. Due to the incorporation of Monte Carlo simulation and the individual treatment of each molecular species, the results from MCell contain realistic stochastic noise based on the spatial arrangement and number of participating molecules.

Unlike metabolic networks, signaling networks can undergo significant temporospatial changes in embryonic development to endow cells specific identities and to fulfill particular functions within them. For example, a fly is different from a mouse because the molecular interactions within cells of the former produce different signals from the molecular interactions within cells of the latter in body plan development. Since recent progress in developmental biology has indicated that the pathways controlling embryonic development are highly conserved in different animals in both composition and function [82–85], to reveal how slightly different pathways, following what rules, lead to distinctively dissimilar morphogenesis is a great challenge. This, therefore, raises issues of modeling parallel, interactive molecular networks. We list some, but not all, issues here.

First, signaling in a cell is not autonomous in cell fate determination. In development, a cell does not know when to divide, when to die, and when to differentiate. It also does not know, in the absence of environmental messages, whether to differentiate into a myocyte or a neuron. Thus, single cell modeling may not be enough to reveal what we want to know.

Second, various variations can occur, which can be normal and abnormal. In fact, cancer has been seen as aberrant developmental events. To simulate only the normal case is insufficient to understand the properties of signaling networks.

Third, relevant to but different from context dependency is gene function polymorphism. Not like enzymes in metabolic networks showing high specificity, genes in signaling networks can produce and transfer different signals. These constitute basic features of tissue scope molecular level signaling modeling. Considering a small $100 \times 100 \times 100$ tissue cube contains 1 million cells, these issues cannot readily be solved by available modeling platforms.

1.5.3 Re-Programming Signaling Process in a Cell

One aspect that signaling modeling can make contribute to is the re-programmability of molecular networks, which has been an important research topic (Tada et al. 2001; Hakelien 2002). Carina Dennis, Natures Australia correspondent, describes the technique of turning an adult human cell back to an embryonic state as cellular alchemy [88]. Usually, from state A, an embryonic state, to state B, a state of a fully differentiated myocyte, more than one network configurations mush be underwent. Among explosive combinatorial conditions, how to find a feasible path, consisting of a series of molecular switches, really make sense for experimenters. A wealth of knowledge on dynamics of molecular interaction is very helpful for correct re-programming.

1.6 Practical Applications of Systems Biology

Systems Biology offers possibility of creating new opportunities for drug target selection based on predictive models. For example, pathway based disease models can be very helpful at the preclinical stage to identify potential toxic effects of lead compounds. If a compound targets network hub, the possibility that such a drug will give rise to a number of side effects is quite high. However, if drug targets turn out to be (a) non-hubs or (b) multiple weak binders in the network collectively bringing about the effect, such lead compounds will be preferred over the rest. Also, the disease and population based drug response models can help lower R&D costs. A prior assessment of side effects/toxic effects can result in speeding up drug discovery, leading to significant savings.

By producing detailed route maps of molecular circuitry in the cell, it is possible, in theory, to develop smarter therapeutic strategies. However, the success of this strategy depends upon completeness and accuracy of relevant data. Systems biology approaches have played a key role in understanding AstraZeneca's Iressa (gefitinib) Lliver abnormalities were identified by Pfizer, and Johnson & Johnson identified a kinase inhibitor mechanism (extracted from Rubenstein 2008). Dr.Rubenstein's recent book also includes examples describing nanosystems studies to construct a predictive model for transcription control, ChIP-on-chip technology for global transcription factor identification, and methylation-specific polymerase chain reaction (PCR) for global DNA methylation detection as an entry point to epigenetics.

Identifying systems, building biologically accurate models, with appropriate parameters, performing sensitivity analysis provides a robust ecosystem for carrying out drug development studies. In our experience, the community will increasingly focus on building virtual cell (e.g., virtual *E. coli*, virtual Pseudomonas) and whole organ (virtual heart, virtual multi-organ diabetic model) in the near future. Professor Dennis Nobel's group already has significant contribution in this direction. Prof. Nobel is one of the pioneers of Systems Biology and developed the first viable mathematical model of the working heart in 1960. His research focuses on using computer models of biological organs and organ systems to interpret function from the molecular level to the whole organism. Together with international collaborators, his team has used supercomputers to create the first virtual organ, the virtual heart.

The impact of systems biology is also visible through the work of Dr. Jasin A. Papin of the University of Virginia. Recently, his group constructed the first Leishmania major metabolic network that accounts for 560 genes, 1,112 reactions,

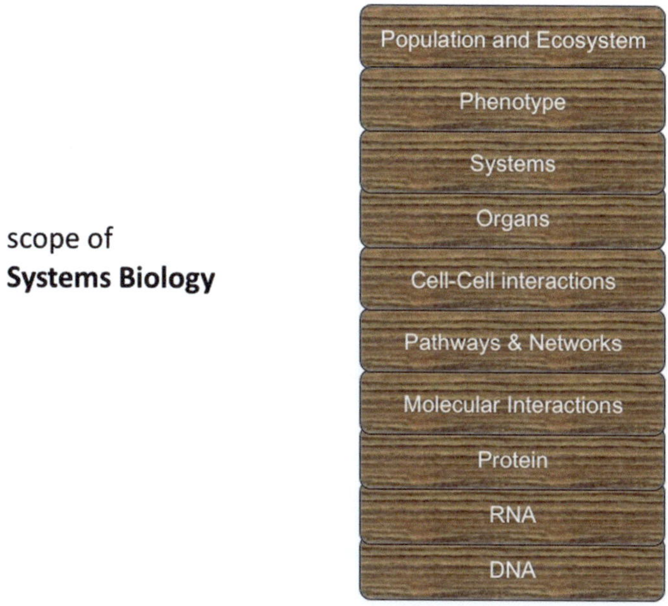

Fig. 1.4 Different levels when connected give a reasonably accurate picture

1,101 metabolites, and eight unique subcellular localizations. Also, the same group was involved in building a genome-scale constraint-based model of the Pseudomonas aeruginosa strain PAO1, mapping 1,056 genes whose products correspond to 833 reactions and connect 879 cellular metabolites.

1.7 Conclusion

A system is not equal to the sum of its components. This is especially true of biological systems that show robustness and emergent properties. Due to dynamic and complex interaction among components within and between different levels (Fig. 1.4), the biophysical and biochemical laws that describe these components cannot explain the collective behavior of a system. A grand challenge in systems biology is to identify these rules at the interface and expand in either direction. It is easy to model energy transactions as the energy transfer reactions have been well studied in physical and chemical systems. The more challenging task is to simulate collaborative interactions among molecules that produce and transfer signals.

As always, new challenges demand new strategies. Signaling pathways, the most difficult to model due to a heterogenous mix of activities involved, can be seen as a kind of molecular body language. We argue that to simulate these molecular activities using a language at a level that matches the molecular body language is a preferable approach. The language should have following minimum features: time-dependent and molecular behavior features, a switchable link between molecules, explicitly defined semantics of interaction, dynamic logging of molecular interaction, hardwiring cellular events with molecular events, and an extension to multicellular modeling capability. We are currently working on building such a language, though its effectiveness hasn't yet been determined.

One of the challenges in Systems Biology is to identify a complete parts list of a cell and tie them by way of equations, conditional statements that are context dependent. The purpose is to move from structural knowledge to functional knowledge of the system. One of the unsolved mysteries of science is how does the behaviour of a cell at different scales relate to the physiological phenomenon. Constructing a cell from its bare components calls for excellent engineering knowledge, not only for integrating small cell parts into pathways and networks, but also for reverse engineering of the parts from experimental data. The construction of a detailed cell map has to be aided by novel experimental and computational approaches. The future of experimental system biology lies in the invention of novel approaches that generate high throughput and noise free data. In addition, advancement of computational systems biology depends on invention of truly integrated algorithms that are adaptive, robust and capable of simulating multi-scale system. The algorithms will fully integrate different levels of abstractions and reconcile the basic assumptions involved in different timescale and time-span involved. Last but not least, algorithms should also model the smooth transition of a model from mesoscopic to macroscopic scale.

Key: Terms Commonly Used in Systems Biology

Modules are subnetworks with a specific function and which connect with other modules often only at one input node and one output node.
Robustness describes how a network is able to maintain its functionality despite environmental perturbations that affect the components. Robustness also reduces the range of network types that researchers must consider, because only certain types of networks are robust.
Network motifs *Patterns of subgraph that recur within a network more often than expected at random.*
Path An unbroken series of linear steps. A path has one entry (input) and one exit (output) point.
Pathway A collection of convergent, divergent and cyclic paths. A pathway may have one entry point and many side branches as exit points. The side branches connect a pathway with other pathways. Often, energy-consuming pathways are coupled to energy generating pathways to maintain the overall energy budget.
Network. A set of interacting pathways. A network has multiple entries and multiple exits. Traditionally, pathway was more used for describing metabolic processes and network for gene regulation and signal transduction. Yet there can

be metabolic networks, signaling networks, and hybrid networks comprising both metabolic and signaling pathways. The topology of networks reflects some fundamental properties of biological systems involved, and it can be reprogrammed in cells in response to external signals.
Module. A module is a relatively independent functional unit in a cell, which may comprise one or several cross-interacting pathways and autonomously performs a specific function. A functional module can have different structural organization in different cells and at different time, reflecting the substitutability and overlap of gene function. Some biological activities like feedback and amplifier can be explained better in terms of module rather than of pathways or molecules.
Modularity describes the extent to which a system is divided into modules.
Complexity. Biological complexity can be gauged in different dimension. It may cover structural and functional interaction among elements, and the evolution of the systems and subsystem they create. Many mathematical concepts and tools, such as self-organization theory, nonlinear equations, cellular automata and chaos, are used to describe complex biological phenomena.
Robustness. The property of system which indicates the resistance to internal errors and external perturbations
Model. A model is a formal or abstract representation of a system, usually in the form of a set of objects and the relations between them.
System. Consisting of more than one component physically that can be sub systems at lower level, a system possesses more attributes and behaves more complex than any of its component.
Systems Biology. An approach to link the constituent elements of a system with its higher level behavior.
Systems Engineering is a methodology developed in engineering areas but applied in biological modeling to build complex systems from a raw material of components.
Forward Engineering follows a bottom-up approach to model a system and its functional process with known information about its elements.
Reverse engineering is a top-down process, inferring the internal structure and components according to systems behavior.
Systems Theory is a mechanical understanding of system structure behavior.

References

Adami C (2002) What is complexity. Bioessays 24:1085–1094
Bhalla U, Iyengar R (1999) Emergent properties of networks of biological signaling pathways. Science 283:381–387

Bornholdt S (2001) Modeling genetic networks and their evolution: a complex dynamical systems perspective. Biol Chem 382:1289–1299

Bradley C, Harris G, Pullan A (2001) The computational performance of a high-order coupled fem/bem procedure in electropotential problems. IEEE Trans Biomed Eng 48:1238–1250

Elowitz MB (2002) Stochastic gene expression in a single cell. Science 297:1183–1186

Fodor SP, Read JL, Pirrung MC, Stryer L, Lu AT, Sola D (1991) Light-directed, spatially addressable parallel chemical synthesis. Science 251:767–773

Gillespie DT (1977) Exact stochastic simulation of coupled chemical reactions. J Phys Chem 81:2340–2361

Gillespie DT (2001) Approximate accelerated stochastic simulation of chemically reacting systems. J Chem Phys 115:1716–1733

Gulrajani R, Savard P, Roberge F (1988) The inverse problem in electrocardiography: solutions in terms of equivalent sources. Crit Rev Biomed Eng 16:171–214

Hakelien AM et al (2002) Reprogramming fibroblasts to express t-cell functions using cell extracts. Nat Biotechnol 20:460–466

Hao Z, Huang S, Dhar P (2003) The next step in systems biology: Simulating temporo-spatial dynamics of the molecular networks. BioEssays 26:68–72

Haseltine EL, Rawlings JB (2002) Approximate simulation of coupled fast and slow reactions for stochastic chemical kinetics. Chem Phys 117:6959–6969

Johnston P, Gulrajani R (1997) A new method for regularization parameter determination in the inverse problem of electrocardiography. IEEE Trans Biomed Eng 44:19–39

Lynch M, Conery JS (2003) The origins of genome complexity. Science 302:1401–1404

Morton-Firth CJ, Bray D (1998) Predicting temporal fluctuations in an intracellular signalling pathway. J Theor Biol 192:117–128

Rubenstein K (2008) Systems biology: a disruptive technology. CHI insight pharma reports. Ed. 156 pages

Savageau P (1971) Parameter sensitivity as a criterion for evaluating and comparing the performance of biochemical systems. Nature 229:542–544

Strogatz S (2001) Exploring complex networks. Nature 410:268–276

Stundzia AB, Lumsden CJ (1996) Stochastic simulation of coupled reaction-diffusion processes. J Comput Phys 127:196–207

Swain PS (2002) Intrinsic and extrinsic contributions to stochasticity in gene expression. Proc Natl Acad Sci 99:12795–12800

Tada M et al (2001) Nuclear reprogramming of somatic cells by *in vitro* hybridization with es cells. Curr Biol 11:1553–1558

Wagner A, Fell D (2001) The small world inside large metabolic networks. Proc R Soc Lond B Biol Sci. 268:1803–1810

Yang J, Lusk R, WH Li (2003) Organismal complexity, protein complexity, and gene duplicability. Proc Natl Acad Sci U S A 100:15661–15665

Further Reading

de Jong H (2002) Modeling and simulation of genetic regulatory systems: a literature review. J Comput Biol 9:67–103

Hasty J and Issacs F (2001) Designer gene networks: towards fundamental cellular control. CHAOS 11:207–220

Hlavacek WS, Faeder JR et al (2006) Rules for modeling signal-transduction systems. Sci STKE 344:re6

Hartwell LH, Hopfield JJ, Leibler S, Murray AW (1999) From molecular to modular cell biology. Nature 409:247–252

Iyengar R (2009) Computational biochemistry: systems biology minireview series. J Biol Chem 284:5425–5426

Nurse P (1997) Reductionism: the ends of understanding. Nature 387:657–657

Sauro HM, Bergmann FT (2008): Standards and ontologies in computational systems biology. Essays Biochem 45:211–222

Schadt EE, Zhang B, Zhu J (2009) Advances in systems biology are enhancing our understanding of disease and moving us closer to novel disease treatments. Genetica 136:259–269

Schulze WX, Deng L et al (2005) Phosphotyrosine interactome of the ErbB-receptor kinase family. Mol Syst Biol 1:0008

Smolen P, Baxter DA, Byrne JH (2000) Modeling transcriptional control in gene networks–methods, recent results, and future directions. Bull Math Biol 62:247–292

Yu RC, Resnekov O, Abola AP et al (2008) The Alpha Project: a model system for systems biology research. IET Syst Biol 2:222–233

Chapter 2
Why Systems Biology Can Promote a New Way of Thinking

Alessandro Giuliani

Abstract This chapter deals with the effect Systems Biology had on the Nature of what we consider 'an explanation' in Biological Science. I try and demonstrate how the most relevant change carried out by Systems Biology approach was the shift from the molecular layer as the definitive place where causative process start to the elucidation of the among elements (at any level of biological organization they are located) interaction network as the main goal of scientific explanations. This change of perspective allows to dissipate a widespread idealistic nightmare looking at the single molecules as Maxwell-demon-like intelligent agents. The recognition that genes work in networks has as consequence the existence of discrete 'allowed global modes' of gene expression. This theoretical expectation was verified by the incredibly narrow space of different tissues (each corresponding to a largely invariant gene expression profile)—around 200 tissue types for all the metazoans emerging from the transfinite number of possible combinations of the expression values of around 30,000 genes. This is a crucial step for generating a scientifically sound framework to address global biological regulation.

Systems Biology approach makes obsolete the debate between 'reductionist' and 'holistic' approach in favor of a 'middle-out' paradygm formally identical to the time honored chemical thought. This is probably the brightest promise of Systems Biology to scientific knowledge.

Keywords Attractor in systems biology · Maxwell's demons · Levinthal paradox · Network · Protein contact network (PCN) · Metabolic network

2.1 Introduction

The classical form in which biological systems are described (being they metabolic charts, gene expression regulation pathways, protein-protein interaction maps, food webs and so forth) corresponds to a set of nodes linked by edges in which the nodes

A. Giuliani (✉)
Environment and Health Department, Istituto Superiore di Sanitá,
Viale Regina Elena 299, 00161 Roma, Italy
e-mail: alessandro.giuliani@iss.it

V. Singh, P. K. Dhar (eds.), *Systems and Synthetic Biology*,
DOI 10.1007/978-94-017-9514-2_2

are the basic elements of the described system (genes, proteins, metabolites and so forth) and the edges connecting them some rules of the kind ‘is transformed into’ or ‘is increased by’.

The figures normally present in books and scientific papers implicitly consider these pathways as linear causative chains in which a signal starting from a molecular perturbation, after a sequence of if-then events, emerge as a biological end-point (Tun et al. 2011). Normally these processes are referred as ‘cascades’ provoking a progressive amplification of the initial stimulus (MacFarlane 1964). On the contrary, biological effectors (being them genes, proteins, hormones.), with only few exceptions, work in networks, and this fact implies a completely different form of biological regulation with respect to the ‘cascade’ model: the *entire* network has, thanks to its wiring structure, few preferred modes corresponding to the stable configuration of the network itself (Tun et al. 2011; Kauffman 1993; Huang et al. 2005), any perturbation, being it pharmacologically induced or coming from a mutation in a crucial gene, ends up into one or the other of these allowed states without any simple relation with the features of the applied perturbation (Tun et al. 2011).

Figure 2.1, taken from (Huang 2009), depicts the change in perspective shifting from pathway to network paradigm.

Without entering in the physical processes instantiating such intermingled (and largely invariant) networks, Systems Biology scholars can make use of a purely phenomenological view on biological regulation adopting some general concepts of dynamics. This is a necessity, if we consider that, thanks to the development of high throughput methodologies the graphs corresponding to the ‘perceived’ regulation networks became larger and larger and ask for some form of global analysis in order to get rid of their wild multiplicity.

The approach considering the graph as a system of differential equations in which an entering stimulus, correspondent to a modification of a peripheral node of the network, is progressively processed according to the wiring architecture and kinetics constraints of the network itself, while being the most potentially exhaustive avenue of research is severely hampered, in the case of biological systems, by a lot of problems. First of all the practical impossibility to attach to the whole set of edges reliable kinetic-like weights for quantifying the entity of the between elements correlation. Only in the case of very small networks this can be done by means of the statistical estimation of the parameters from experimental data, but it is well known that in physiological settings these weights can vary of orders of magnitude (Laughlin et al. 2000). Moreover in many cases we cannot rely on the complete knowledge of the wiring diagram of the network. For these (and other) reasons many authors preferred a purely topological approach to the analysis of biological networks, considering the presence of a link between two nodes as a pure yes/no binary relation and limiting themselves to statistical descriptions making use of the so called graph-invariants, i.e. a collection of indexes that, relying on the simple count of nodes and edges, enable the analyst to identify crucial elements of the network (like the so-called hubs, nodes engaged in a very large amount of relations) or to highlight specific features of the entire network architecture responsible for some aspects of the studied system

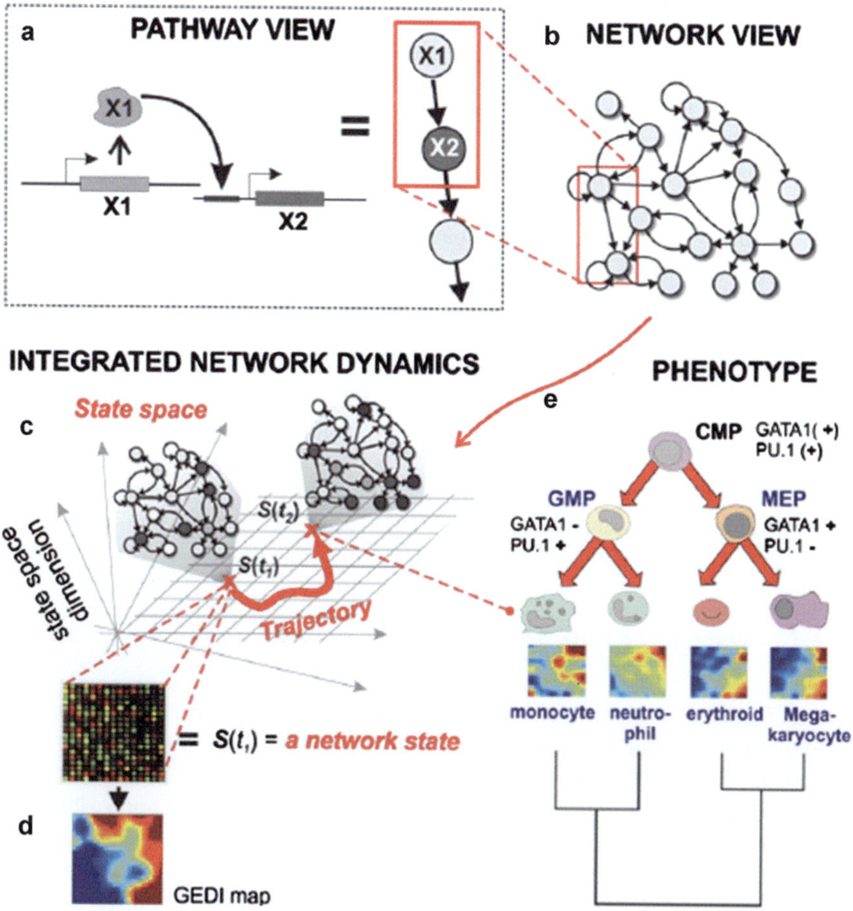

Fig. 2.1 Panel **a** reports the usual pathway view: a molecule produced by gene X1 acts as modulator (positive or negative) of gene X2 that in turn acts on another gene and so forth. If we consider these linear pathways are part of a network (panel **b**) we understand how the only allowable states are those corresponding to the network configurations that occupy energy minima in the state space having as dimensions the actual values of the nodes (panel **c**). The presence of a strong correlation structure among the nodes (in panel **d** the gene expression network is reported in which the node values correspond to the expression values of different ORFs) creates a 'rugged energy landscape' over the state space with only few valleys correspondent to differentiated tissues having a strongly invariant gene expression profile expressed as GEDI (Gene Expression Dynamics Inspector) map in which each pixel corresponds to a gene whose expression value is paralleled by a different color. These maps, collectively correspond to observed phenotypes (panel **e**)

behaviour (this is the case of the so called 'scale-free' architecture that was demonstrated to be at the basis of the huge resilience of biological systems) (Watts and Strogatz 2004).

The consideration of biological systems at the coarse-grain level of the graph topological approach is, in my opinion, a very important first step for the development of a sort of biological statistical mechanics in which the actual behaviour of the global system can be predicted by a convenient statistics over its constituentparts (Giuliani 2010).

In the case of statistical mechanics of inanimate systems this was the case with the Boltzmann microscopic definition of entropy as a statistical index computed over the microstates frequency distribution of the studied system (Giuliani 2010). This deliberate coarse-grain approach that abandoned the dream of following the trajectories of the single elements for a population level view, enabled scientists to get a link between microscopic and macroscopic physical descriptions (Laughlin et al. 2000; Watts and Strogatz 2004; Giuliani 2010; Karsenti 2008).

In the following I will try and describe the search for a Boltzmann-like approach to biology by the critical analysis of different regulation network-like systems, in the same time I hope it will be clear how this effort is strictly consistent with very fertile lines of epistemological lines of thought, mainly chemical research tradition and multidimensional statistics (Di Paola et al. 2012; Benigni and Giuliani 1994).

2.1.1 The Concept of Attractor in Systems Biology

The concept of attractor was developed in dynamical systems theory, where the whole system is thought as evolving towards a preferred (minimal energy) state called an attractor set, and represented such as a point, a curve, and a manifold in the state space. The study of folding process in proteins, where the impossibility for the linear chain of amino-acids to randomly explore all the possible configurations in biologically plausible times before settling down in the native 3D structure (the so called Levinthal paradox taking its name by the crucial observation made by Cyrus Levinthal that in his 1969 paper (Levinthal 1969) computes in the order of millions of years the duration of protein folding process as compared by the seconds to minutes effective actual time) is the field of biomolecular science where a 'goal-oriented' trajectory driven by the existence of a preferred configuration 'attracting' the system trajectories in the state space was studied more in depth.

Figure 2.2 reports a simplified view of the folding process of a protein in energetic terms: this representation is named 'folding funnel' (Dill and Chan 1997) and stresses the fact that different initial states corresponding to different positions in the upper part of the funnel converge on the same potential well at the bottom of the funnel.

The fact the potential well bottom (equilibrium state) does not correspond to a fixed configuration but to a set of possible states makes it possible protein dynamics that is crucial for exerting its physiological role. The fact protein molecule behavior can be fully interpreted as the dynamics of a network in which the amino-acid residues are the nodes and the between-residues contacts the edges (Di Paola et al. 2012) makes the folding funnel metaphor perfectly suited for gene expression network, the only difference being the knowledge we have of the physical forces shaping

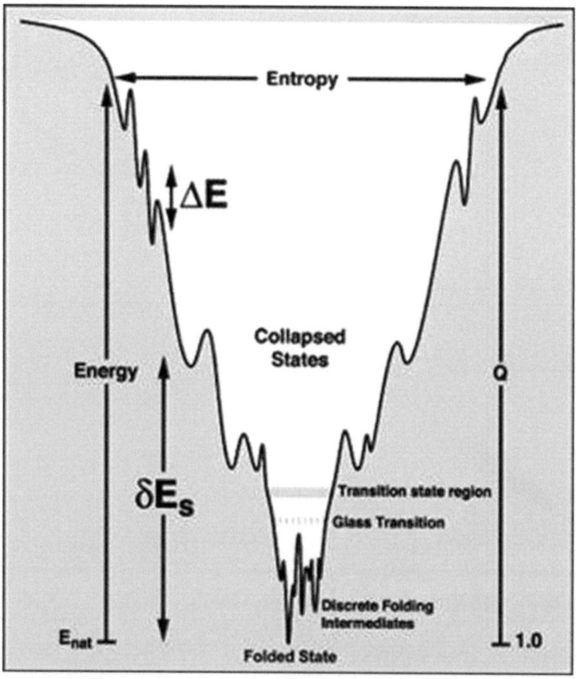

Fig. 2.2 The protein folding trajectory can be represented as a descent along a potential well. At the beginning, the fact the protein molecule is in a fully disordered state, corresponds to the width of the well, measured by the entropy of the system (many equally probable configurations of the polymer at the *top*). The vertical axis of the figure corresponds to the total energy of the system, the folding trajectory is driven by an energy gradient going down the funnel. Thanks to thermal agitation, the molecule can escape local minima and reach the bottom of the funnel that in turn is not a single point but a 'rugged landscape' corresponding to slightly different configurations of the molecule allowing for protein dynamics that in turn is essential for its physiological role

protein folding behavior (Hydrophobic interaction, Van der Walls forces, Hydrogen bonding.) and the almost complete ignorance about the force fields in which gene expression networks are embedded.

The common experience of any experimentalist dealing with microarray data is the fact that any two independent samples of the same cell kind when correlated over the expression of more than 20,000 different gene products display a near to unity correlation (see Fig. 2.3).

This marked invariance is normally 'given for granted' by biologists that historically focused on the (small) deviations from the native tissue profile in gene expression space looking for specific altered genes without having any perception of the 'elephant-in-the-room' correspondent to the invariance. Systems Biology is starting to look at the nature and origin of this elephant not only for knowledge reasons, but for incredibly urgent and dramatic practical motivations. As a matter

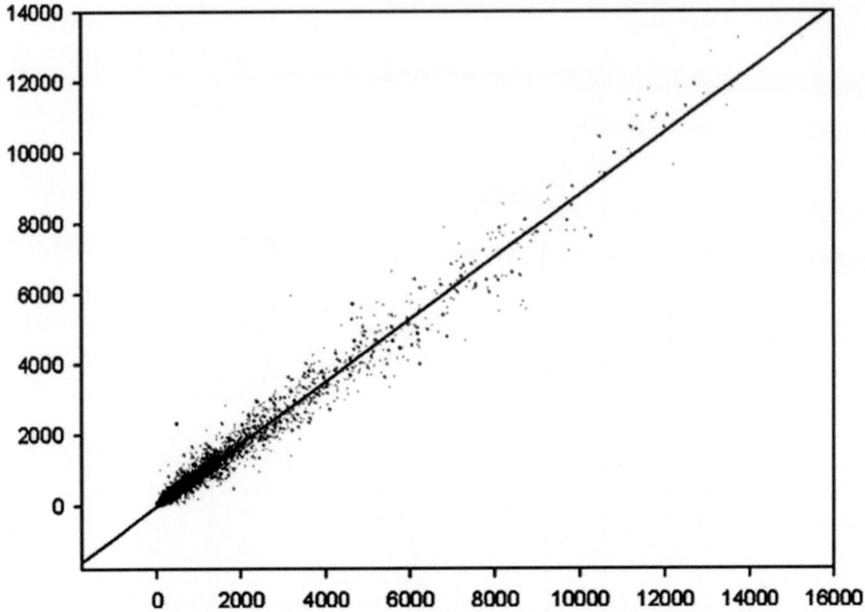

Fig. 2.3 The X and Y axes of the figure correspond to two independent samples of the same tissue type (in this case macrophages). The approximately 23,000 vector points correspond to different gene expressions coming from a microarray experiment. Notwithstanding the fact the graph spans four order of magnitudes there is a remarkable order of gene expression level corresponding to a Pearson correlation coefficient $r = 0.99$ between the profiles. This invariance comes from the fact each tissue is an attractor in the gene expression space, the (relatively small) scattering around the identity *line* corresponds to the motions 'inside the attractor', these motions are analogous to the dynamics of a protein molecule around its native state and are the only ones eventually affected by disease states or pharmacological perturbations. (Giuliani 2010)

of fact in two very important papers (Overington et al. 2006; Hopkins 2008) Overington, Hopkins and colleagues gave a very crude (but statistically clear) picture of the state of pharmacology research and development: the number of new drugs arriving at the market stage dramatically decreased starting from the 80's of the last century and the classes of receptors they are supposed to bind were already known since 50 years, the concept of a 'druggable genome' with myriads of new drug targets supposed to be revealed by genome project simply does not exist or, better, the targets are not 'druggable'. The same basic idea of network stable states allows to understand what happened: the only 'simple targets' whose modification is expected to give rise to a macroscopic, organism-scale observable effect are those located at the extreme periphery of the interaction network, while the modification of a node located in the internal position of the network is immediately buffered by the feedback relations so that the system cannot be modified by pharmacological intervention (Tun et al. 2011).

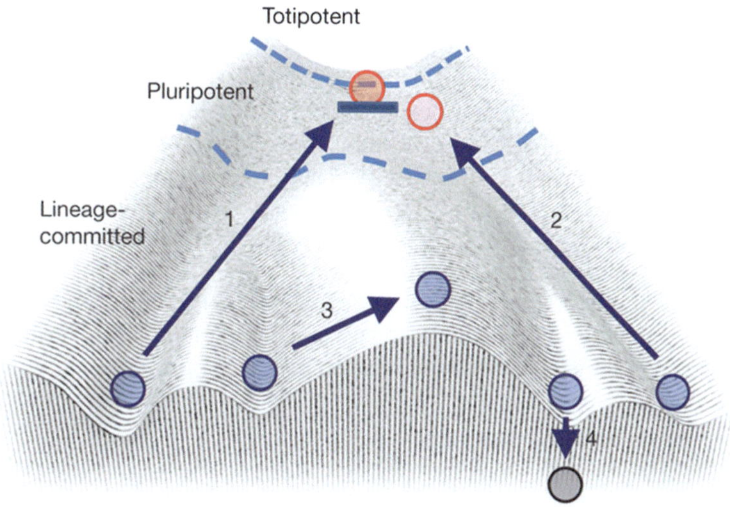

Fig. 2.4 At the cell population level, the ensemble of cells' potency in iPS reprogramming process can change in a probabilistic manner like rolling up and down on the epigenetic energy landscape towards a specific valley having definite potency (attractor state). A lower potency can be pushed up by a competent stimulus to another allowed potency level on the landscape corresponding to another (less stable) equilibrium endowed with a higher differentiation potential energy. Figure cited from (Yamanaka 2009)

On the contrary, the recent Nobel prize to Yamanaka and Gurdon tells us a completely different story (Yamanaka 2009; Yamanaka and Blau 2010): a bunch of effectors, in a still largely unknown way, is able to transmit to the 'system as a whole' an effective stimulus able to push the entire network along a 'counter-gradient' trajectory going back to an higher energy state corresponding to an undifferentiated, totipotent state. It is remarkable that Yamanaka explains this effect using the Conrad Waddington epigenetic landscape (Waddington 1957) a precursor of energy landscapes, in which unstable, 'high energy' states (and thus states in which the system can be modified by an external modification) are represented as ridges and stable states (attractors) by valleys, as depicted in Fig. 2.4, coming from a Yamanaka paper (Yamanaka 2009).

The attractor view allows scientists to eliminate the need to impose the presence of 'intelligent agents' (Maxwell's demons) in order to get rid of specific and finely tuned behaviors: the system 'lives' in a non-uniform state space (the ensemble of all the possible system configurations) characterized by a so called 'rugged landscape' (Frauenfelder 1991) where the energy minima (valleys of the landscape, quasi-equilibrium configurations) correspond to attractor states (Frauenfelder 1991).

Each system accommodates towards the energy minimum nearest to it, consistently with the marked 'context dependence' (e.g. sensitivity to microenvironment) of biological regulation. Metaphorically, C. H. Waddington (Waddington 1957) suggested that cell fate would be determined by a trajectory toward a local minimum (attractor) on epigenetic energy landscape, where a series of "valleys" and "ridges" describe stable cellular states (local minima) and barriers (local maxima) between those states, respectively. The epigenetic landscape is a proposal for the existence of global molecular regulation in cell fate decision. The word 'global' underlines the fact 'energy' is computed over the entire state space (in the case of transcription dynamics, the genome-wide expression) and not over few specific genes.

Clearly, as above stated, this is a pure phenomenological proposal that does not enter the molecular mechanisms supporting it even if cytoskeleton organization (Ingber 1999) or confinement by phase transitions (Hyman and Simons 2012) allowing the selection of specific pathway in complex microenvironments are very plausible candidates. Limiting ourselves to data analysis coming from actual biological experimentation, it is sufficient to imagine these discrete states corresponding to 'allowed positions in the transcriptome space', coming from the presence of a still unknown origin field sensed by the entire genome and driving its collective behavior (Huang et al. 2005). It is worth noting that a very basic 'toolbox' made of principal component analysis (principal components being the coordinated fluxes of variations of many different genes), network invariant descriptors (with the assignment to each node a set of measures related to its role in the network wiring), cluster analysis (very dense clusters in the phase space correspond to attractors) are sufficient to undergo such 'biological dynamics' avenue of research (Huang et al. 2005; Huang 2009; Benigni and Giuliani 1994).

In (Huang et al. 2005), the authors offer a very thorough proof-of-concept of the relevance of considering a differentiated state of a cell population as an attractor in the proper dynamical sense. They demonstrate that, after perturbation induced by two completely different chemical stimuli (atRA and DMSO respectively) initially inducing a completely different response in terms of gene expression, the system returns back to the same attractor point in the genome expression phase space (see Fig. 2.5).

The above behavior corresponds exactly to the basic definition of an attractor as a state 'attracting' the perturbation trajectories of the system. This stems from the fact attractor states are stable and, if the perturbation is not sufficiently strong to push the system outside its 'basin of attraction', soon or later the system will come back to its original attractor state loosing memory of the nature of the initial perturbation. Thus the specific differences in mechanism of action of the two effectors are not so relevant in terms of the resulting effect that is largely dependent on the affected system modes.

A very important consequence of the presence of an attractor-like regulation is the impossibility to maintain the classical discrimination of house-keeping vs. specifically regulated genes and the importance of low-variance genes with the consequent need to re-cast the idea of what a 'pathway' (or a 'gene signature') is (Tsuchiya et al. 2010; Venet et al. 2011).

Clearly it is for sure that different genes have different discrimination ability for specific diseases, and again these specific genes could be more useful than other for diagnostic purposes, but this has only to do with our specific discrimination goals: the system as it is works in a self-coherent way on the whole-genome scale.

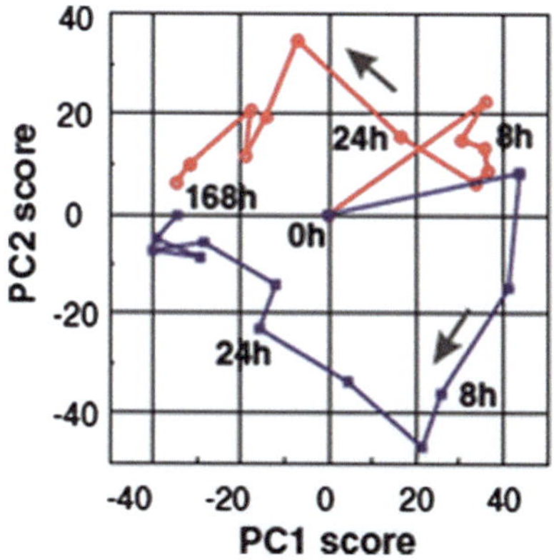

Fig. 2.5 Attractor approaching coming from two different trajectories: principal component analysis for highly expressed 2773 genes following atRA and DMSO stimulus shows two different trajectories on the space spanned by the first two principal components (PC1 and PC2; *red circles* for atRA; *blue squares* for DMSO). Figure comes from (Huang et al. 2005) and allows to appreciate how two different stimuli initially make the system to go away from the same initial state (0 h) toward two different directions, when the transients settle down and the system reaches a new stable configuration, the two trajectories converge to the same attractor state (168 h) losing memory of their different paths. The strong between genes correlation allows to collapse the entire multidimensional expression space into a bi-dimensional component space

Especially relevant is the fact that the genome dynamics involve both highly and lowly expressed genes, which are generally considered noisy and insignificant in microarray experiments.

These findings give an immediate explanation to the recent 'iconoclastic' results obtained by Venet et al. (2011) thoroughly commented in (Jordan 2012) demonstrating the practical equivalence between random collection of genes and specific signatures for breast cancer prognosis. Along the same line is the finding of the complete equivalence of different random gene selections for tracking hematopoietic differentiation demonstrated by Felli et al. (2010). In the attractor model, lowly expressed genes are effective players in global gene regulation, given they are integral part of collective expression modes elicited by the perturbation (treatment, mutation, differentiation stimulus, etc.); this implies that any sufficiently dense sampling of genetic probes gives us a relevant picture of the collective mode (Felli et al. 2010; Censi et al. 2011).

A very recent work describing the architecture of whole genome regulation as emerging from results coming from ENCODE project (Gerstein et al. 2012) is consistent with the view of a dense interconnected network working as a whole and thus asking for system-level description of gene expression dynamics.

2.1.2 'Bottom-up', 'Top-down' or (Better) 'Middle-out'?

If we assume a classical molecular approach, we make the implicit assumption that the 'effective flux' of causation starts at the most microscopic level of the biological matter and progressively emerges at more macroscopic levels by an interaction chain.

We refer to this approach as 'bottom-up' and it was at the basis of molecular biology research in these last 50 years: each disease, each general condition is approached by looking for its molecular determinants.

On the other hand, a physician adopts a 'top-down' diagnostic approach, even if he is convinced the basic causative layer of the still unknown disease he suspects a given patient is affected is located at some fundamental level, he must orient the search for a proof of his conjectures in a top-down way by looking for objective data (biomarkers from blood or urine analysis, image analysis as NMR or X-rays, biopsies.) collected starting from the goal, and then driven by the global state of the patient that implicitly is supposed to influence the microscopic findings. The same 'top-down' approach is implicitly assumed in 'goal-driven' phenomena like development even if embryologists actively look for the way to turn development into a bottom-up explanation, being the top-down approach considered as a constraint arising from the lack of a sufficiently accurate knowledge of development.

All in all, the choice of one approach or the other is often not-decidable: from a certain point of view the 'ultra-reductionist' exclusively bottom-up perspective is totally unrealistic for the obvious reason any organism is subjected to a huge number of top-down constraints coming from the fact they live in a physical world (gravity (Ingber 1999), electromagnetism (Sebastian et al. 2001), thermodynamics (Shakhnovic 2006)) as well as from higher order perturbations affecting molecular targets (synaptic modifications induced by learning (Malenka and Nicoll 1993), hormonal changes due to psychological and social stress (Catalani et al. 2011)). All these phenomena ask for a top-down causation complementing the bottom-up mechanisms.

On the other hand, a purely top-down approach will end up into a pure descriptive/diagnostic and mainly tautological body of knowledge in which any knowledge element is at its best a 'diagnostic marker' of something else or, worst, a necessary consequence of a global principle. This happens for example in some misconceptions of evolutionary theory that virtually inhibit any fundamental research on the causes of observed phenomena by simple stating 'if it is there it means that it must be there because it is convenient' that is in some cases is nothing more nothing less than tautological 'just-so-stories'.

Network (or better graph) paradigm are located half-way between these two extremes and for their very basic nature make these two opposite epistemological approaches obsolete.

The classic Konigsberg bridge problem introduced graph theory in eighteenth century. The problem had the following formulation: does there exist a walk crossing each of the seven bridges of Konigsberg exactly once? The solution to this problem appeared in 'Solutio Problematis ad geometriam situs pertinentis' in 1736 by Euler

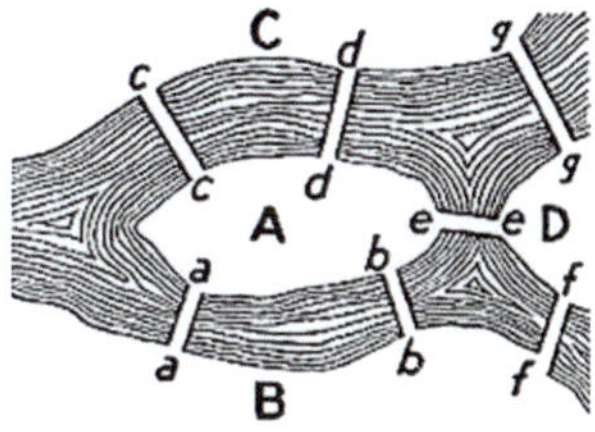

Fig. 2.6 The Konigsberg bridge problem: the seven bridges (edges) extremities are indicated by letters (nodes)

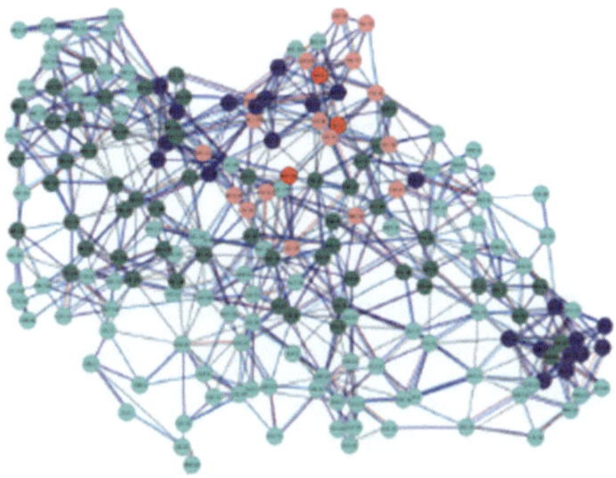

Fig. 2.7 A Protein Contact Network (PCN) this is a complex graph in which each node corresponds to an aminoacid residue and each edge to a physical contact between two residues. The nodes are variously colored according to amionacid chemico-physical features. (Di Paola et al. 2012)

(Di Paola et al. 2012). This structure was called a graph and this was the first time a problem was codified in terms of nodes and edges linking nodes (Fig. 2.6).

A graph G is a mathematical object used to model complex structures and it is made of a finite set of vertices (or nodes) V and a collection of edges E connecting two vertices (Fig. 2.7).

It is relatively easy to extract from graphs many descriptors located at local (single nodes), global (entire network) and mesoscale (clusters of nodes, optimal paths) levels. Thus we can compute the degree of each node (how many links are attached to a given node) that is a local, microscopic characteristic by which we can in principle locate the most important elements in a complex system (bottom-up approach) or we can compute the so called 'average shortest path' or 'characteristic length' of a graph corresponding to the average length of minimal paths connecting all the node pairs (this corresponds to a mesoscopic feature of the system) or the general connectivity of the network (this allows for a top-down study of the network as a whole) (Watts and Strogatz 2004; Di Paola et al. 2012).

It is important to stress all these different view are strictly intermingled among them, given they derive from the same basic representation (the graph) so that any view influences (and in turn is influenced) by all the others. The necessary (and natural) interaction of different level view is called 'middle-out' approach to stress the fact the interest is focused on the mesoscopic level, i.e. on the pattern of between elements relation and not on the fundamental features of the constituting elements (Csermely et al. 2005).

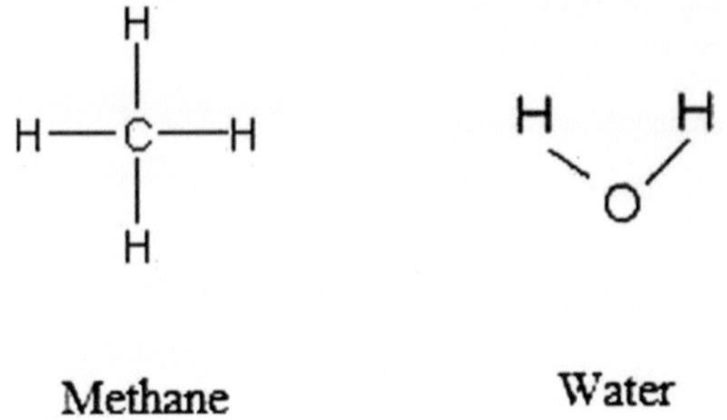

Fig. 2.8 Structural formulas are graphs in which the edges are covalent bonds between atoms (nodes)

The science that was mostly influenced by this 'naturally systemic' view is chemistry that uses since decades the most widespread (and effective) graph formalization of all: the structural formula (Di Paola et al. 2012). Figure 2.8 reports the structural formulas of methane and water.

Every chemistry student knows very well that an hydrogen atom embedded into methane molecule has different features than *the same* hydrogen atom of a water molecule: e.g. the hydrogen in the water molecule has a partial positive charge much greater than the methane hydrogen for the greater electronegativity of oxygen with respect to carbon atom. This is a clear example of top-down causation: the properties of the most basic level (atom) depends on the features of the entire system (molecule). In the same time both methane and water molecules derive their features from the constituent atoms (bottom-up causation). Stressing the two 'directions of causality' is in any case out-of-scope, because the chemical graph incorporates both into a global systemic reasoning made it possible by the formula. If we shift to more complex organic molecules formulas we can appreciate the richness of the possibilities offered by this approach (by the way thousands of different quantitative features of the molecules can be directly derived from structural formulas so that, strictly speaking, properties like solubility, melting point, molar refractivity, partition coefficients can be considered as graph descriptors (Fredenslund et al. 1979)).

What is important to stress here is that the network paradigm introduces a unique synthesis between reductionist (all is in the molecules) and holistic (all is in the whole) approaches. A clear example of the efficiency of this 'graph-based' reasoning more linked to Systems Biology problems is the prediction of lethal mutants in yeast by the graph analysis of metabolicnetwork (Palumbo et al. 2005; Palumbo et al. 2007).

From a purely topological point of view, each node of a network is uniquely defined by its position in the graph. Obviously, when dealing with experimentally derived and not abstract networks, each node has a name (a particular gene, protein, metabolite) pointing to a rich basin of knowledge and evoking cognition resonance to the specialist mind and the same is true for the edges. However, if we are interested in discovering what can be inferred solely from topological information (so acquiring a Boltzmann-like statistical attitude sacrificing the unique personality of the element to the search of a mesoscopic principle), we should try and predict some relevant features of the studied system without relying on the particular 'nature' of nodes and edges, but only taking into consideration their connectivity pattern. In other terms all the properties relative to each node (edge) must be derived only by its pattern of relations and thus by its peculiar location in the complete graph. In (Palumbo et al. 2005; Palumbo et al. 2007) the authors checked for the possibility to derive, from purely topological information on the metabolic network of yeast (Saccharomyces Cerevisiae), the lethal character of genetic mutations. The metabolic network of microorganisms is very well understood: it can be considered as a graph having enzymatic reactions as edges and metabolites as nodes. Since an enzymatic reaction is catalysed by one or more enzymes, an edge can also represent the enzymes involved in the reaction. This opens the way to a straightforward analysis of the possibility to derive biologically meaningful features at a macroscopic scale (entire organism) from network topology: the elimination of an enzyme by a knock-out experiment implies the elimination from the network of the edge (or edges since the same enzyme can catalyze different reactions) corresponding to that particular enzyme (Palumbo et al. 2005). If it is possible to pick up a connectivity descriptor able to unequivocally define essential enzymes (those enzymes whose lack provoke the yeast death) we can safely assume the biological relevance of the metabolism 'wiring structure', irrespective of the specific nature of the involved enzymes, and consequently deriving a mesoscopic biological principle (Giuliani 2010).

In the considered case of yeast metabolic network, the analysis of 36 lethal mutations out of the 412 relative to enzymes involved in metabolism, reported in the Stanford repository (http://www-sequence.stanford.edu/group/yeast_deletion_project /deletions3.html href) and in Jeong and colleagues (Giuliani 2010; Palumbo et al. 2005; Palumbo et al. 2007), allowed the authors to discover that all of the enzymes corresponding to lethal mutations, when deleted, prevent the connections between the separate nodes (Palumbo et al. 2005; Palumbo et al. 2007). No alternative path is available to connect the separate nodes and this mesoscopic features based on paths along the network explains the essential character of each specific mutation on a pure topological basis (Fig. 2.9).

This 'essentiality-by-location' mesoscopic principle equating the lethal character of a mutation to the lack of an alternative path in the network, was confirmed (Palumbo et al. 2007) demonstrating that a double mutation involving two enzymes that per se are not essential acquires essentiality and then causes the death of the organism, if the double knock-out provokes the 'lack of alternative path' condition. This illustrates the emergent character of the 'essentiality by location' principle: the arising of lethality by the summation of two non lethal events derives from the existence of a global metabolism architecture and thus cannot be inferred by going in depth into the nature of the two enzymes, in other words is a collective emergent property of the network system (Giuliani 2010).

It is worth noting the authors (Palumbo et al. 2005; Palumbo et al. 2007) did not find any exception to this rule: if an alternative pathway does exist then the mutation is not lethal. These data suggest the lack of 'purely kinetic' lethal mutations, i.e. situations in which the poor kinetic properties of alternative paths do not allow the yeast to survive. This points to a remarkable difference between metabolic and artificial networks: if we think of a road map, an accident causing a block of a

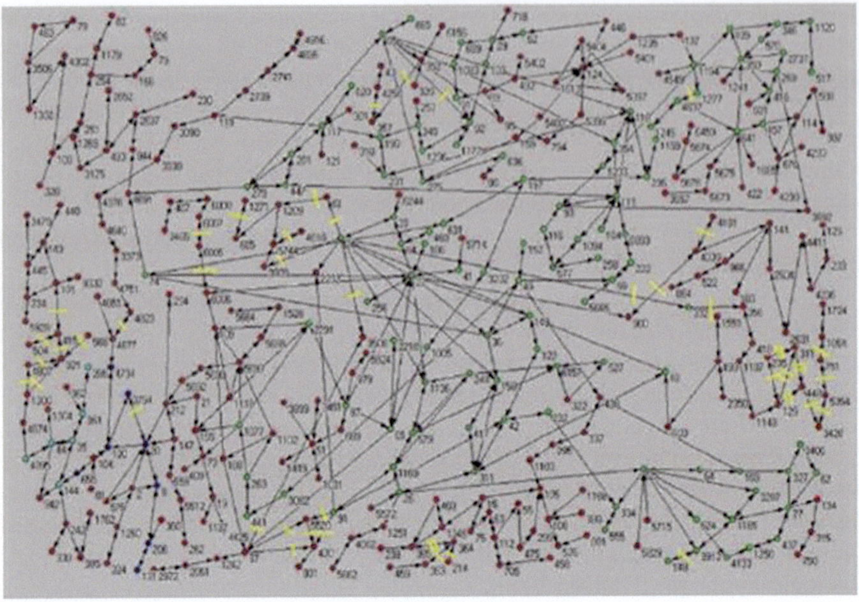

Fig. 2.9 The metabolic network of yeast is depicted in the figure, the enzymatic reactions correspond to edges, while the nodes point to the metabolites. The *yellow* signs indicate the analyzed mutations. (Giuliani 2010)

huge highway (kinetically optimal path) causes the traffic flow to shift into much narrower alternative roads (kinetically non-optimal paths), this will provoke soon or later a traffic jam that will make impossible a normal traffic flux with a consequent detrimental condition for the entire system. The fact such a situation was never observed, allows for the speculation that kinetic constraints in biological networks are not hard-wired in the network architecture and can be relatively easily circumvented. There are in fact some experimental data demonstrating the possibility of many orders of magnitudes variations of kinetic parameters of biochemical reactions (Von Dassow et al. 2000; Russell et al. 2009).

All in all this case showed the existence of a mesoscopic level (the network) whose behaviour cannot be simply derived by the knowledge of the constituting elements while, in an apparently counterintuitive manner with respect to the reductionist paradigm, influencing the microscopic level.

2.2 Conclusions

If Systems Biology holds a promise to provoke a big advancement in Biology this is not for an ancillary work of 'intelligent data mining and storage' enabling the scientists to pick up what they are more interested to from those 'Definitive Libraries' big relational data sets are becoming. This is certainly a useful work to be done (even if this is becoming a very risky business in terms of deterioration of the possibilities of falsification and of generation of self-sustained mythologies as aptly pointed out by Rzhetsky and co-workers (Rzhetsky et al. 2006)) but this is not the peculiarity of Systems Biology.

The specific role of Systems Biology is, in my opinion, to contaminate mainly mechanistic biological thinking with a relational paradigm analogue to chemical thought. This contamination can be hardly underestimated, mechanistic approach already gave clear signs of having ended its possibilities to say something new and to help to discover efficient therapies (Csermely et al. 2005). This shift of paradigm can be symbolized by the shifting from linear chain of events (pathway, the way of reasoning of mechanistic approach) to complex graph (network, with the corollary of the presence of few ideal forms or stable configuration is the seal of system approach). We made clear how this shift can by no means referred to the old 'bottom-up' vs. 'holistic' opposition but, on the contrary proposes a 'middle-out' approach focusing on relational structures. These relational structures can be analyzed at any scale of definition, given the conceptual and mathematical (very simple indeed) tools for analyzing networks are identical whatsoever the character of the network from between amino-acid residues contacts inside a protein, to gene expression network and food-webs.

Acquiring this paradigm will force Biology to abandon some vitalistic concepts in which molecules are considered as intelligent agents, one of them being the separation between 'house-keeping' and 'differentiative' genes that implies a super-natural controller that decides which activities are good for the cell as it is (house-keeping) and which ones are the price the cell pays for being part of a tissue, of an organ, and of an entire organism. The existence of global regulations of gene expressions driven by the existence of 'attractors', i.e. general configurations of the transcriptome state that are more stable than others allows to insert gene regulation in the realm of physical world. The point is that this description is till purely phenomenological: we have no idea of the physical basis of collective regulations.

Therefore, elucidation the physical origin of collective behaviors might provide novel insights for cell fate decisions, especially considering how well-known master instructive genes, such as Yamanaka factors (Yamanaka 2009; Yamanaka and Blau 2010) can drive genomes in differentiation of pluri-potent stem cells. The recent Nobel prize to Shinya Yamanaka and sir John Gurdon, for their demonstration of the possibility to reprogramming a mature cell population back to their stem state, goes along the same 'system' approach to biological regulation: the simple fact that the mature cell population can 'go backward' implies the presence of a collective transcriptome state that can be pushed back by a competent stimulus to another allowed location in the phase space corresponding to another (less stable) equilibrium endowed with an higher differentiation potential energy. This kind of behavior can be rationalized by the same methodological tools routinely used by chemical-physics where the 'going backward' of a system to another equilibrium state can be achieved by the flow of energy. This release of energy being maximally efficient when the system occupies what we call 'inflection points' so opening very attractive (even if largely futuristic) scenarios to a state-dependent (and thus maximally effective) therapeutic intervention on biological systems. But we remain with the huge curiosity about the 'material instantiation' of such dynamics, e.g. what 'piece of matter' must be maintained at a mostly invariant state (such as invariance of gene expression profiles for a specific tissue) thus driving the entire regulation machinery.

In an enlightening work (Tompa and Rose 2011) Peter Tompa and George Rose drove our attention to what they call the 'central biological question of the twenty first century', i.e. 'how does a viable cell emerge from the bewildering complexity of its molecular components?'. They use as analogy the Levinthal paradox (Levinthal 1969), pointing to another (still more drastic) paradox arising from the so called interactome, i.e. the necessity to maintain an extremely ordered pattern of relative spatial positions (and relative concentrations) of proteins mutually interacting in the cell. The authors (Tompa and Rose 2011) estimate a transfinite number of alternative orderings equal to 10^{7200} for the relatively small yeast proteome made up of 4500 protein species. The maintaining of a strict order in protein-protein interaction pattern is mandatory for an efficient metabolism, given no ordered reaction pathways can be achieved in a diffusive regimen. The maintaining of the interactome in its 'native state' is a natural candidate to act as primary driving force influencing gene expression regulation and generating the incredible invariance of gene expression profiles of a given tissue.

Beside these fascinating speculations, what is really important is the re-opening of basic science frontiers in biology after a period in which the basic dogmas were considered already known and firmly established.

References

Benigni R, Giuliani A (1994) Quantitative modeling and biology: the multivariate approach. Am J Physiol 266(35):R1697–R1704

Catalani A, Alemà GS et al (2011) Maternal corticosterone effects on hypothalamus-pituitary-adrenal axis regulation and behavior of the offspring in rodents. Neurosci Biobehav Rev 7:1502–1517

Censi F, Giuliani A, Bartolini P et al (2011) A multiscale graph theoretical approach to gene regulation networks: a case study in atrial fibrillation. IEEE Trans Biomed Eng 58(10):2943–2946

Csermely P, Agoston V, Pongor S (2005) The efficiency of multi-target drugs: the network approach might help drug design. Trends Phramacol Sci 26:178–182

Dill KA, Chan HS (1997) From Levinthal to pathways to funnels. Nat Struct Biol 4:10–19

Di Paola L, De Ruvo M, Paci P, Santoni D, Giuliani A (2012) Protein contact networks: an emerging paradigm in chemistry. Chem Rev (in press)

Felli N, Cianetti L, Pelosi E et al (2010) Hematopoietic differentiation: a coordinated dynamical process towards attractor stable states. BMC Syst Biol 4:85

Frauenfelder H, Sligar SG, Wolynes P (1991) The energy landscapes and motions of proteins. Science 254:1598–1603

Fredenslund A, Gmehling J, Rasmussen P (1979) Vapor-liquid equilibria using UNIFAC: a group contribution method. Elsevier Scientific, New York

Gerstein MB, Kundaje A, Hariharan M et al (2012) Architecture of the human regulatory network derived from ENCODE data. Nature 489:91–100

Giuliani A (2010) Collective motions and specific effectors: a statistical mechanics perspective on biological regulation. BMC Genomics 11(Suppl 1):S2

Hopkins AL (2008) Network pharmacology: the next paradigm in drug discovery. Nat Chem Biol 11:682–690
Huang S (2009) Reprogramming cell fates: reconciling rarity with robustness. Bioessays 31:546–560
Huang S, Eichler G, Bar-Yam Y et al (2005) Cell fates as high-dimensional attractor states of a complex gene regulatory network. Phys Rev Lett 94:128701–128705
Hyman AA, Simons K (2012) Beyond oil and water: phase transitions in cells. Science 337:1047–1049
Ingber D (1999) How cells (might) sense microgravity. FASEB J 13 (Suppl):S13–S15
Jordan B (2012) Are expression profiles meaningless for cancer studies? Bioessays 34:730–733
Karsenti E (2008) Self organization in cell biology, a brief history. Nat Rev Mol Cell Biol 9:255
Kauffman SA (1993) The origins of order. Oxford University, New York
Laughlin RB, Pines D et al (2000) The middle way. Proc Natl Acad Sci U S A 97:32–37
Levinthal C (1969) How to fold graciously. In: De Brunner JTP, Munch E (eds) Mossbauer spectroscopy in biological systems. University of Illinois press, Illinois, pp 22–24
MacFarlane RG (1964) An enzyme cascade in the blood clotting mechanism, and its function as a biochemical amplifier. Nature 202:498–499
Malenka RC, Nicoll RA (1993) NMDA-receptor-dependent synaptic plasticity: multiple forms and mechanisms. Trends Neurosci 16(12):521–527
Overington JP, Al-Lazikani B, Hopkins JL (2006) How many drug targets are there? Nat Rev Drug Discov 5:993–996
Palumbo MC, Colosimo A, Giuliani A, Farina L (2005) Functional essentiality from topology features in metabolic networks: a case study in yeast. FEBS Lett 579:4642–4646
Palumbo MC, Colosimo A et al (2007) Essentiality is an emergent property of metabolic network wiring. FEBS Lett 581(13):2485–2489
Russell D, Lasker K et al (2009) The structural dynamics of macromolecular processes. Curr Opin Cell Biol 21:97–108
Rzhetsky A, Iossifov I et al (2006) Microparadigms: chains of collective reasoning in publication about molecular interactions. Proc Natl Acad Sci U S A 103:4940–4945
Sebastian JL, Munoz S et al (2001) Analysis of the influence of the cell geometry, orientation and cell proximity effects on the electric field distribution from direct RF exposure. Phys Med Biol 46:213–219
Shakhnovic E (2006) Protein folding thermodynamics and dynamics: where physics, chemistry and biology meet. Chem Rev 106(5):1559–1588
Tompa P, Rose G (2011) The Levinthal paradox of interactome. Protein Sci 20:2074–2079
Tsuchiya M, Piras V, Giuliani A et al (2010) Collective dynamics of specific gene ensembles crucial for neutrophil differentiation: the existence of genome vehicles revealed. PLoS ONE 5(8):e12116
Tun K, Menghini M, D'Andrea L, Tanaka H, Dhar P, Giuliani A (2011) Why so few drug targets: a mathematical explanation? Curr Comput Aided Drug Des 7(3):206–213
Venet D, Dumont JE, Detours V (2011) Most random gene expression signatures are significantly associated with breast cancer outcome. PLoS Comput Biol 7(10):e1002240
Von Dassow G, Meir E et al (2000) The segment polarity network is a robust developmental module. Nature 406:188–192
Waddington CH (1957) The strategy of the genes: a discussion of some aspects of theoretical biology. Macmillan, New York
Watts DJ, Strogatz SH (2004) Collective dynamics of 'small world' networks. Nat Rev Genet 5:101–113
Yamanaka S (2009) Elite and stochastic models for induced pluripotent stem cell generation. Nature 460:49–52
Yamanaka S, Blau HM (2010) Nuclear reprogramming to a pluri-potent state by three approaches. Nature 465:704–710

Chapter 3
Modelling Methodologies for Systems Biology

Vikram Singh

Abstract This chapter intends to introduce various strategies for simulating the biological systems. We start by presenting simple biological systems based on elementary mono- and bimolecular chemical reactions and explain the concepts of chemical kinetics via the Michaelis–Menten mechanism. In most of the cases, the time evolution of a biological system can be assumed to be a continuous and deterministic one. By evolving the chemical reaction system, using ordinary differential equations (ODEs), one can reproduce the underlying dynamics of the biological processes. We describe the essential methods of solving ODEs like, Euler, Runge Kutta, and their application in some models. How and under what circumstances these methods should be used in the Systems Biology is illustrated. It is important to note that for small systems where intrinsic fluctuations are large, the connection between the macroscopic description of dissipative processes and the corresponding microscopic description is not straightforward. We discuss stochasticity in the biological systems and give an outline of the Gillespie's stochastic simulation algorithm (SSA). By applying it to some biological systems, we show when and why it is important to use this method over the continuous approximation.

Keywords Chemical kinetics · Michaelis–Menten kinetics · Cooperativity · Hill equation · Deterministic modelling · Euler method · RK4 method · Stochastic modelling · Gillespie's Stochastic Simulation Algorithm (SSA)· Brusselator · Repressilator · SBML · SBGN · BioPAX

3.1 Introduction

Studies in systems biology consist of four-fold path: (i) System structure, (ii) System dynamics, (iii) The control method, and (iv) The design method (Kitano 2001). Advances in biotechnology and high throughput data techniques are enabling us to build large-scale biological networks that describe the structure of the system under

V. Singh (✉)
Centre for Computational Biology and Bioinformatics,
Central University of Himachal Pradesh, Dharamshala, India-176215
e-mail: vikramsingh.jnu@gmail.com

V. Singh, P. K. Dhar (eds.), *Systems and Synthetic Biology*,
DOI 10.1007/978-94-017-9514-2_3

study. A network representing any biological system can be characterised by states that evolve over time, dynamically. These states are comprised of a set of interacting chemical species that react via various reaction channels.

Mathematical modelling of biological systems has become an essential and integral part in the studies of biological systems and plays an important role in the study of various levels of systems biology. Though, one need to integrate the experimentation with the theoretical framework to completely understand the interrelationships of different components of any system, computer simulations of mathematical models and the concepts of nonlinear dynamical systems theory provide a systematic way to describe various events of a biological process in a simple manner.

There has been a long tradition of studying the evolution of dynamical system states as the reaction rate equations (in the form of ordinary differential equations), but in the presence of very low number of molecules, these reactions do not follow a continuous route but are discrete in nature. This chapter tries to review the key concepts in modelling a biological system in both approaches, deterministic as well as stochastic.

In this chapter we discuss the design aspects and numerical simulation techniques of the models of biochemical reaction networks, which is organised as follows: In Sect. 3.2, the basics of chemical kinetics is presented, and in Sect. 3.3, Michaelis–Menten kinetics and Hill equation formalisms are described. Deterministic methods of solving coupled nonlinear equations are explained in Sect. 3.4. Section 3.5 provides an outline of stochasticity in biological systems and Gillespie's method. Chapter concludes with a brief discussion on various standards and softwares used in systems biology in Sect. 3.6.

3.2 Chemical Kinetics

Chemical kinetics provides a formal way to study, *how fast* the amount of reactant and product change during a reaction. **Law of Mass Action** states that the rate of reaction depends upon the molecular concentrations of the reactants. Since concentrations fall with time, therefore rate changes. Rate of a reaction is defined as

$$\frac{dX}{dt} = \lim_{\delta t \to 0} \frac{X(t + \delta t) - X(t)}{\delta t}$$

where, $X(t)$ is the concentration of reactant at time t.

3.2.1 Zero-order Reaction

In this type of a reaction, product is formed without any change in the concentration of the reactants, like:

$$\phi \xrightarrow{k_0} X$$

Rate of the zeroth order reaction is constant $\frac{dX}{dt} = k_0$. Examples in biological modelling include synthesis of mRNA molecules from the fixed pool of DNA.

3.2.2 *First-order Reaction*

Reaction mechanisms of irreversible first order reaction are given as following:

$$X \xrightarrow{k_1} \phi$$

$$X \xrightarrow{k_1} Y$$

$$X \xrightarrow{k_1} Y + Z$$

Rate of the reaction is directly proportional to the substrate concentration and is given by $\frac{dX}{dt} = k_1 X$. Examples from biological processes that can be modelled as this type of reaction mechanism are molecular degradation, mRNA translation into a protein, decomposition of a complex into its constituents. While molecular kinetics is zeroth order and stable, cellular kinetics is first order and unstable (Harta et al. 2001).

For a reversible reaction of first order:

$$X \underset{k_{-1}}{\overset{k_1}{\rightleftharpoons}} Y$$

the rate equation is given as,

$$\frac{dX}{dt} = -k_1[X] + k_{-1}[Y]$$

where the first term on the right is the rate of consumption of X, and the second term is the rate of formation of Y. Example can be a reversible conformational change of a protein from one confirmation to another.

3.2.3 *Second-order Reaction*

Second-order reaction is one in which either two molecules of one species or one molecule of two different species interact to form a product. Few examples are substrate binding to enzyme, ligand binding to receptors, and protein binding to other proteins or nucleic acids.

Second-order irreversible reaction is given by

$$X + Y \xrightarrow{k_2} Z$$

with the reaction rate

$$\frac{dX}{dt} = -k_2[X][Y]$$

In most cases, these reactions are reversible, so the net rate of the reaction is given by the difference between rates of forward and reverse reactions that may be first or second order reactions.

The reaction mechanism for a reversible second-order reaction is

$$X + Y \underset{k_{-2}}{\overset{k_2}{\rightleftharpoons}} Z$$

with the elementary reaction rate

$$\frac{dX}{dt} = -k_2[X][Y] + k_{-2}[Z].$$

This type of reactions are quite common in biological systems and may be observed in two scenarios: (i) two compounds X and Y synthesize a macromolecule by the breakage and synthesis of covalent bonds, or (ii) both the interacting molecules are held-together, by hydrogen bonds or other physio-chemical forces, to form a complex that has specific functionality. For example binding of a transcription factor to a DNA site, thereby forming a complex to activate or repress the transcription process.

3.2.4 Modelling of Gene Expression

A simple model of protein translation can be given by the following set of equations (Thattai and van Oudenaarden 2001). As the DNA sequence transcribing for mRNA is in fixed copy numbers so it can be assumed as a constant.

$$DNA \xrightarrow{k_1} DNA + mRNA$$

$$mRNA \xrightarrow{k_2} mRNA + Protein$$

$$mRNA \xrightarrow{k_3} \phi$$

$$Protein \xrightarrow{k_4} \phi$$

Considering that the above system follows linear kinetics, rate equations in mRNA and Protein can be given as,

$$\frac{d[mRNA]}{dt} = k_1 - k_3[mRNA]$$

$$\frac{d[Protein]}{dt} = k_2[mRNA] - k_4[Protein]$$

3.3 Michaelis–Menten Kinetics

Consider a reaction of the type

$$E + S \xrightarrow{k_0} E + P$$

If treated with the ordinary chemical kinetics scheme, velocity of this reaction will turn out to be linear, implying that one can get as much product as much substrate is increased. It is true only for the limit when number of enzyme molecules are larger than the substrate molecules. If substrates are more in numbers, they would have occupied all the active sites of the enzyme molecules. Now reaction rate will remain constant whatever the amount of substrates is added to the system.

In 1913, German biochemist Leonor Michaelis and Canadian physician Maud Menten provided a way to solve it. They proposed that each molecule of substrate will first form a complex with enzyme via a second order reversible reaction. This complex will, then, decompose into the final product and the enzyme itself.

$$S + E \underset{k_{-1}}{\overset{k_1}{\rightleftharpoons}} ES \xrightarrow{k_2} P + E$$

We want an expression for the forward rate, i.e. $k_2[ES]$, but this cannot be solved analytically.

Assume that $k_2 << k_1$ and k_{-1}, namely that there is a quasi-equilibrium of the enzyme-substrate complex ES. It assumes that the concentration of this reaction intermediate is constant, i.e. its derivative is zero.

$$d[ES]/dt = 0$$

If initially $[E] << [S]$, then this means that $E_t = [E] + [ES]$. (E_t stands for the total enzyme present in the system)

From the full ordinary differential equations, we get

$$d[ES]/dt = k_1[E][S] - (k_{-1} + k_2)[ES]$$

Substituting assumptions (1) and (2), we get

$$0 = k_1\big(Et - [ES]\big)[S] - \big(k_{-1} + k_2\big)[ES]$$

solving for [ES], we get,

$$[ES] = \frac{E_t[S]}{\frac{k_{-1}+k_2}{k_1} + [S]}$$

assuming $\frac{k_{-1}+k_2}{k_1} = K_M$, it becomes,

$$[ES] = \frac{E_t[S]}{K_M + [S]}$$

K_M is known as the Michaelis constant of this reaction.

As the velocity of a composite reaction depends upon the slow reaction, we get reaction velocity as,

$$v = k_2[ES] = \frac{k_2 E_t[S]}{K_M + [S]}$$

For the case $[S] = K_M$, velocity will be maximum $v_{max} = k_2 E_t$

$$v = \frac{v_{max}[S]}{K_M + [S]}$$

3.3.1 *Cooperativity*

Enzyme can bind more than one substrate molecules at different binding sites. In general, the binding of first substrate to the enzyme changes the rate at which second substrate will bind to it. If the binding rate of second substrate increases it is called the positive co-operativity. If the binding rate decreases, it is called as negative co-operativity.

$$S + E \underset{k_{-1}}{\overset{k_1}{\rightleftharpoons}} ES_1 \overset{k_2}{\rightarrow} P + E$$

$$S + ES_1 \underset{k_{-3}}{\overset{k_3}{\rightleftharpoons}} ES_2 \overset{k_4}{\rightarrow} P + E$$

By making the same pseudo state assumptions, one gets equations for complexes ES_1 and ES_2 of the following form

$$[ES_1] = \frac{K_2 E_t[S]}{K_1 K_2 + K_2[S] + [S]^2} \qquad and \qquad [ES_2] = \frac{E_t[S]^2}{K_1 K_2 + K_2[S] + [S]^2}$$

That gives the velocity function as following,

$$v = k_2[ES_1] + k_4[ES_2] = \frac{(k_2 K_2 + k_4[S]) E_t[S]}{K_1 K_2 + K_2[S] + [S]^2}.$$

For independent binding sites, above equation gives the binding rate that is just the twice of single binding rate.

$$v = 2\frac{k_2 E_t[S]}{K + [S]}$$

In the limit, when the binding of second S becomes infinitely fast, velocity equation gives rise to Hill equation.

Consider the case when $k_3 \rightarrow \infty$ and $k_1 \rightarrow 0$, keeping $k_1 k_3$ constant, then velocity function will be given by

$$v = \frac{\left(k_2 K_2 + k_4[S]\right) E_t[S]}{K_1 K_2 K_2[S] + [S]^2} \rightarrow \frac{v_{max} S^2}{K_m^2 + S^2}$$

This is the Hill equation with Hill coefficient 2. This heuristic equation is used to describe a co-operative reaction.

For an enzyme having n binding sites

$$v = \frac{v_{max} S^n}{K_m^2 + S^n} \tag{3.1}$$

Hill coefficient "n" provides a quantitative measure for characterising binding co-operativity. $n > 1$ corresponds to positive co-operativity, $n < 1$ corresponds to negative co-operativity and $n = 1$ means there is no co-operation.

3.4 Deterministic Modelling

Associate a single state variable X(t) with each species of the system. At any time t, collection of population of all these state variables $X_1, X_2, \ldots, X_N$ represents the state (or configuration) of this system. With the progress of time, due to interactions amongst themselves, population of constituent species will change and system will reach to another state. To understand the evolution of dynamics, one need to write a differential equation corresponding to each species, describing its change in concentration over time. If there are N species in the system, one gets a set of N coupled differential equations, like following.

$$\frac{dX_1}{dt} = f_1\left(X_1, X_2, \ldots, X_N\right)$$
$$\frac{dX_2}{dt} = f_2\left(X_1, X_2, \ldots, X_N\right)$$
$$\ldots\ldots$$
$$\frac{dX_N}{dt} = f_N\left(X_1, X_2, \ldots, X_N\right)$$

For small systems involving one or two species that follow simple functions, differential equations can be solved analytically. e. g. constant synthesis, radioactive decay, autocatalytic production of single species etc.. As the number of variables increase, functions take non-linear complex form and finding analytical solutions become difficult. For these systems, to know the dynamics of different variables over time, one need to use the methods of numerical simulations to find the approximate dynamical behaviour. In this section, we describe two widely used numerical methods for deterministic simulation of coupled differential equations.

3.4.1 Euler's Method

Consider a simple case of one variable.

$$\frac{dX}{dt} = f(X)$$

with the initial condition at time $t = 0$, $X(0) = X_0$.

Discretization using Taylor series expansion gives the following Euler's approximation

$$X(t + dt) = X(t) + dt * f\big(X(t + dt)\big) \tag{3.2}$$

When time step dt is sufficiently small, this method provides a fairly good approximation to the exact analytical solution. By using the value X_0 at t_0, one can find the value X_1 at time $t_1 = t_0 + dt$. By iterating the process over time, one can slowly build the dynamics X_2, X_3 ... etc. (Press et al. 2005).

In Euler's method the local error term is of the order of h^2 while global error is of the order h.

3.4.2 Runge Kutta Method

The most widely used method for the deterministic simulations is the Runge-Kutta fourth-order method, popularly known as RK4 method.

Unlike the Euler method this method calculates four slopes values, at the initial point, two times at the mid point and at the end point of the time step. Updates in the variable is then made, for the next time step, by adding their weighted average to their previous values (Press et al. 2005).

$$\begin{aligned}
k_1 &= h * f\big(t_n, x_n\big) \\
k_2 &= h * f\left(t_n + \frac{h}{2}, x_n + \frac{k_1}{2}\right) \\
k_3 &= h * f\left(t_n + \frac{h}{2}, x_n + \frac{k_2}{2}\right) \\
k_4 &= h * f\big(t_n + h, x_n + k_3\big) \\
x_{n+1} &= x_n + \frac{k_1 + k_2 + k_3 + k_4}{6} + O\big(h^5\big)
\end{aligned} \tag{3.3}$$

As most of the biological systems are autonomous in nature, time term does not appear in the differential equations representing change in the concentration of species in the given biological system. Therefore, while calculating various slopes in equation, including time in the argument of the function is not required. However for the system of more than one variable, one must calculate all the four slopes corresponding to every variable of the system.

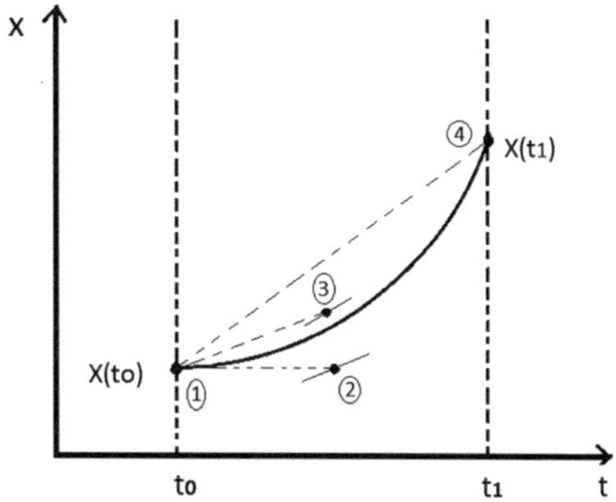

Fig. 3.1 In RK4 method, four slopes are calculated to decrease the error

In the RK4 method the local error term is of the order of h^5 while global error is of the order h^4. Due to significant decrease in errors, with respect to Euler's method, RK4 method provides a better approximation of solution Fig. 3.1.

For the rest of this section, Brusselator and Repressilator are discussed as model examples and their modelling aspects are elaborated.

3.4.3 The Brusselator

Brusselator is a theoretical model, proposed by Ilya Prigogine and Rene Lefver in 1968, to study the systems involving autocatalytic reactions and showing oscillatory dynamics. Being conceptualized at the Free University of Brussels, this oscillator model was named as Brusselator by JJ Tyson in 1976. The Belousov–Zhabotinsky reaction (BZ reaction) is one of the examples of a chemical system exhibiting oscillations due to autocatalytic reactions.

The brusselator system is given by,

$$\begin{aligned} A &\xrightarrow{k_1} X \\ 2X + Y &\xrightarrow{k_2} 3X \\ B + X &\xrightarrow{k_3} Y + C \\ X &\xrightarrow{k_4} D \end{aligned} \tag{3.4}$$

The concentrations of reactants A and B, and of products C and D are assumed to be constant. The intermediate species X and Y, whose concentrations vary over time, are of interest in this system. Second reaction is an autocatalytic reaction, in which

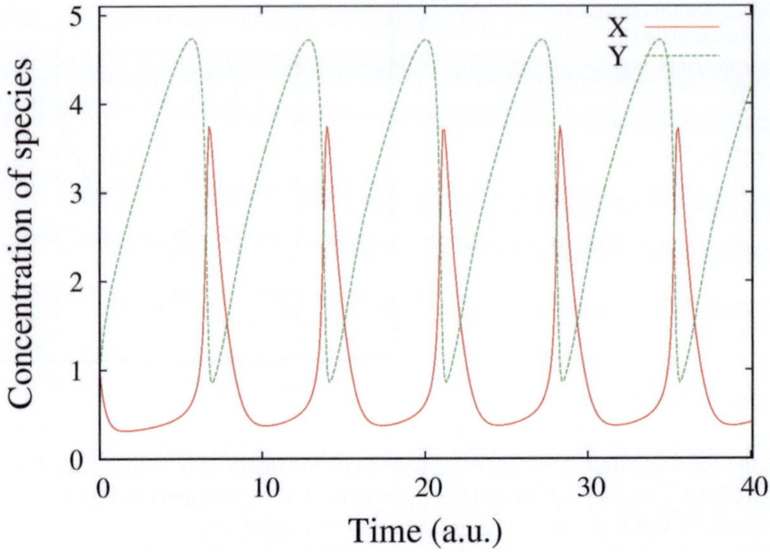

Fig. 3.2 Dynamics of brusselator system modelled via RK4 method of deterministic simulation. Initial values of X and Y are taken as 1 and all the reaction rates are also set to 1. Concentrations of reactants A and B are fixed at 1 and 3 respectively

two molecules of X produce three X molecules. The reaction rates for X and Y are given as

$$\frac{dX}{dt} = k_1[A] + k_2[X]^2[Y] - k_3[B][X] - k_4[X]$$
$$\frac{dY}{dt} = -k_2[X]^2[Y] + k_3[B][X] \tag{3.5}$$

Figure 3.2 depicts the oscillatory dynamics of this system when simulated deterministically, using RK4 method.

3.4.4 *The Repressilator*

The repressilator is a synthetic genetic oscillator (Elowitz and Leibler 2000) that uses three genes (*lacI* from *E. coli*, *tetR* from tetracycline-resistance transposon Tn10 and *cI* from λ phase) as shown in Fig. 3.3. The gene *lacI* expresses repressor protein LacI which inhibits the transcription of *tetR*. TetR represses the transcription4 of *cI* whose protein product CI in turn represses the transcription of lacI, completing the negative feedback cycle.

The deterministic model based on reaction rate equations for this system consists of six dynamical variables, three corresponding to mRNA concentrations and

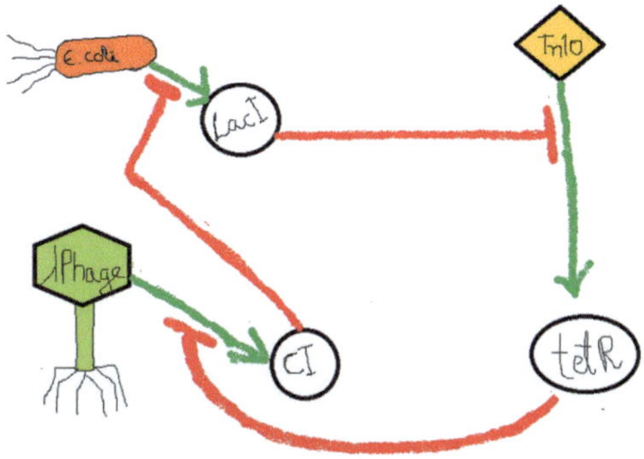

Fig. 3.3 Repressilator Circuit. *Bars* show the repression in transcription. Each gene can be in either of the two states, active or inactive, depending upon the unbinding or binding of the repressor protein to its promotor region

other three corresponding to protein concentrations. To deterministically evolve the system, following six coupled first-order differential equations have to be solved. As discussed in the previous section about the chemical kinetics, co-operativity and the Hill equation, reaction rate equations for the mRNAs and the proteins can be written as following. (For details, refer to BIOMD0000000012 hosted on Biomodels database, Li et al. 2010)

$$\frac{d[mRNA_{tetR}]}{dt} = -k_d * mRNA_{tetR} + \frac{a_1 * K_M^n}{K_M^n + LacI^n} + a_0$$
$$\frac{d[mRNA_{cI}]}{dt} = -k_d * mRNA_{cI} + \frac{a_1 * K_M^n}{K_M^n + TetR^n} + a_0$$
$$\frac{d[mRNA_{lacI}]}{dt} = -k_d * mRNA_{lacI} + \frac{a_1 * K_M^n}{K_M^n + CI^n} + a_0$$
$$\frac{d[TetR]}{dt} = k_1 * mRNA_{tetR} - k_2 * TetR$$
$$\frac{d[CI]}{dt} = k_1 * mRNA_{cI} - k_2 * CI$$
$$\frac{d[LacI]}{dt} = k_1 * mRNA_{lacI} - k_2 * LacI \tag{3.6}$$

Where mRNAi represent the mRNA and LacI, TetR and CI are representing the proteins from the corresponding gene. n is the Hill coefficient, k_d is the degradation rate of the mRNAs, k_1 is the rate of formation for protein from the corresponding mRNA and the k_2 is the degradation rate of the proteins. a_0 is a constant that determines the gene expression in the presence of saturating amount of repressor proteins, and the constant a_1 corresponds to varying amount of gene expression. Deterministic simulation of the above reaction set is shown in the Fig. 3.4. Following values of the various constants were used in the modelling:

$n = 2$, $k_d = 0.347$, $k_1 = 6.931$, $k_2 = 0.069$, $K_M = 40$, $a_0 = 0.03$, $a_1 = 29.97$

In the paper, Elowitz and Leibler used the following scaled versions of the above set of detailed equations. i and j vary according to the repressilator design, such that

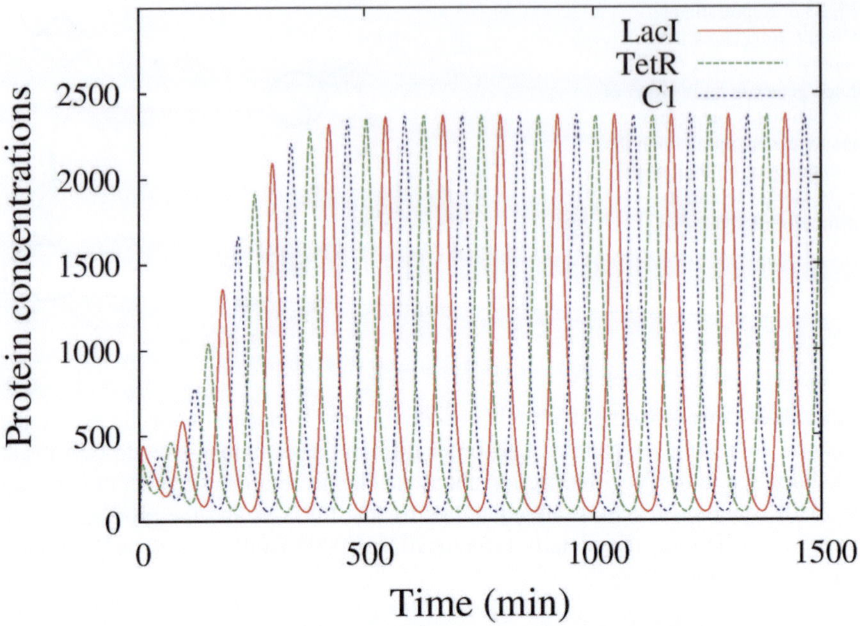

Fig. 3.4 Deterministic dynamics of the constituent proteins of the repressilator

$Protein_j$ represses the transcription of $mRNA_i$.

$$\frac{d[mRNA_i]}{dt} = -mRNA_i + \frac{\alpha}{1 + Protein_j^n} + \alpha_0$$
$$\frac{d[Protein_i]}{dt} = -\beta(Protein_i - mRNA_i) \tag{3.7}$$

3.5 Stochastic Simulation

3.5.1 Noise in Biological Systems

Gene regulation is inherently a noisy process due to stochasticity present at various steps such as promoter binding, transcription, translation, diffusion, protein degradation and so on.

Intracellular noise or intrinsic noise is the one that results because of probabilistic nature of biochemical reactions. It may arise from stochastic events during the process of gene expression, from the level of promoter-binding to mRNA translation to protein degradation. Intrinsic noise may play a deciding role in governing the dynamics of the system, if the number of reacting molecules is low. If the number of these reacting molecules are high i.e. in the thermodynamic limit (N and V tends to infinite, while N/V remains finite), dynamics of the system can very well approximated by the deterministic method.

Extracellular noise or extrinsic noise arises due to the differences between cells, either in local environment (e.g. pH, temperature etc.) or in the concentration or activity of any factor that affects gene expression (e.g. number of RNA polymerase, ribosomes etc.).

Both intracellular and extracellular noise lead to the fluctuations in the dynamcis of a single cell that causes the cell-to-cell variability (Elowitz et al.2002; Rosenfeld et al. 2005). While Noise is the main reason for various imprecisions in the genetic events, phenotype variability etc., it plays an important role in the context of regulation (Eldar and Elowitz 2010). If the noise is below some critical value, it may result in stochastic resonance (Hanggi 2002). Under certain circumstances it may cause the phenotypic switching (intrinsic transition from one state to another) (Acar et al. 2008), can induce synchrony amongst a group of cells (Zhou and Kurths 2003; Singh et al. 2010) and may even be the guiding factor for the self-organization at sub-network level (Fange and Elf 2006).

As these stochastic effects play a crucial role in regulatory networks, it is important to study the time evolution of the biological system as a discrete, stochastic process. Algorithm proposed by D. T. Gillespie in 1977 provides an exact stochastic simulation of a spatially homogeneous chemical system.

3.5.2 *Gillespie's SSA*

While the differential reaction-rates equations modelling schema to chemical kinetics assumes the time evolution of the system as the deterministic and continuous, in nature this dynamics is neither a deterministic process nor is continuous. Population changes in the species are always in discrete numbers and the occurrence of reactions is probabilistic.

The dynamics of a system (e.g. a cell) consisting N chemical species that react via M reaction channels, can be specified like following (Nandi et al. 2007),

$$X_{i1} + X_{i2} + \cdots \xrightarrow{c_j} X_{i3} + X_{i4} + \cdots$$

where the X's denote the various chemical species present in the system, and c_j's are the corresponding rate of the jth reaction channel (say R_j).

As the state of the system evolve stochastically, fluctuations originate in the species population. These fluctuations are generally termed as internal noise (van Kämpen 1981) and depend on the volume of the system and the reaction propensities of various reactions. For systems having small population of species, the strength of this noise can not be treated perturbatively.

Master equation provides a formal description of the evolution of this system in terms of configurational probabilities (Oppenheim et al. 1977). One can define the configuration, C, of a system by the number of molecules of various chemical species present, namely $C = X_1, X_2, \ldots$ ..., where X_i represents the number of molecules of ith chemical species. If $P(C,t)$ represents the probability of configuration C at time t and $\{W\}$ are the transition probabilities between two configurations, then master equation is written as,

$$\frac{dP(C,t)}{dt} = -\sum_{C'} P(C,t)W_{C \to C'} + \sum_{C} P(C',t)W_{C' \to C} \tag{3.8}$$

Gillespie's stochastic simulation algorithm (SSA) (Gillespie 1977) provides an exact method to simulate the above master equation. This algorithm assumes that the chemical system under consideration is spatially homogeneous, system volume is fixed and temperature is constant (i.e. thermal fluctuations are not considered). For any given system that consists of N chemical species reacting via M reaction channels, Gillespie's SSA attempts to estimate the answers to following two questions:

(i) When the next reaction will occur?
(ii) Which reaction channel will it follow?

To simulate a chemical reaction system, one need to characterize each reaction channel of the system and it can be done using following two quantities,

1. State-change vector, ν_{ij}, is defined as the change in X_i molecular population due to R_j reaction event.
2. Propensity function, a_j, is defined as $a_j(\mathbf{c})dt \equiv$ probability that one reaction event, R_j, will occur in next infinitesimal time interval (t, t+dt) if the system's configuration $C = \mathbf{c}$ at time t.

Without going into the detailed mathematical description of the algorithm, in the following, we present the essential steps to implement the Gillespie's stochastic simulation algorithm (SSA) (Gillespie 1977).

1. Initialize the time $t = t_0$ and the system's configuration state $C = \mathbf{c}_0$. Also initialize uniform random number generator.
2. Given the system is in state $\mathbf{c}$ at time t, calculate all $a_j(\mathbf{c})$ and also their sum $a_0(\mathbf{c})$.
3. Generate two random numbers r_1 and r_2 and using following equations, get the values of τ, time after which next reaction will occur, and j, the channel next reaction will follow.
   - $\tau = (1/a_0(\mathbf{c}))\ln(1/r_1)$
   - j = smallest value satisfying $\sum_{j'=1}^{j} a_{j'}(\mathbf{c}) > r_2 a_0(\mathbf{c})$
4. Change the system's configuration and increment the time
5. Record ($\mathbf{c}$,t). Return to step 2 or else stop.

3.5.2.1 Brusselator Dynamics Using Gillespie's SSA

Following the above described algorithm, Brusselator system as described in Eq. set 3.4 is simulated using the initial conditions as $X = 1000, Y = 2000$. Stochastic reaction constants used are $c_1 = 5, c_2 = 0.025, c_3 = 0.00005, c_4 = 5$. Figure 3.5 shows one such simulation.

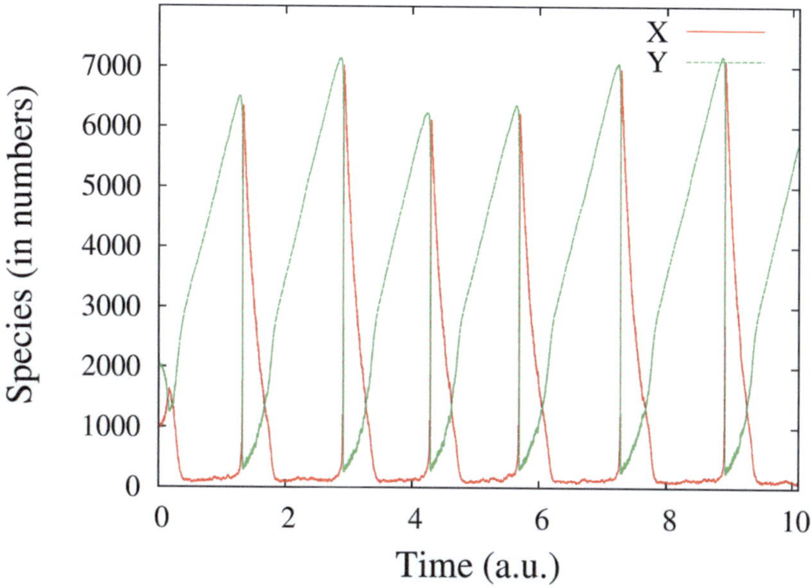

Fig. 3.5 Stochastic dynamics of two species X and Y of the Brusselator

3.5.2.2 Repressilator Dynamics Using Gillespie's SSA

To evolve the repressilator stochastically one need to consider all the reactions occuring within the system. Elowitz and Leibler considered two operator sites present in each promoter (Elowitz and Leibler 2000). In the following, we describe all the reaction steps necessary to stochastically simulate repressilator dynamics using Gillespie's SSA. UOP represents un-occupied promoter, SOP represents a promoter in which a single operator site is occupied and BOP is for the promoter in which both the operator sites are occupied.

$$UOP_{tetR} + LacI \underset{c_7}{\overset{c_1}{\rightleftharpoons}} SOP_{tetR} \ : \ SOP_{tetR} + LacI \underset{c_{10}}{\overset{c_4}{\rightleftharpoons}} BOP_{tetR}$$

$$UOP_{cI} + TetR \underset{c_8}{\overset{c_2}{\rightleftharpoons}} SOP_{cI} \ : \ SOP_{cI} + TetR \underset{c_{11}}{\overset{c_5}{\rightleftharpoons}} BOP_{cI}$$

$$UOP_{lacI} + CI \underset{c_9}{\overset{c_3}{\rightleftharpoons}} SOP_{lacI} \ : \ SOP_{lacI} + CI \underset{c_{12}}{\overset{c_6}{\rightleftharpoons}} BOP_{lacI}$$

$$UOP_{tetR} \xrightarrow{c_{13}} mRNA_{tetR} \ : \ SOP_{tetR} \xrightarrow{c_{14}} mRNA_{tetR} \ : \ BOP_{tetR} \xrightarrow{c_{15}} mRNA_{tetR}$$

$$UOP_{cI} \xrightarrow{c_{16}} mRNA_{cI} \ : \ SOP_{cI} \xrightarrow{c_{17}} mRNA_{cI} \ : \ BOP_{cI} \xrightarrow{c_{18}} mRNA_{cI}$$

$$UOP_{lacI} \xrightarrow{c_{19}} mRNA_{lacI} \ : \ SOP_{lacI} \xrightarrow{c_{20}} mRNA_{lacI} \ : \ BOP_{lacI} \xrightarrow{c_{21}} mRNA_{lacI}$$

$$mRNA_{tetR} \xrightarrow{c_{22}} TetR$$

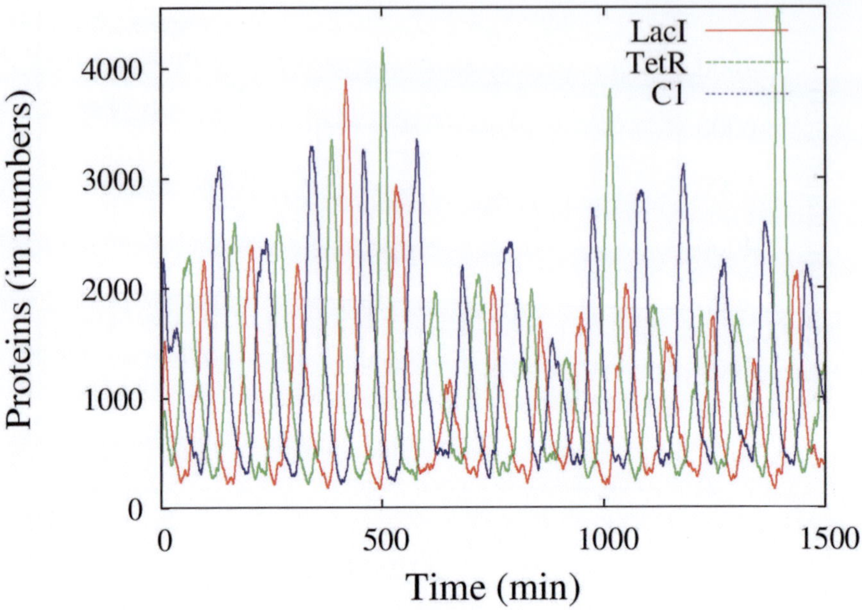

Fig. 3.6 Stochastic dynamics of three constituent proteins of repressilator

$$mRNA_{cI} \xrightarrow{c_{23}} CI$$

$$mRNA_{lacI} \xrightarrow{c_{24}} LacI$$

$$mRNA_{tetR} \xrightarrow{c_{25}} \Phi \; : \; TetR \xrightarrow{c_{28}} \Phi$$

$$mRNA_{cI} \xrightarrow{c_{26}} \Phi \; : \; CI \xrightarrow{c_{29}} \Phi$$

$$mRNA_{lacI} \xrightarrow{c_{27}} \Phi \; : \; LacI \xrightarrow{c_{30}} \Phi \qquad (3.9)$$

One simulation of the stochastic simulation of repressilator is shown in Fig. 3.6. Reaction constants used in the simulation are as following,

$c_1 = c_2 = c_3 = c_4 = c_5 = c_6 = 1\ \text{nM}^{-1}\text{s}^{-1}$
$c_7 = c_8 = c_9 = 224\ \text{s}^{-1}$
$c_{10} = c_{11} = c_{12} = 9\ \text{s}^{-1}$
$c_{13} = c_{16} = c_{19} = 0.5\ \text{s}^{-1}$
$c_{14} = c_{15} = c_{17} = c_{18} = c_{20} = c_{21} = 0.0005\ \text{s}^{-1}$
$c_{22} = c_{23} = c_{24} = 0.167\ \text{mRNA}^{-1}\text{s}^{-1}$
$c_{25} = c_{26} = c_{27} = \ln(2.0)/120\ \text{s}^{-1}$
$c_{28} = c_{29} = c_{30} = \ln(2.0)/600\ \text{s}^{-1}$

3.5.3 *Improvements and Alternative Methods for Gillespie's SSA*

One of the major limitation of the of the Gillepsie's algorithm is that it is highly computationally expensive. The Next-Reaction method proposed by M. A. Gibson and J. Bruck (2000) and the Tau-Leap method proposed by D.T. Gillespie 2001 are the two algorithm that provide improvements in the computational speed with minimal loss in the accuracy of the Gillespie's SSA.

StochSim algorithm, proposed by Morton-Firth and Bray in 1998, attempts to provide an alternative method for stochastic simulation of chemical reaction system (Morton-Firth and Bray 1998). In this algorithm every interacting particle is represented as an individual object. These objects react with another such objects according to probability distribution function that are derived from concentrations and rate constants, usually known from the experimental data.

Transcription, translation, export and other biochemical processes are not instantaneous inside a cell. It takes typically 10–20 min from transcription factor binding to the actualization of mRNA and similarly around 1–3 min in the translation of mRNA into the protein (Barrio et al. 2006). These delays can be upto 40–50 min for transcription and 8–10 min for long eukaryotic genes (Cai 2007). If delays in biochemical processes are comparable to the time scales characterizing the genetic system, these should be incorporated in the mathematical model describing that system. Readers are referred to (Barrio et al. 2006) and (Cai 2007) for the algorithms to incorporate the delay in the stochastic simulation of the biological reaction system.

3.6 Standards and Tools for Systems Modelling

Systems Biology Markup Language (SBML) has become a standard for representing biochemical reaction networks (Huck et al. 2003). It is an open-source, free to use, XML based language that provides a framework to exchange models of biological systems between different tools for simulation and analysis. Although it is developed as a language researchers working in the area of Systems Biology generally use it for exchange of data rather than writing codes for systems modelling. According to sbml.org, currently more that 250 softwares support the SBML.

CellML (http://www.cellml.org/) is another standard language for storing and exchanging the mathematical models (Cuellar et al. 2003). It is also XML based and is free to use. CellML tends to be more modular in nature while SBML is hierarchical.

Systems Biology Graphical Notation (SBGN) is a standard format for representing biological interactions graphically (Novère et al. 2009). It has now become a standard in the drawing of network diagrams. It is also unrestricted in use and is independent of the underlying operating system. Three types of diagrams cover various aspects of biological systems: (i) process description, (ii) entity relationship, and (iii) activity flow.

Biological Pathways Exchange (BioPAX) is a language being developed with an aim to provide a standard exchange format for the biological pathway data (Demir

Table 3.1 A brief overview of the frequently used softwares in the modelling of biological systems

Software	Web-address	Summary
XPPAut	http://www.math.pitt.edu / bard/xpp/xpp.html	For solving ordinary differential equations, delay differential equations, plotting phase planes and bifurcation analysis (Ermentrout 2002)
CellDesigner	http://www.celldesigner.org/	A structured diagram editor in which networks are drawn as per SBGN and are stored using SBML. Models can be simulated using SBML ODE solver or Copasi (Funahashi et al. 2003)
Copasi	http://www.copasi.org/ tiki-view_articles.php	A stand-alone program that can be used to simulate the models in SBML format using ODEs or Gillespie's SSA (Hoops et al. 2006)
E-Cell	http://www.e-cell.org/	Allows user to define functions of various proteins and interactions and then simulates the cell behaviour by numerically integrating the implicit differential equations (Tomita et al. 1999)
MCell	http://www.mcell.cnl.salk .edu/	A modelling tool for stochastic simulations of cellular signalling processes in 3-D subcellular environment using Monte-Carlo algorithms in space and time (Stiles and Bartol 2001)
StochSim	http://www.pdn.cam.ac.uk /groups/comp-cell/StochSim. html	Provides the implementation of StochSim algorithm (Morton-Firth and Bray 1998)
StochKit	http://www.engineering.ucsb .edu/ cse/StochKit/	Allows simulations via Gillespie's SSA, tau-leap method etc. Also, provides statistical analysis for the verification of stochastic behaviour (Sanft et al. 2011)
MATLAB	http://www.mathworks.in /products/matlab/	ode23 for solving ordinary differential equations and direct Method, SSA_constitutive etc. for Gillespie's SSA. Several other packages for various applications
R package	http://www.r-project.org/	deSolve package for solving ordinary differential equations and GillespieSSA package for stochastic simulation using Gillespie's direct method and many other applications

et al. 2010). Biological pathways can be represented at cellular as well as molecular level.

Table 3.1 provides a list of softwares that are most-often used in the modelling studies of biological systems.

References

Acar M, Mettetal JT, van Oudenaarden A (2008) Stochastic switching as a survival strategy in fluctuating environments. Nature Genet 40:471–475

Barrio M, Burrage K, Leier A, Tian T (2006) Oscillatory regulation of Hes1: discrete stochastic delay modelling and simulation. PLoS Comput Biol 2:1017–1030

Cai X (2007) Exact stochastic simulation of coupled chemical reactions with delays. J Chem Phys 126:124108

Cuellar AA, Lloyd CM, Nielsen PF, Bullivant DP, Nickerson DP, Hunter PJ (2003) An overview of CellML 1.1, a biological model description language. SIMULATION: Transactions of The Society for Modeling and Simulation International 79(12):740–747

Demir E et al (2010) The BioPAX community standard for pathway data sharing. Nat Biotechnol 28:935–942

Eldar A, Elowitz MB (2010) Functional roles for noise in genetic circuits. Nature 467:167–173

Elowitz MB, Leibler S (2000) A synthetic oscillatory network of transcriptional regulators. Nature 403:335–338

Elowitz MB, Levine AJ, Siggia ED, Swain PS (2002) Stochastic gene expression in a single cell. Science 297:1183–1186

Ermentrout B (2002) Simulating, analyzing and animating dynamical systems: a guide to XPPAUT for researchers and students. SIAMP, Philadelphia

Fange D, Elf J (2006) Noise-induced min phenotypes in *E. coli*. PLoS Comp Bio 2(6):637–648

Funahashi A, Tanimura N, Morohashi M, Kitano H (2003) CellDesigner: a process diagram editor for gene-regulatory and biochemical networks. BIOSILICO 1:159–162

Gibson MA, Bruck J (2000) Efficient exact stochastic simulation of chemical systems with many species and many channels. J Chem Phys 104:1876–1889

Gillespie DT (1977) Exact stochastic simulation of coupled chemical reactions. J Phys Chem 81:2340–2361

Gillespie DT (2001) Approximate accelerated stochastic simulation of chemically reacting systems. J Chem Phys 115:1716–1733

Hanggi P (2002) Stochastic resonance in biology. ChemPhysChem 3:285

Harta Y, Antebib YE, Mayoa AE, Friedmanb N, Uri Alon, U (2001) Design principles of cell circuits with paradoxical components. Proc Natl Acad Sci U S A 109:8346–8351

Hoops S, Sahle S, Gauges R, Lee C, Pahle J, Simus N, Singhal M, Xu L, Mendes P, Kummer U (2006) COPASI: a COmplex PAthway SImulator. Bioinformatics 22:3067–3074

Huck M, Finney A et al (2003) The Systems Biology Markup Language (SBML): a medium for representation and exchange of biochemical network models. Bioinformatics 19:524–531

Kitano H (2001) Foundations of systems biology. The MIT, Cambridge

Le Novère N et al (2009) Systems biology graphical notation. Nat Biotechnol 27:735–741

Li C, Donizelli M, Rodriguez N, Dharuri H, Endler L, Chelliah V, Li L, He E, Henry A, Stefan MI, Snoep JL, Hucka M, Le Novére N, Laibe C (2010) BioModels Database: an enhanced, curated and annotated resource for published quantitative kinetic models. BMC Syst Biol 4:92

Morton-Firth CJ, Bray D (1998) Predicting temporal fluctuations in an intracellular signalling pathway. J Theor Biol 192:117–128

Nandi A, Santhosh G, Singh R, Ramaswamy R (2007). Effective mechanisms for the synchronization of stochastic oscillators. Phys Rev E 76, 041136

Oppenheim I, Schuler KE, Weiss GH (1977) Stochastic processes in chemical physics: the master equation. MIT, Cambridge

Press WH, Flannery BP, Teukolsky SA, Vetterling WT (2005) Numerical recipes in C: the art of scientific computing (second edition). Cambridge university press, New Delhi

Prigogine I, Lefver R (1968) Symmetry breaking instabilities in dissipative systems. J Chem Phys 48:1695

Rosenfeld N, Young JW, Alon U, Swain PS, Elowitz MB (2005) Gene regulation at the single-cell level. Science 307:1962–1965

Sanft KR, Wu S, Roh, M, Fu J, Lim RK, Petzold LR (2011) StochKit2: software for discrete stochastic simulation of biochemical systems with events. Bioinformatics 27(17), 2457–2458 (2011)

Singh RKB, Singh V, Ramaswamy R (2010) Stochastic synchronization of circadian rhythms. J Syst Sci Complex 23(5):978–988

Stiles JR, Bartol TM (2001) Monte Carlo methods for simulating realistic synaptic microphysiology using MCell. In: De Schutter E (ed) Computational neuroscience: realistic modeling for experimentalists. CRC, Boca Raton, pp 87–127

Thattai M, van Oudenaarden A (2001) Intrinsic noise in gene regulatory networks. Proc Natl Acad Sci U S A 98:8614–8619

Tomita M et al (1999) E-CELL: software environment for whole-cell simulation. Bioinformatics 15(1):72–84

Tyson JJ (1976) The Belousov-Zhabotinskii reaction, lecture notes in biomathematics 10. Springer-Verlag, Heidelberg

van Kämpen NG (1981) Itô versus Stratonovich. J Stat Phys 24:175–187

Zhou C, Kurths J (2003) Noise-induced synchronization and coherence resonance of a Hodgkin-Huxley model of thermally sensitive neurons. Chaos 13(1):401–409

Chapter 4
In silico Identification of Eukaryotic Promoters

Venkata Rajesh Yella and Manju Bansal

Abstract The identification of promoters is essential for complete annotation of genomes and better understanding of gene regulatory networks. Experimental methods for promoter identification are costly, time-consuming and labor intensive. Hence, *in silico* methods are an attractive alternative. Computational methods for promoter prediction methods are easy, fast and can provide reliable results. A promoter prediction algorithm identifies promoter regions based on the idea that, promoter regions are different from other genomic regions in their features (sequence, context and structure). Promoter prediction algorithms are broadly classified as *ab initio*, hybrid and homology-based, depending on the information used for model design. The different approaches used in promoter prediction are briefly described here.

Keywords Promoter prediction programs · FirstEF · CpGProD · Eponine · PromoterInspector · PromPredict · EP3 · PromH

4.1 Introduction

Recent advances in genome sequencing techniques have provided a wealth of base sequence information, from which the coding and regulatory sequences need to be identified. While experimental as well as *in silico* tools are available for identifying coding sequences, locating regulatory sequences like promoters is a great challenge and the currently available methods are not very efficient. Promoter identification is essential for several reasons: annotating genomic regions for understanding genome architecture and understanding gene regulatory networks. Promoters are identified on the whole genome scale, using experimental techniques like binding assays, ChiP-chip, ChiP-seq, etc, which are costly, labor intensive and time consuming. Hence,

M. Bansal (✉) · V. R. Yella
Molecular Biophysics Unit, Indian Institute of Science,
Bengaluru, Karnataka, India
e-mail: mb@mbu.iisc.ernet.in

V. R. Yella
e-mail: yvrajesh@mbu.iisc.ernet.in

V. Singh, P. K. Dhar (eds.), *Systems and Synthetic Biology*,
DOI 10.1007/978-94-017-9514-2_4

it may not be feasible to characterize all genomes in detail experimentally. Alternatively, computational methods are available to identify promoters, as well as coding regions. There are several Promoter Prediction Programs (PPPs) available, which use different features or statistical models and identify either transcription start sites (TSSs) or promoter regions. In this chapter, we briefly describe the architecture of Eukaryotic promoters and the different kinds of promoter prediction algorithms currently available.

4.2 Eukaryotic Promoter Architecture

A promoter region is generally defined as any genomic DNA where the transcription machinery assembles and initiates transcription. The promoter region consists of protein binding regions along with the transcription start site (TSS). Promoter architecture in Prokaryotes and Eukaryotes differs in complexity. In Prokaryotes, a single RNA polymerase transcribes all types of RNAs and the promoter regions are characterized by the presence of −35 and −10 elements and in some cases the UP element as well. Overall, in the Prokaryotes, the regulatory region is located within 100 base pairs relative to the TSS. In Eukaryotes, promoter structure is more complex, with the complexity increasing from single celled yeast to mammals. Eukaryotes have several different types of RNA polymerases (usually three), with each one responsible for the production of different subsets of RNA. RNA polymerase II is responsible for synthesis of all mRNAs and is well studied compared to other RNA polymerases. Hence, only features corresponding to promoters of genes transcribed by RNA polymerase II are discussed below.

In Eukaryotes, the promoter regions are broadly classified as core promoters, proximal promoters and distal promoters. The core promoter region, where the actual basal transcription machinery assembles, is 30–100 nucleotides in length. These regions are characterized by the presence of sequence motifs such as the TATA box and the Inr element. They may also contain downstream elements like DPE, MTE (in humans) along with the associated TSS (Juven-Gershon et al. 2008; Thomas and Chiang 2006). The proximal promoter regions are the sequences located within 500 base pairs relative to the TSS and contain certain proximal promoter elements, which include the GC box, the CAAT box, *cis*-regulatory modules (CRM) (Lenhard and Sandelin 2012), etc. Distal promoter elements include enhancers, insulators and silencers. The distal promoter region does not have a well-defined length and can extend up to 10 kb from the TSS in upstream as well as downstream regions. Distal promoters interact with transcription activators to increase the rate of transcription. In vertebrates, it is known that 5 % of the genes code for specific transcription activators, which interact with proximal and distal promoter regions.

Along with the transcription factor binding elements, mammalian promoter regions also contain CpG islands. In humans, it is known that 60 % of promoters belong to the CpG island-containing class. Figure 4.1 shows a schematic representation of different promoter elements and their activators in Eukaryotes. Recent studies have

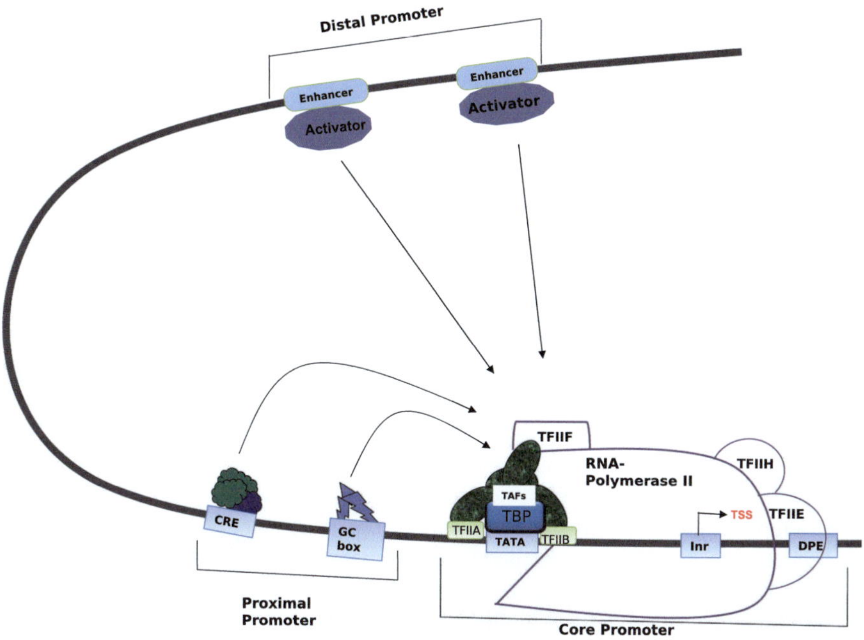

Fig. 4.1 A schematic representation of Eukaryotic RNA polymerase II promoter elements and basal transcription machinery. Promoter regions are divided into three classes, namely, *core promoters*, *proximal promoters* and *distal promoters*. Core promoter elements bind to basal transcription factors like TFIID. Proximal and distal promoter elements bind to transcription activators and increase the rate of transcription

shown that in Eukaryotes, especially in humans, each promoter is associated with many TSSs, which are spread over 50–100 nucleotides (referred to as transcriptionally active regions) (Carninci et al. 2006). Promoters can also be bidirectional (Xu et al. 2009). For detailed reviews on Eukaryotic promoters refer to Juven-Gershon et al. (2008), Lenhard and Sandelin (2012), Sandelin et al. (2007), Thomas and Chiang (2006). Recent understanding of vertebrate promoters is that though promoters differ in their motif content (with most of them lacking a consensus motifs), GC content (with lower Eukaryotes being AT rich and mammals being GC rich), some properties such as nucleosome free region and epigenetic features around TSSs are quite common (Valen and Sandelin 2011).

4.3 Experimental Methods of Promoter Identification

Experimental methods for promoter identification and characterization generally identify TSSs or DNA sequences that bind to proteins such as TFs and RNAPII (Lenhard and Sandelin 2012; Sandelin et al. 2007). Earlier methods such as nuclease protection and primer extension carry out promoter identification on a gene-by-gene basis and cannot be used for whole genome promoter identification. Current high-throughput methods measure either products from transcription (mRNA) or promoter activity in whole genome. They provide a snapshot of all transcribed regions or DNA-protein interactions in the genome for given experimental conditions. Recent advancements in promoter region identification consist of sequencing methods and hybridization methods (Sandelin et al. 2007). Sequencing methods such as RACE, 5‘-tag sequencing and 5‘-3‘ paired-end sequencing provide information about the mRNA or cDNA sequences. All these methods use reverse transcription to get cDNA. Then the cDNA is fragmented and the fragments amplified and sequenced from the 5‘-end. The sequenced fragments are mapped to the genomic DNA sequence to get information about TSS location. Hybridization methods, instead of sequencing, use short oligonucleotides to hybridize with target DNA. Two widely used methods are tiling arrays and ChiP-chip, which characterize TSSs and promoter elements respectively. Oligonucleotide tiling arrays are designed with parts of contiguous regions of sequenced genome or some times even whole genomes. They can provide information about the whole transcriptome along with the location of TSSs. The ChiP-chip method is an application of tiling arrays to identify protein bound regions of genomic DNA. ChiP-chip method uses chromatin immunoprecipitation (ChiP) to isolate DNA-bound promoter-associated proteins and then bound DNA is identified using tiling arrays (Sandelin et al. 2007).

4.4 *In silico* Methods for Promoter Identification

The computational methods for identification of promoter regions are mostly based on the basic premise that promoter regions have distinct sequences when compared to other genomic regions. Promoter Prediction Programs (PPPs) use experimentally identified promoter regions aligned with respect to TSSs, or transcription factor binding site information from databases (TRANSFAC (Wingender et al. 2000), EPD (Schmid et al. 2004) and DBTSS (Suzuki et al. 2002)) as a training dataset, to derive principles that differentiate promoters from non-promoter regions. PPPs can be broadly classified into three types based on the information used for promoter characterization. They are *ab initio*, hybrid and homology based algorithms.

Ab initio or *de novo* methods use only DNA sequence information for promoter identification. *Ab initio* methods are further classified (as shown in Fig. 4.2) as search-by-signal, search-by-content and search-by-structure algorithms based on features used for modeling (Zeng et al. 2009). Some current algorithms integrate two or more features for efficient promoter prediction.

Hybrid methods use sequence information with other accessory information such as epigenetic features, nucleosome occupancy and gene expression data. Homology based PPPs use orthologous gene information to identify promoter elements. Here, we will focus on *ab initio* PPPs in detail and also provide an introduction to other methods. Detailed information on the history, feature selection, model design and performance assessment of these PPPs is available in several excellent reviews (Abeel

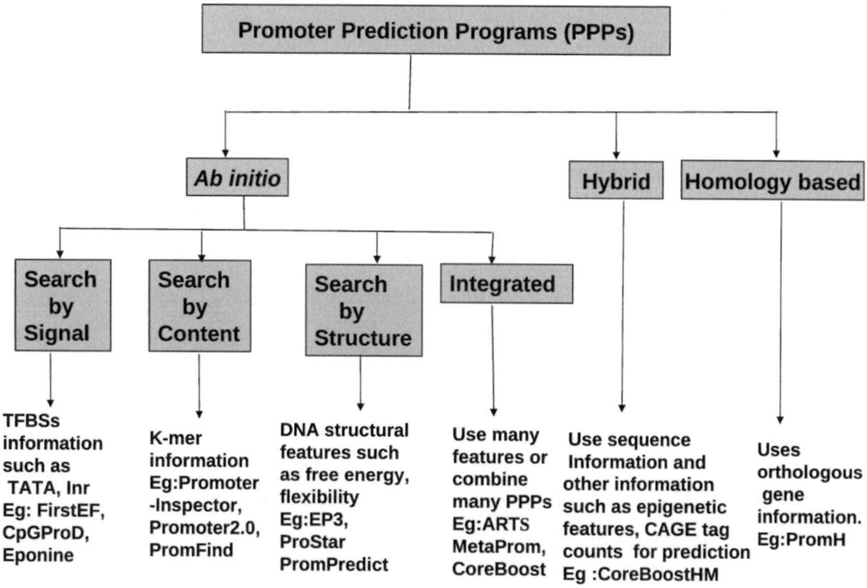

Fig. 4.2 Classification of Promoter Prediction Programs (PPPs) based on the information used for prediction

et al. 2009; Bajic et al. 2004; Bajic et al. 2006; Fickett and Hatzigeorgiou 1997; Ohler and Niemann 2001; Pedersen 1999; Zeng et al. 2009; Zeng 2011).

4.4.1 *Ab initio* Methods

Ab initio algorithms use only DNA sequence information to predict promoter regions. They identify either putative TSSs or promoter regions or in some cases, both. *Ab initio* methods may use three different kinds of features: biological signals such as core promoter elements, TFBSs or sequence context information like oligonucleotide composition or DNA structural features. Along with feature selection, they use different statistical and machine learning methods such as weight matrices (Bucher 1990), artificial neural networks (Reese 2001; Wang and Ungar 2007), Markov chains (Audic and Claverie 1997), quadratic discriminant analysis (Davuluri and Grosse 2001), genetic algorithms (Levitsky and Katokhin 2003), principle component analysis (Li et al. 2008) and kernel methods which employ support vector machines (Abeel et al. 2008b; Gangal and Sharma 2005), etc.

These algorithms search for biological signal features of core promoter elements, for example, the TATA box, initiator element (Inr), DPE (Downstream promoter Element), specific TFBSs and CpG islands (in mammals). Generally, these algorithms either predict core promoter elements or, in some cases, give the TSS position along with the distance between the binding site and the TSS. These models first derive consensus signals from experimentally identified TSSs or promoter elements. They then use different statistical methods like weight matrices, artificial neural networks and discriminant models to discriminate between promoter regions and their neighbouring sequences. Typical examples of this class of PPPs include PWMs (Bucher 1990), NNPP (Reese 2001), CpGProD (Ponger and Mouchiroud 2002), CpG-promoter (Ioshikhes and Zhang 2000), FirstEF (Davuluri and Grosse 2001) and Eponine (Down and Hubbard 2002). Search-by-signal PPPs are considered to be first generation methods. Earlier published PPPs did not use CpG-islands and their prediction efficiency was low, where as recent improved algorithms to predict promoters in mammalian genomes include use of CpG islands (Ioshikhes and Zhang 2000; Ponger and Mouchiroud 2002).

1. ***FirstEF***: FirstEF (Davuluri and Grosse 2001), which uses CpG islands, is not a pure promoter prediction program. It identifies first exons along with putative promoter regions (Bucher 1990). The developers of this PPP observed that CpG distribution in the vicinity of TSSs is bimodal, so there are two classes of first exons that exist, such as CpG containing and non-CpG containing ones. It uses a probabilistic model to identify potential first exons (splice donor sites) for both classes of promoter regions. It considers upstream promoter region and downstream splice donor sites (GT) and checks whether the intermediate region is an exon or not. The algorithm is optimized to find potential first donor sites along with CpG-related and non-CpG-related promoter regions.
2. ***CpGProD***: CpGProD (CpG Island Promoter Detection) uses CpG islands to identify mammalian promoter regions in large genomic sequences (Pedersen 1998). Although it is strictly dedicated to this particular promoter class, which corresponds to 50 % of the genes in humans, it exhibits a higher sensitivity and specificity than the other tools used for promoter prediction.
3. ***Eponine***: Eponine (Down and Hubbard 2002) is one of the best algorithms and uses sequence motif signals for locating the TSS. It combines weight matrices with discrete probability distributions of differently positioned constraints. The Eponine DNA weight matrix model for any signal is represented by the following equation.

$$\phi(i;S) = \log \sum_{j=-\infty}^{+\infty} P(j).W(a+i+j;S) \tag{4.1}$$

P(j) is a discrete probability distribution; W(x;S) is the weight matrix score, aligning the first column to position x on sequence S; a is the center position of the distribution, relative to the TSS; and i is the position of the true TSS. These PWM models were chosen for a set of four constraint elements in 599 mammalian promoter regions. They are

i. a diffuse preference for CpG enrichment downstream of the TSS.

ii. a TATAAA motif with focused distribution centered at position -30 relative to the TSS.

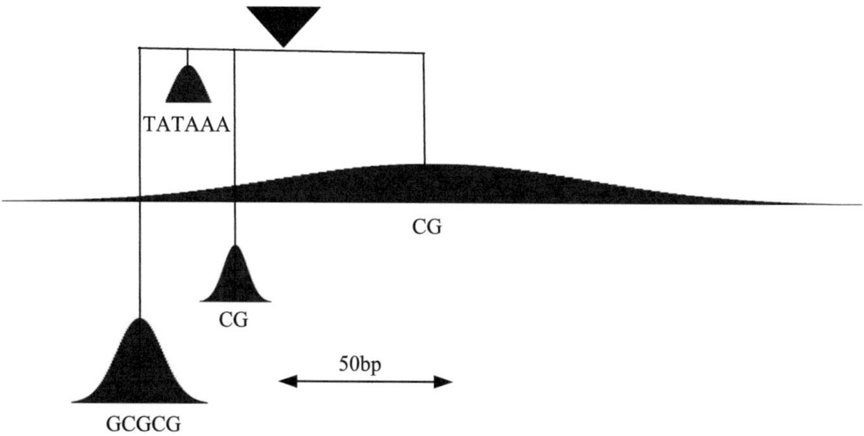

Fig. 4.3 A schematic representation of the Eponine core promoter model, showing four constraint element distributions, which were used for a weight-matrix consensus. (Down and Hubbard 2002)

iii & iv. two GC-rich matrices (GCGCG and GC) closely flanking the TATA box and positioned upstream and downstream respectively (Fig. 4.3).
To derive an efficient model, the data was trained using a relevant vector machine (RVM) algorithm with a Monte Carlo sampling process.

4.4.1.1 Search-by-content Algorithms

Search-by-content algorithms are considered to be more advanced compared to earlier approaches, as they achieve greater sensitivity and specificity. These algorithms are inspired by linguistics. The basic principle underlying all search-by-content methods is that promoter and non-promoter regions differ in their grammar and can be differentiated using certain threshold values. Context features are generally oligonucleotides represented by a set of k-tuples (or k-mers). Promoters and non-promoter regions are different in their tuple statistics. This characteristic statistical property of oligonucleotide composition can be used to discriminate promoter from non-promoter regions. Typical examples of PPPs, which use this feature, include PromFind (Hutchinson 1996), Promoter2.0 (Knudsen 1999), PromoterInspector (Scherf et al. 2000) and PCAHPR (Li et al. 2008). These classes of algorithms were shown to be more discriminative compared to search by signal algorithms. All these PPPs may differ in their statistical models but discriminate promoters from non-promoters using k-mer ($k = 2, 3, ..6$) frequencies.

1. ***PromoterInspector***: PromoterInspector uses discriminant functions to identify promoters and was considered the best PPP at one time (Scherf et al. 2000). This was trained using a brute-force algorithm to discover a set of sequence motifs overrepresented in promoter regions. Their models introduce IUPAC words by incorporating wildcards in multiple positions of an oligomer, except at the

start and end of words (AGCNGCA, AGCNNGCA). Using a certain threshold, it classifies IUPAC words into promoter related and non-promoter related candidates. From these pre-derived threshold values, PromoterInspector scans target the genome through a sliding window to identify promoter regions. The predictions are not strand-specific and do not provide information about the TSS. This tool was developed for mammalian genomes.

4.4.1.2 Search-by-property Algorithms

It is known that DNA structural features play a role in DNA-protein recognition (Pedersen 1998). The biological significance of different DNA structural properties in promoter regions is described in the accompanying chapter 13. These structural features are more conserved compared to sequence features. Search-by-property based algorithms use DNA structural features such as flexibility/bendability, curvature, base stacking and free energy to predict promoter regions. These algorithms are more recent compared to the methods described above and are based on one or more structural features to derive principles of learning. Generally, these kinds of models use simple statistical methods (Abeel et al. 2009); Rangannan and Bansal 2010) or advanced machine-learning approaches such as support vector machines (Abeel et al. 2008b) and are applicable across genomes, though genome based cut-offs may have to be specified. McPromoter (Ohler 2000), Prostar (Goni et al. 2007), EP3 (Abeel et al. 2008a), PromPredict (Rangannan and Bansal 2010) and ProSOM (Abeel et al. 2008b) are examples of these types of methods. Some of these algorithms (Abeel et al. 2008b) cluster sequences using structural profiles and use these clusters to classify unknown sequence into different promoter classes. Others use derived threshold property values to distinguish promoters from non-promoter regions (Abeel et al. 2009; Rangannan and Bansal 2010). If a given genomic sequence has a feature score in a defined window which is greater or smaller (depending on the property) than the pre-derived threshold, then it is classified as a promoter. These algorithms generally identify promoter regions rather than giving TSS positions.

1. ***PromPredict***: PromPredict (Rangannan and Bansal 2010) uses the dinucleotide free energy values obtained from differential melting stability of DNA duplex as a predictor of promoters (SantaLucia 1998). The idea behind using DNA duplex stability is that promoter regions should be less stable than neighbouring regions for easy melting at the time of transcription initiation. Compared to other structural features, stability (or base stacking) is found to be the most prevalent feature in the promoter region (Abeel et al. 2008a). Although it was developed for bacterial promoter prediction, it also works well for Eukaryotes (Morey et al. 2011). The program takes an input genome or a fragment of a sequence along with a defined window (100 or 50) and gives the start and end of predicted promoter regions as well as least stable nucleotide position. PromPredict can be applied to any genome and also to fragments of genomic sequences, independent of their size or GC composition.

2. ***EP3***:EP3 (Abeel et al. 2008a) is similar to PromPredict; it uses a base-stacking property to distinguish promoter regions from other regions. For a given sequence of DNA, it calculates inverted base-stacking values over a window size of 400 base pairs in non-overlapping fashion and calls a region as promoter when the structural feature value crosses the threshold score, which is genome specific.

4.4.1.3 Integrated Algorithms

For *ab initio* promoter prediction, it is important to choose the most discriminatory features along with the discriminative model (statistical model). Some programs integrate different features to achieve better prediction (Zeng et al. 2010). ARTS (Sonnenburg et al. 2006), CoreBoost (Zhao et al. 2007), PromoterExplorer (Xie et al. 2006) and SCS (Zeng et al. 2010) are a few examples of such new-generation algorithms. which use two or more features to predict promoters. PPPs, such as MetaProm (Wang and Ungar 2007), integrate many algorithms to predict promoters. The integrated algorithms are generally better discriminators of promoter regions, compared to the algorithms described earlier.

4.4.2 Hybrid Methods

Hybrid PPPs have been developed very recently. Along with the intrinsic features of promoter sequences, they use experimental information such as gene expression and histone modification data (Wang et al. 2012). CoreBoost_HM (Wang et al. 2009) and a method using ChIP-seq Pol-II enrichment data (Gupta et al. 2010) belong to the class of hybrid PPPs. CoreBoost_HM integrates specific histone modification profiles and DNA sequence features (core promoter elements, TFBSs, flexibility) to predict human Pol II promoters. Similarly another recent method integrates gene expression data from Chip-seq and CAGE methods (average and maximum tag counts per million) as well as DNA sequence features (10 sequence composition variables and 22 property variables) to predict promoter regions in humans. Both these methods have outperformed earlier methods in terms of sensitivity and specificity.

4.4.3 Homology Based

The idea behind using DNA sequence homology for promoter prediction is that, like coding regions, regulatory regions are also evolutionarily under selective pressure and are free of mutations, whereas non-regulatory, non-coding regions can accumulate mutations. Phylogenetic foot printing (Fickett and Wasserman 2000) is one of the methods used in this type of PPP. These methods are only applicable to identify promoter regions of orthologous genes. PromH (Solovyev and Shahmuradov 2003) is one PPP which uses orthologous gene information to predict promoter regions. PromH checks the conservation of TATA boxes in the upstream region, the conservation of nucleotide sequences around the TSS and the conservation of regulatory motifs in the upstream and downstream regions of the TSS and then uses a discriminator function to identify conserved promoter regions in pairs of orthologous genes. The program was developed specifically for testing human and rodent orthologous pairs. These kinds of algorithms are not applicable to whole genome promoter identification.

4.5 Conclusions and Future Perspectives

In silico identification of promoters is a great challenge in computational biology. A large number of promoter prediction programs are available and they differ in terms of the feature used for discriminating promoter regions from the large mass of genome sequence information. Search-by-structure or integrated algorithms appear to be promising as they are applicable to different model systems, whereas hybrid algorithms are generally efficient but are restricted to the systems for which accessory experimental information is available (such as epigenetic features and CAGE tag counts). With the rapid development of high-throughput technologies, which provide genome wide information about transcription, our understanding of promoter features is changing.

Current notion about vertebrate promoters is that while promoter regions differ in their GC and motif content, some common properties are present, such as the nucleosome free region near the TSS and epigenetic features. So, future algorithms can use this information along with other features to design new PPPs. There is always scope for the development of better algorithms based on new features and high throughput data. Most of the current PPPs are focused on promoter regions of protein coding genes. Now, with the increasing importance of non-coding RNAs in gene regulation, it is essential to analyze them. New algorithms are needed to identify promoter regions of these non-coding genes. Promoter prediction is required even if we have experimental promoter data, as we need statistical models to understand and explain promoter architecture. Up and down regulation of genes and interaction between genes is carried out through the inherent features of promoter regions. So, promoter identification and its characterization as weak or strong can serve as an important input for better understanding of systems biology of diverse organisms.

Acknowledgement MB is a recipient of the J. C. Bose National Fellowship of DST, India. We thank Rajasekaran for assistance in the preparation of Fig. 4.1.

References

Abeel T, Saeys Y, Bonnet E, Rouze P, Van de Peer Y (2008a) Generic eukaryotic core promoter prediction using structural features of DNA. Genome Res 18(2):310–323

Abeel T, Saeys Y, Rouze P, Van de Peer Y (2008b) ProSOM: core promoter prediction based on unsupervised clustering of DNA physical profiles. Bioinformatics 24(13):24–31

Abeel T, Van de Peer Y, Saeys Y (2009) Toward a gold standard for promoter prediction evaluation. Bioinformatics 25(12):i313–i320

Audic S, Claverie JM (1997) Detection of eukaryotic promoters using Markov transition matrices. Comput Chem 21(4):223–227

Bajic VB, Seah SH (2003) Dragon gene start finder: an advanced system for finding approximate locations of the start of gene transcriptional units. Genome Res 13(8):1923–1929

Bajic VB, Seah SH, Chong A, Zhang G, Koh JL, Brusic V (2002) Dragon Promoter Finder: recognition of vertebrate RNA polymerase II promoters. Bioinformatics 18(1):198–199

Bajic VB, Tan SL, Suzuki Y, Sugano S (2004) Promoter prediction analysis on the whole human genome. Nat Biotechnol 22(11):1467–1473

Bajic VB, Brent MR, Brown RH, Frankish A, Harrow J, Ohler U, Solovyev VV, Tan SL (2006) Performance assessment of promoter predictions on ENCODE regions in the EGASP experiment. Genome Biol 7(Suppl 1):1–13

Bucher P (1990) Weight matrix descriptions of four eukaryotic RNA polymerase II promoter elements derived from 502 unrelated promoter sequences. J Mol Biol 212(4):563–578

Carninci P, Sandelin A, Lenhard B, Katayama S, Shimokawa K, Ponjavic J, Semple CA, Taylor MS, Engstrom PG, Frith MC, Forrest AR, Alkema WB, Tan SL, Plessy C, Kodzius R, Ravasi T, Kasukawa T, Fukuda S, Kanamori-Katayama M, Kitazume Y, Kawaji H, Kai C, Nakamura M, Konno H, Nakano K, Mottagui-Tabar S, Arner P, Chesi A, Gustincich S, Persichetti F, Suzuki H, Grimmond SM, Wells CA, Orlando V, Wahlestedt C, Liu ET, Harbers M, Kawai J, Bajic VB, Hume DA, Hayashizaki Y (2006) Genome-wide analysis of mammalian promoter architecture and evolution. Nat Genet 38(6):626–635

Davuluri RV, Grosse I, Zhang MQ (2001) Computational identification of promoters and first exons in the human genome. Nat Genet 29(4):412–417

Down TA, Hubbard TJ (2002) Computational detection and location of transcription start sites in mammalian genomic DNA. Genome Res 12(3):458–461

Fickett JW, Hatzigeorgiou AG (1997) Eukaryotic promoter recognition. Genome Res 7(9):861–878

Fickett JW, Wasserman WW (2000) Discovery and modeling of transcriptional regulatory regions. Curr Opin Biotechnol 11(1):19–24

Gangal R, Sharma P (2005) Human pol II promoter prediction: time series descriptors and machine learning. Nucleic Acids Res 33(4):1332–1336

Goni JR, Perez A, Torrents D, Orozco M (2007) Determining promoter location based on DNA structure first-principles calculations. Genome Biol 8(12):R263

Gupta R, Wikramasinghe P, Bhattacharyya A, Perez FA, Pal S, Davuluri RV (2010) Annotation of gene promoters by integrative data-mining of ChIP-seq Pol-II enrichment data. BMC Bioinformatics 11Suppl 1:S65

Hutchinson GB (1996) The prediction of vertebrate promoter regions using differential hexamer frequency analysis. Comput Appl Biosci 12(5):391–398

Ioshikhes IP, Zhang MQ (2000) Large-scale human promoter mapping using CpG islands. Nat Genet 26(1):61–63

Juven-Gershon T, Hsu JY, Theisen JW, Kadonaga JT (2008) The RNA polymerase II core promoter—the gateway to transcription. Curr Opin Cell Biol 20(3):253–259

Knudsen S (1999) Promoter2.0: for the recognition of PolII promoter sequences. Bioinformatics 15(5):356–361

Lenhard B, Sandelin A, Carninci P (2012) Metazoan promoters: emerging characteristics and insights into transcriptional regulation. Nat Rev Genet 13(4):233–245

Levitsky VG, Katokhin AV (2003) Recognition of eukaryotic promoters using a genetic algorithm based on iterative discriminant analysis. In Silico Biol 3(1-2):81–87

Li X, Zeng J, Yan H (2008) PCA-HPR: a principle component analysis model for human promoter recognition. Bioinformation 2(9):373–378

Morey C, Mookherjee S, Rajasekaran G, Bansal M (2011) DNA free energy-based promoter prediction and comparative analysis of Arabidopsis and rice genomes. Plant Physiol 156(3):1300–1315

Ohler U (2000) Promoter prediction on a genomic scale—the Adh experience. Genome Res 10(4):539–542

Ohler U, Niemann H (2001) Identification and analysis of eukaryotic promoters: recent computational approaches. Trends Genet 17(2):56–60

Ohler U, Liao GC, Niemann H, Rubin GM (2002) Computational analysis of core promoters in the Drosophila genome. Genome Biol 3(12):RESEARCH0087

Pedersen AG, Baldi P, Chauvin Y, Brunak S (1998) DNA structure in human RNA polymerase II promoters. J Mol Biol 281(4):663–673

Pedersen AG, Baldi P, Chauvin Y, Brunak S (1999) The biology of eukaryotic promoter prediction—a review. Comput Chem 23(3–4):191–207

Ponger L, Mouchiroud D (2002) CpGProD: identifying CpG islands associated with transcription start sites in large genomic mammalian sequences. Bioinformatics 18(4):631–633

Prestridge DS (1995) Predicting Pol II promoter sequences using transcription factor binding sites. J Mol Biol 249(5):923–932

Rangannan V, Bansal M (2010) High-quality annotation of promoter regions for 913 bacterial genomes. Bioinformatics 26(24):3043–3050

Reese MG (2001) Application of a time-delay neural network to promoter annotation in the Drosophila melanogaster genome. Comput Chem 26(1):51–56

Sandelin A, Carninci P, Lenhard B, Ponjavic J, Hayashizaki Y, Hume DA (2007) Mammalian RNA polymerase II core promoters: insights from genome-wide studies. Nat Rev Genet 8(6):424–436

SantaLucia J (1998) A unified view of polymer, dumbbell, and oligonucleotide DNA nearest-neighbor thermodynamics. Proc Natl Acad Sci USA 95(4):1460–1465

Scherf M, Klingenhoff A, Werner T (2000) Highly specific localization of promoter regions in large genomic sequences by PromoterInspector: a novel context analysis approach. J Mol Biol 297(3):599–606

Schmid CD, Praz V, Delorenzi M, Perier R, Bucher P (2004) The Eukaryotic Promoter Database EPD: the impact of in silico primer extension. Nucleic Acids Res 32(Database issue):D82–D85

Solovyev VV, Shahmuradov IA (2003) PromH: Promoters identification using orthologous genomic sequences. Nucleic Acids Res 31(13):3540–3545

Sonnenburg S, Zien A, Ratsch G (2006) ARTS: accurate recognition of transcription starts in human. Bioinformatics 22(14):e472–e480

Suzuki Y, Yamashita R, Nakai K, Sugano S (2002) DBTSS: dataBase of human Transcriptional Start Sites and full-length cDNAs. Nucleic Acids Res 30(1):328–331

Thomas MC, Chiang CM (2006) The general transcription machinery and general cofactors. Crit Rev Biochem Mol Biol 41(3):105–178

Valen E, Sandelin A (2011) Genomic and chromatin signals underlying transcription start-site selection. Trends Genet 27(11):475–485

Wang J, Ungar LH, Tseng H, Hannenhalli S (2007) MetaProm: a neural network based meta-predictor for alternative human promoter prediction. BMC Genomics 8:374

Wang J, Ma C, Zhou D, Zhang L, Zhou Y (2012) Accurately predicting transcription start sites using logitlinear model and local oligonucleotide frequencies. In: Bio-Inspired Computing and Applications, pp 107–114

Wang X, Xuan Z, Zhao X, Li Y, Zhang MQ (2009) High-resolution human core-promoter prediction with CoreBoost_HM. Genome Res 19(2):266–275

Wingender E, Chen X, Hehl R, Karas H, Liebich I, Matys V, Meinhardt T, Pruss M, Reuter I, Schacherer F (2000) TRANSFAC: an integrated system for gene expression regulation. Nucleic Acids Res 28(1):316–319

Xie X, Wu S, Lam KM, Yan H (2006) PromoterExplorer: an effective promoter identification method based on the AdaBoost algorithm. Bioinformatics 22(22):2722–2728

Xu Z, Wei W, Gagneur J, Perocchi F, Clauder-Munster S, Camblong J, Guffanti E, Stutz F, Huber W, Steinmetz LM (2009) Bidirectional promoters generate pervasive transcription in yeast. Nature 457(7232):1033–1037

Zeng J, Zhu S, Yan H (2009) Towards accurate human promoter recognition: a review of currently used sequence features and classification methods. Brief Bioinformatics 10(5): 498–508

Zeng J, Zhao XY, Cao XQ, Yan H (2010) SCS: signal, context, and structure features for genome-wide human promoter recognition. IEEE/ACM Trans Comput Biol Bioinform 7(3):550–562

Zhang MQ (2011) Computational promoter prediction in a vertebrate genome. In: Handbook of Statistical Bioinformatics, pp 73–85

Zhao X, Xuan Z, Zhang MQ (2007) Boosting with stumps for predicting transcription start sites. Genome Biol 8(2):R17

Chapter 5
Hill Equation in Modeling Transcriptional Regulation

Silpa Bhaskaran, Umesh P. and Achuthsankar S. Nair

Abstract Quantitative analysis of the dynamics in cellular systems is a key aspect of systems biology. Gene regulatory networks, especially transcriptional regulatory network are studied widely by the community with such a focus. Mathematical models of gene regulatory networks are developed for understanding the dynamics by quantifying the interaction between the regulatory components. Hill equation is accepted as a quite useful by means of modeling the regulatory functions of transcriptional regulatory network. Even though its application in this scenario is constrained, the foundation upon the basic enzyme kinetics and simplicity makes Hill equation a well-accepted model for transcriptional regulatory interactions. In this chapter we give an account on the role of Hill equation in modeling transcriptional regulatory interactions mediated by the transcription factors. The Law of Mass Action and the Michaelis- Menten Kinetics is illustrated to provide a background picture. The chapter sketches out the modeling of the gene input functions in transcriptional regulatory network based on the actions of transcription factors. The feasibilities and limitations of Hill equation for modeling the transcriptional regulatory interactions is also discussed in the chapter.

Keywords Hill equation · Transcriptional regulatory network · Transcription factors · Cooperativity

5.1 Introduction

Transcriptional regulatory network can be considered as the principal gene regulatory network as gene expression is regulated mainly at the level of transcription (Barberis and Petrascheck 2003). Initiation of the transcription process is regulated by this network which ultimately influences the production of basic building blocks of life, the proteins. Transcriptional regulatory network determines the rate of production of each protein required by the cell. Mathematical models of transcriptional regulatory network often describe the effect of the binding of transcription factors

Umesh P. (✉) · S. Bhaskaran · A. S. Nair
Dept. of Computational Biology and Bioinformatics, University of Kerala, Kerala, India
e-mail: toumesh@gmail.com

V. Singh, P. K. Dhar (eds.), *Systems and Synthetic Biology*,
DOI 10.1007/978-94-017-9514-2_5

(regulatory proteins- activator and repressor) upon the regulatory region of the gene, which in turn directs the rate of transcription of that particular gene. One of the simple and popular model among them is the Hill equation. Hill equation accounts for the cooperative binding of the transcription factors which is often observed in biological systems for achieving maximal binding affinity. Hill equation is considered as a suitable formalism for modeling the functions of transcriptional regulatory network as it exhibits many required characteristics that are experimentally observed (Santillan 2008). In this chapter a brief account on Hill equation is given, based on the transcriptional regulation.

The chapter is organized as follows: Section 2 gives an overview of the transcriptional regulatory network and the regulation mechanism carried out by the transcription factors. Section 3 discusses the modeling of gene input functions using Hill equation based on the law of mass action and Michelis- Menten kinetics. A detailed illustration on the derivation of Hill equation for modeling the binding of transcription factors to the regulatory region is also included. The final section is a discussion on the role of Hill equation in modeling the transcriptional regulatory network along with its capabilities and limitations.

5.2 Transcriptional Regulation and Transcriptional Regulatory Network

Transcription is the process of synthesis of mRNA from the DNA. The process is initiated by the binding of RNA Polymerase to specific region in DNA called promoter, which generates its complementary, single stranded mRNA. Promoter lies in the upstream of the gene region that codes for the protein. Binding of the enzyme to the promoter is regulated by specific proteins called transcription factors. The gene that encodes this transcription factor proteins will be regulated by other transcription factors which are encoded by other genes which in turn is regulated by some other transcription factors and so on. This chain of regulatory interactions together constitutes the transcriptional regulatory network (Fig. 5.1).

5.2.1 Transcription Factorand Binding Site

The transcription factor and its binding site assembled with the gene constitute the elements or components of the transcriptional regulatory network. Genes transcribe mRNA while the transcription factors or the regulatory proteins (regulators) regulate the protein synthesis by binding to the regulatory regions in DNA. Transcription factors are of two kinds: activators and repressors. The binding of the transcription factors to the regulatory region influences (promotes if transcription factor is activator, blocks if it is repressor) the binding of RNA Polymerase enzyme to the initiation site in promoter and thereby the gene expression also.

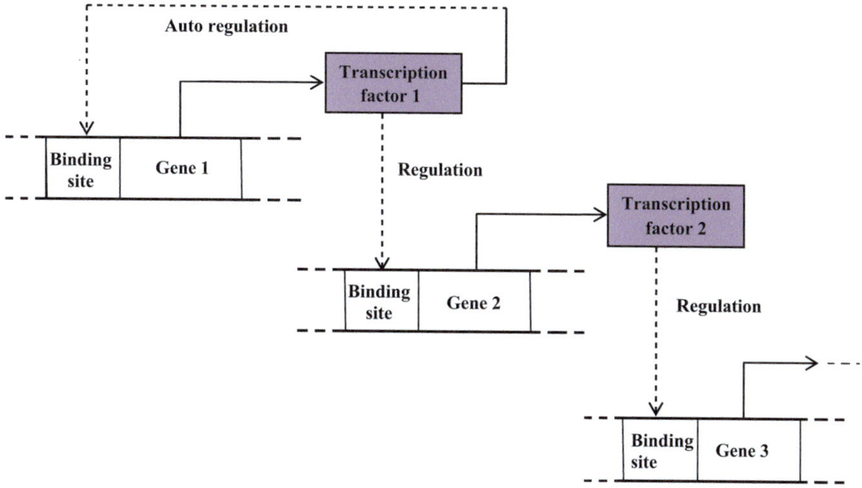

Fig. 5.1 Representation of transcriptional regulatory network

5.2.1.1 Binding of Repressor

Repressor binds to the operator sequence and prevents the RNA Polymerase from binding to the transcription initiation site in the promoter. Thus the transcription process is repressed, and also the gene expression. In certain cases, a specific ligand molecule called inducer binds to the repressor which prevents the binding of repressor or causes the bound repressor to release from the regulatory region. Thus the gene is expressed and this process of increased expression is called induction (Slonczewski and Foster 2009).

Figure 5.2 shows the effect of an inducer in the transcription process. In Figure 5.2a, the inducer inhibits the repressor from being bound to the binding site and so transcription is activated. In Fig. 5.2b, the inducer left the repressor, so it could bind to the site and thus transcription is repressed.

Some other repressors require a small ligand called co-repressor for making efficient binding to the regulatory site (Slonczewski and Foster 2009). With the influence of co-repressor, the repressor protein is able to bind the operator effectively and repress gene expression (Fig. 5.3a). When this co-repressor is released from the repressor, the repressor will not be able to bind to its binding site. Thus the gene gets expressed. This process is called de-repression (Fig. 5.3b).

5.2.1.2 Binding of Activator

Activator protein binds to the operator sequence and activates the transcription process, thereby enabling gene expression. Unlike the case of repressor, activators bind

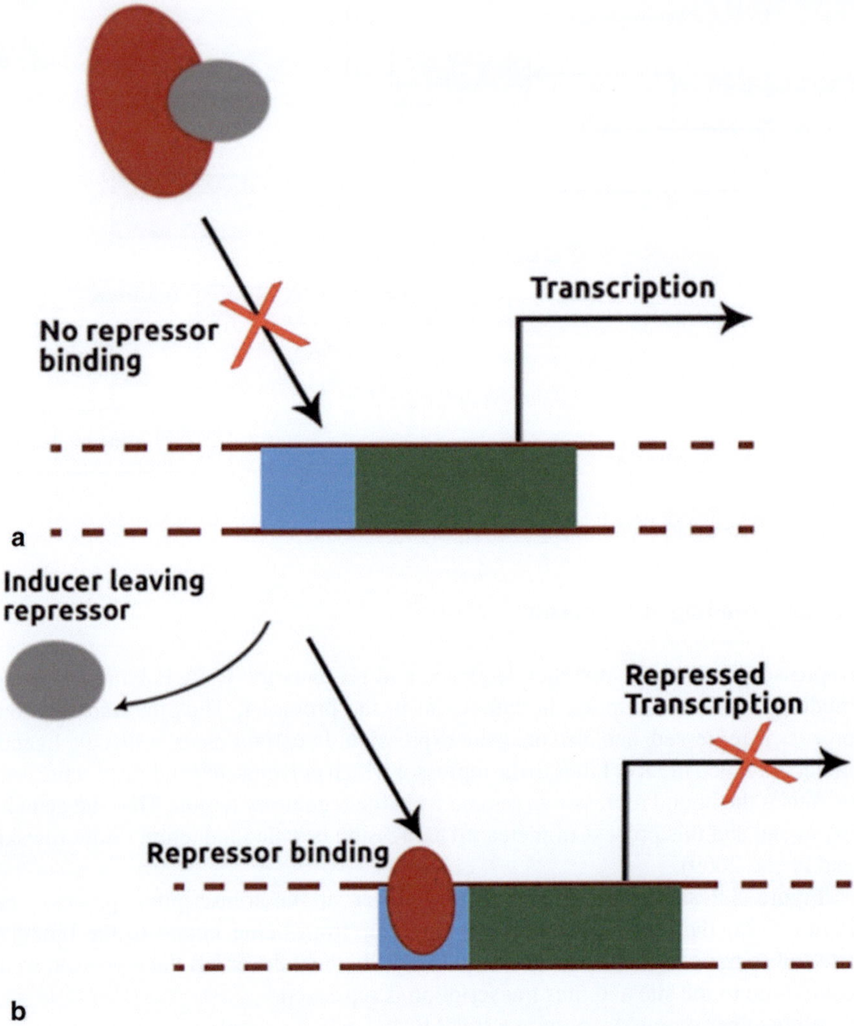

Fig. 5.2 **a** When inducer bounds to repressor. **b** When inducer leaves repressor

efficiently to the binding site only in the presence of the inducer molecule (Slonczewski and Foster 2009). The inducer- activator complex binds to the respective site and transcription is initiated (Fig. 5.4a). When the inducer leaves the activator or if inducer is absent, the activator cannot bind to the regulatory region which blocks the RNA Polymerase from initiating the transcription. So the gene expression is inhibited (Fig. 5.4b).

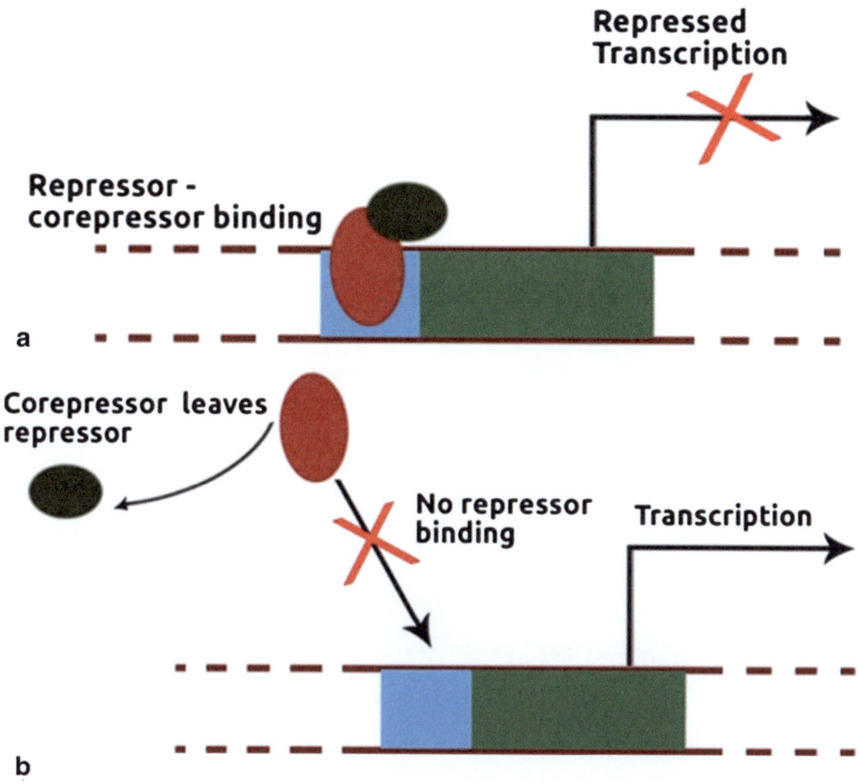

Fig. 5.3 **a** When co-repressor bounds to repressor. **b** When co-repressor leaves the repressor

5.3 Hill Equation

The Hill equation was originally formulated in 1910 by Archibald Vivian Hill in order to describe the binding of oxygen to hemoglobin based on experimental findings (Barcroft and Hill 1910). Later it was used to describe the ligand-receptor interactions in the field of biochemistry, pharmacology etc. and were also applied in mathematical modeling of gene expression in 1960s (Griffth 1968). Hill equation was derived from the Michaelis-Menten kinetics which describes the enzyme reaction mechanism based on the law of mass-action. Michaelis-Menten kinetics failed to explain the cooperativity shown by the ligands during their binding to the respective sites (Wikibooks 2013). In a protein with several binding sites, the affinity for further ligands to get bound to the protein may vary if there are already bound ligands. This happens because of the cooperativity or the interaction among the sites. Hill equation counts this cooperativity by adding one coefficient to the Michaelis- Menten kinetics, i.e. the Hill coefficient.

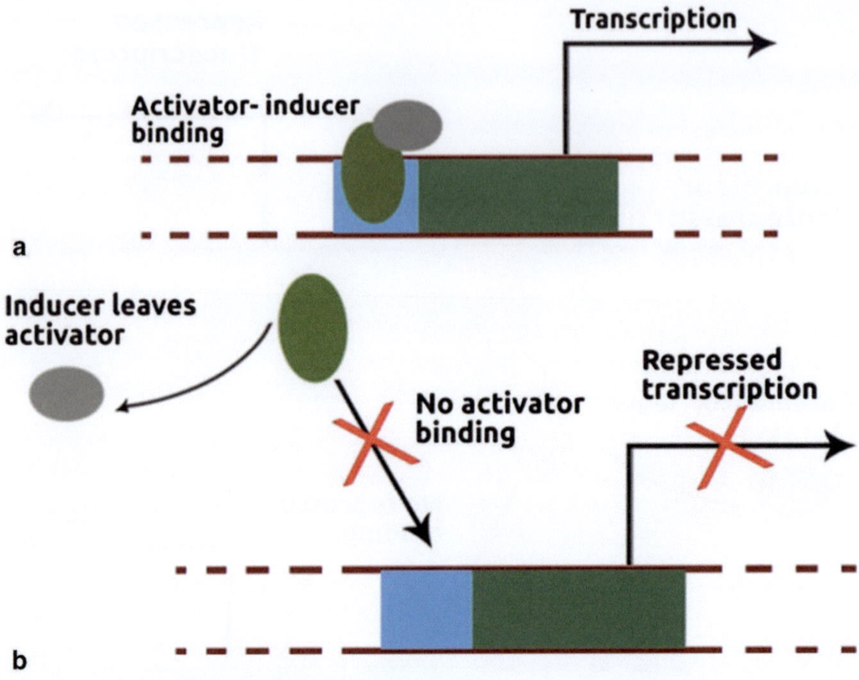

Fig. 5.4 **a** When inducer bounds the activator. **b** When inducer leaves the activator

A brief explanation on the law of mass action and the Michaelis-Menten Kinetics is explained in the subsequent sections.

5.3.1 Law of Mass Action

The mass-action law was introduced by Cato M. Guldberg and Peter Waage during the period 1864–1879 (Guldberg and Waage 1864). It states that the rate of any given chemical reaction is proportional to the product of the concentrations of the reactants. Consider reactants R1 and R2 react together to give the products P1 and P2,

$$\mathrm{aR1} + \mathrm{bR2} \leftrightharpoons \mathrm{cP1} + \mathrm{dP2}$$

where a, b, c, and d are the number of moles of the corresponding reactants and products. Then according to the law of mass action,

$$\text{Rate of forward reaction } \alpha [\mathrm{R1}]^{\mathrm{a}}.[\mathrm{R2}]^{\mathrm{b}}$$

$$\text{Rate of forward reaction } = \mathrm{K_{f.}}[\mathrm{R1}]^{\mathrm{a}}.[\mathrm{R2}]^{\mathrm{b}}$$

where Kf is the rate constant for forward reaction.

Similarly, rate of backward reaction $= K_b.\ [P1]^c.\ [P2]^d$ where K_b is the rate constant for backward reaction.

At equilibrium state, the rate of forward reaction becomes equal to the rate of backward reaction.

i.e. $K_{f.}[R1]^a.[R2]^b = K_b.[P1]^c.[P2]^d$ or $Kf/Kb = ([P1]^c.[P2]^d)/([R1]^a.[R2]^b)$

$$\text{Therefore, } K = \frac{[P1]^c[P2]^d}{[R1]^a +]R2]^b} \text{ where } K = Kf/Kb \tag{5.1}$$

K is called as the equilibrium constant or more specifically the dissociation constant. Dissociation constant measures the tendency of a larger object to fall apart into its separate subunits or components. Its value is determined by experimental data and gives an indication on the degree to which dissociation occurs. If the dissociation constant is small, then there is a high affinity between the components.

5.3.2 *Michaelis—Menten Kinetics*

Michaelis—Menten Kinetics is one of the simplest models of enzyme kinetics. The kinetics is named after German biochemist Leonor Michaelis and Canadian physician Maud Menten. They studied the enzymatic reaction mechanism in invertase that catalyzes the hydrolysis of sucrose into glucose and fructose and proposed this model (Michaelis and Menten 1913). According to Michaelis—Menten Kinetics, all enzyme-single substrate reaction mechanism can be generalized as,

$$E + S \underset{K_{-1}}{\overset{K_1}{\leftrightharpoons}} ES \overset{K_2}{\rightarrow} E + P$$

where E is the enzyme and S is the substrate. E and S react together to form the complex ES at the rate K_1 which dissociates at the rate K_{-1}. At the rate $K_{2,}$ the complex ES form the product P.

Applying the Law of Mass Action we can derive the rate of change of each of the reactants and products. Thus the rate equation for the dynamics of this reaction can be derived as (Klipp et al 2009) a set of differential equations,

$$\frac{dS}{dt} = -k_1 E.S + k_{-1} ES \tag{5.2}$$

$$\frac{dES}{dt} = k_1 E.S - (k_{-1} + k_2)ES \tag{5.3}$$

$$\frac{dE}{dt} = -k_1 E.S + (k_{-1} + k_2)ES \tag{5.4}$$

$$\frac{dP}{dt} = k_2 ES \tag{5.5}$$

If the initial concentration of the substrate is much larger than the enzyme concentration, then the concentration of the ES complex remains constant at certain state. At this state we have to consider the conservation of enzyme only.

$$\text{So, } \mathrm{E_{total}} = \mathrm{E} + \mathrm{ES} = \text{a constant or } \mathrm{E} = \mathrm{E_{total}} - \mathrm{ES} \tag{5.6}$$

At this steady state, $\frac{dES}{dt} = 0$. Substituting this in Eq. (5.2) gives,

$$K_{1.}E.S = (k_{-1} + k_2)ES \tag{5.7}$$

Substituting Eq. (5.6) will give,

$$K_1(E_{total} - ES)S = (k_{-1} + k_2)ES \tag{5.8}$$

$$K_1.E_{total}.S = K_1.\ ES.\ S + (k_{-1} + k_2)ES \tag{5.9}$$

$$K_{1.}\, E_{total.}S = ES.\ (K_1 S + K_{-1} + K_2) \tag{5.10}$$

$$ES = \frac{K_1.\ E_{total}.S}{K_1 S + K_{-1} + K_2} \tag{5.11}$$

$$ES = \frac{E_{total}.S}{S + \left(\frac{K_{-1}+K_2}{K_1}\right)} \tag{5.12}$$

The reaction rate is equal to the dissociation rate of the substrate or the formation rate of the product. So the reaction rate,

$$\frac{dP}{dt} = \frac{k_2.\ E_{total.}S}{S + \left(\frac{K_{-1}+K_2}{K_1}\right)} \tag{5.13}$$

$$\text{More Simply, } \frac{dP}{dt} = \frac{V_{max}.S}{S + K_m} \tag{5.14}$$

$$where \quad V_{\max} = k_2.\ E_{total} \quad \text{and} \quad K_m = \frac{K_{-1} + K_2}{K_1}$$

This is the expression for Michaelis- Menten kinetics. Here V_{max} is called the maximal velocity which is the maximum reaction rate that can be achieved when the enzyme is completely saturated with substrate. K_m is called the Michaelis- Menten constant which is equal to the substrate concentration at which the reaction rate is half maximal.

The binding of transcription factors to the regulatory region in promoter for regulating transcription can be considered as an enzyme-substrate reaction. The rate of the transcription process is determined by how effectively the transcription factors

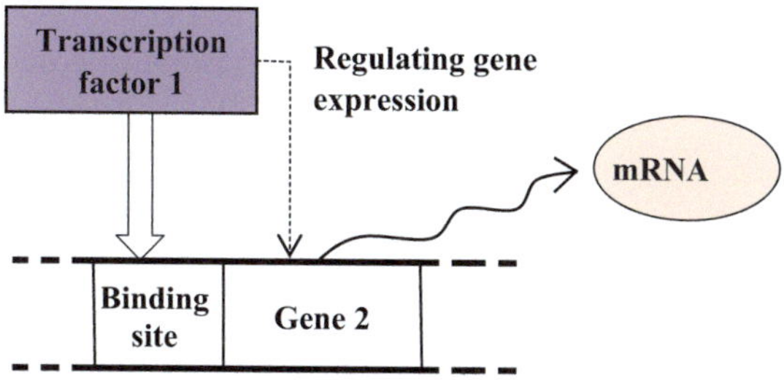

Fig. 5.5 Transcription factor activity

attach to the binding site of the gene. So when transcriptional regulation is mathematically modeled, we describe the input of the transcription factors on the transcription process. This can be explained using a mathematical function which is called as the gene input function. A gene input function describes the strength of the effect of a transcription factor on the transcription rate of that particular gene. The input function relates the input signals and the transcription rate. This input function will be an increasing function, if the transcription factor is an activator and will be a decreasing function if it is a repressor. Hill equation is regarded as a useful function that describes any real gene input functions (Alon 2007).

5.3.3 *Modeling Gene Input Functions using Hill Equation*

Now let us see how Hill equation efficiently models the gene input functions in transcriptional regulatory network. The given diagram (Fig. 5.5) depicts the transcription factor binding activity that we are explaining using Hill equation. The transcription factor (either activator or repressor) binds to the site in promoter and regulates the gene expression which eventually determines the transcription rate of mRNA.

We illustrate how Hill equation models transcription regulation in three cases: when (i) an activator is bound (ii) repressor is bound and (iii) an inducer is bound to repressor.

5.3.3.1 Binding of Repressor to Promoter

Consider the reaction,

$$X + D \underset{K_b}{\overset{K_f}{\leftrightharpoons}} [XD]$$

Here, in the forward reaction, the transcription factor protein, X binds to the binding site, D of the promoter to form the complex XD at the rate K_f. In the backward reaction XD is dissociating into X and D at the rate K_b. Transcription of the gene occurs only when X is not bound or when D is free. According to the conservation equation, the total concentration of the DNA site,

$$[D_{Tot}] = [\mathrm{D}] + [\mathrm{XD}]$$

Or

$$[\mathrm{XD}] = [\mathrm{D_{Tot}}] - [\mathrm{D}] \tag{5.15}$$

According to the law of mass action, the rate of change of concentration of [XD] can be defined as,

$$\frac{d[XD]}{dt} = K_f[X][\mathrm{D}] - K_b[XD] \tag{5.16}$$

As steady state,
$$\frac{d[XD]}{dt} = 0 \tag{5.17}$$

$$\text{So,}\quad K_f[X][\mathrm{D}] = K_b[XD] \tag{5.18}$$

$$[\mathrm{X}][\mathrm{D}] = \frac{K_b}{K_f}[XD] \tag{5.19}$$

$$[\mathrm{X}][\mathrm{D}] = K_d\,[XD] \tag{5.20}$$

where, $K_d = \frac{K_b}{K_f}$ and is called as the equilibrium constant or the dissociation constant.

$$\therefore K_d = \frac{[X][D]}{[XD]}$$

The constant K_d has units of concentration and measures the tendency for [XD] to fall apart into its two separate subunits. If K_d is small, then there is a high affinity between X and D.

Substituting Eq. (5.15) in Eq. (5.20):

$$[\mathrm{X}][\mathrm{D}] = \mathrm{K_d}([\mathrm{D_{Tot}}] - [\mathrm{D}])$$

$$\mathrm{K_d}[\mathrm{D}] + [\mathrm{X}][\mathrm{D}] = \mathrm{K_d}[\mathrm{D_{Tot}}]$$

$$[\mathrm{D}](\mathrm{K_{d+}}[\mathrm{X}]) = \mathrm{K_d}[\mathrm{D_{Tot}}]$$

$$\mathrm{K_{d+}}[\mathrm{X}] = \frac{\mathrm{K_d}[\mathrm{D_{Tot}}]}{[D]} \tag{5.21}$$

$$\frac{[D_{Tot}]}{[D]} = \frac{K_d + [X]}{K_d}$$

$$\frac{[D_{Tot}]}{[D]} = 1 + \frac{[X]}{K_d}$$

$$\frac{[D]}{[D_{Tot}]} = \frac{1}{1 + \frac{[X]}{K_d}} \tag{5.22}$$

$\frac{[D]}{[D_{Tot}]}$ is the probability that the site D is free and is a decreasing function of the concentration of the repressor X. If $[X] = K_d$, the probability for the site being free is $\frac{1}{2}$ or 50 %. Also when there is no repressor, [X] is 0 and the site will be always free and $\frac{[D]}{[D_{Tot}]}$ is 1 indicating high probability for the site being free. In such a case, an RNA polymerase can bind the site and transcribe at maximum rate. This rate of transcription from a free binding site is called maximal transcription rate, β. The rate of mRNA production or the promoter activity is β times the probability that the binding site is free (value of β ranges from approximately 10^{-4} to 1 mRNA/s).

$$\text{Promoter activity or Rate of mRNA production} = \beta . \frac{1}{1 + \frac{[X]}{K_d}} \tag{5.23}$$

So if $[X] = K_d$, the promoter activity is reduced to 50 % of its maximal transcription rate. i.e. promoter activity $= \frac{\beta}{2}$. This value of [X] required for repressing the promoter activity by 50 % of its maximal transcription rate is called the repression coefficient.

In reality, most transcription factors are composed of multiple subunits and in order to achieve maximum activity these multiple subunits cooperatively bind the binding site. Suppose there are n subunits, then the Hill equation of input function of the gene bound with repressor is,

$$\text{Promoter activity} = \beta . \frac{1}{1 + \left[\frac{[X]}{K_d}\right]^n} \tag{5.24}$$

and n is called as the Hill coefficient.

The plot (Fig. 5.6) shows the behavior of hill equation model for repressor with-varying values for n. It is clear that as the value of n goes higher the steeper the plot becomes. Also as theconcentration of the repressor is increased, the rate of transcription is decreased.

5.3.3.2 Binding of Activator to Promoter

Consider the transcription factor, X be an activator protein. As mentioned in the previous section, the total concentration of the DNA site is,

$$[D_{Tot}] = [D] + [XD]$$

Or

$$[D] = [D_{Tot}] - [XD] \tag{5.25}$$

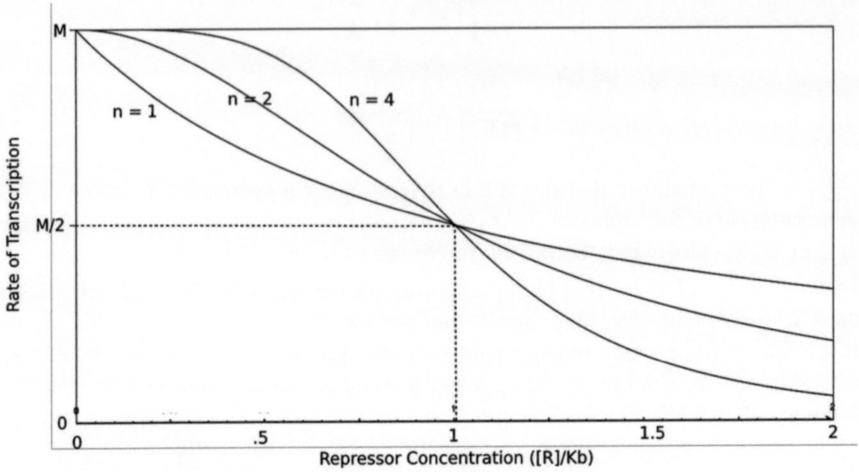

Fig. 5.6 Hill equation model of repressor with varying values for *n*. (Alon 2007)

Substituting (5.25) in (5.20):

$$[\mathrm{X}]([D_{Tot}] - [XD]) = K_d\,[XD]$$

$$[\mathrm{X}]\,[D_{Tot}] - [\mathrm{X}]\,[\mathrm{XD}] = K_d\,[XD]$$

$$[\mathrm{X}][D_{Tot}] = K_d[\mathrm{XD}] + [\mathrm{X}][\mathrm{XD}]$$

$$[\mathrm{X}][D_{Tot}] = [\mathrm{XD}](K_d + [X])$$

$$\frac{[X][D_{Tot}]}{[XD]} = K_d + [X]$$

$$\frac{[D_{Tot}]}{[XD]} = \frac{K_d + [X]}{[X]}$$

$$\frac{[XD]}{[D_{Tot}]} = \frac{[X]}{K_d + [X]}$$

$$\therefore \text{Promoter activity or Rate of mRNA production} = \beta.\ \frac{[\mathrm{X}]}{K_d + [X]} \tag{5.26}$$

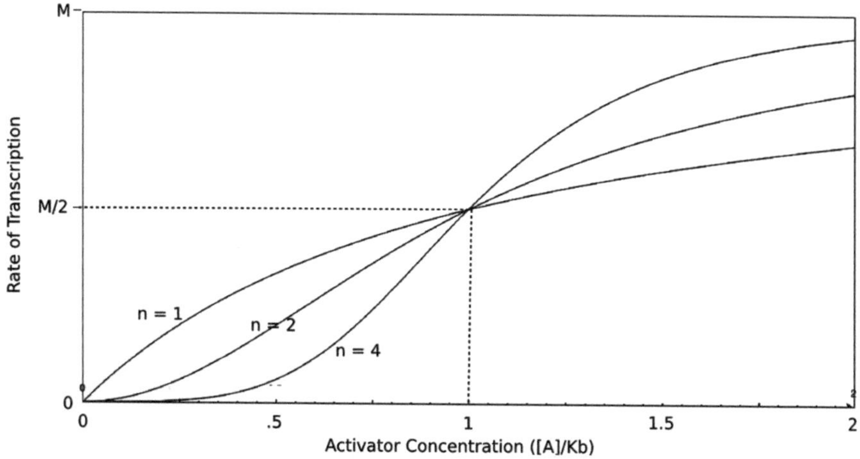

Fig. 5.7 Hill equation model of activator with varying values for n. (Alon 2007)

Thus when an activator with n subunits is bound, the Hill equation for the input function of the geneis,

$$\text{Promoter activity} = \beta . \frac{[X]^n}{K_{d+}[X]^n} \tag{5.27}$$

where n is called the Hill coefficient.

The plot (Fig. 5.7) shows the behavior of hill equation model for activator with varying values for n. Here also it is clear that as the value of n determines the steepness of the function.

5.3.3.3 Binding of Repressor to an Inducer

Let X be the repressor protein which binds with an inducer molecule, S_x to form the complex $[XS_x]$. So the total concentration of the repressor protein,

$$X_T = X + [XS_x]$$

Or

$$X = X_T - [XS_x] \tag{5.28}$$

where X is the repressor in free form. If the formation of $[XS_x]$ is at a rate k_{on} and its dissociation is in the rate k_{off}, then according to the law of mass action,

$$\frac{d[XS_x]}{dt} = K_{on}\, X.\ S_x - K_{off}[XS_x] \tag{5.29}$$

$$\text{At steady state,} \quad \frac{d[XS_x]}{dt} = 0, \text{which implies,}$$

$$\mathrm{XS_x = K_x[XS_x]} \tag{5.30}$$

$$\text{Where } \mathrm{K_x} = \frac{d_{off}}{d_{on}}$$

Substituting X with $\mathrm{X_T - [XS_x]}$ from Eq. (5.28) in Eq. (5.30):

$$[XS_x] = \frac{X_{Tot} S_x}{K_x + S_x} \tag{5.31}$$

But as $\mathrm{S_x}$ is an inducer, only the X unbound to $\mathrm{S_x}$ is active since only the unbound X can bind to the promoter binding site and repress the transcription process. So this active repressor,

$$\mathrm{X^* = X_{Tot} - [XS_x]}$$

$$\mathrm{X^*} = \mathrm{X_{Tot}} - \frac{X_{Tot} \cdot S_x}{K_x + S_x}$$

$$\mathrm{X^*} = \frac{X_{Tot}}{1 + \frac{S_x}{K_x}} \tag{5.32}$$

where $\mathrm{X^*}$ is the active repressor or the repressor unbound to inducer. Hill input equation explain this as,

$$X* = \frac{X_{Tot}}{1 + \left[\frac{[S_x]}{K_x}\right]^n} \tag{5.33}$$

5.4 Discussion

In order to achieve maximum binding affinity while binding to the promoter binding site, transcription factors usually exhibit cooperativity among themselves. Cooperativity is a biological characteristic which can be described as a variation in the binding affinity for other binding sites, caused by the binding of a ligand to its corresponding binding site in the same molecule. This cooperativity is achieved by binding as multiple subunits and not as monomers. It requires the interactions between multiple binding sites also. A remarkable feature of Hill equation model is that it takes this cooperativity into account through the Hill coefficient. Hill coefficient (here, n) is the measure of the cooperativity among the transcription factors binding to the regulatory region in the promoter. It determines the steepness of the gene input function. The larger is the value of n, the more steep-like the input function (Alon 2007). It thus represents the response of regulatory network to the transcriptional input. According to the mathematical derivation, Hill coefficient denotes the number of binding sites, but it is not often true. The fact is that Hill coefficient will always be equal to or less than the number of binding sites (Weiss 1997).

If Hill coefficient > 1 positive cooperativity results, meaning that the binding of a particular transcription factor to its binding site increases the binding affinity of other transcription factors for simultaneous binding. If Hill coefficient < 1 negative cooperativity results, meaning that the binding of that particular transcription factor to its binding site decreases the binding affinity of other transcription factors for simultaneous binding. If Hill coefficient $= 1$, it is non-cooperative, that is, the binding of the transcription factor to its corresponding site will not alter the binding affinity of other transcription factors.

Hill coefficient, n is introduced as a term that gives the number of binding sites which should always be an integer. But while fitting Hill equation to experimental data, this is a rare case. It would be accurate only when extreme positive cooperativity is present. This indicates the inability of Hill equation in providing a proper model of the real biological system. Adding regulator molecules to the equation is also a challenge. The Hill equation we discussed here considers irreversible reactions only (Reversible Hill equation has also been put forwarded recently (Hofmeyr and Cornish-Bowden 1997; Westermark et al. 2004)). Also it couldn't easily model the multi-reactant systems. Mechanistically, Hill equation is based on the concept that all binding sites for a given ligand are bound at once. This is unrealistic (Sauro 2011). It is impossible to infer all the details of DNA-transcription factor interactions from this model. Even if the regulatory interaction details are available it is difficult to derive the best fitting Hill function parameters simply from it. However Hill equation is used widely for modeling the transcriptional regulation as it effectively describes its sigmoid behavior and often fit to experimental data quite well (Sauro 2011).

Acknowledgment First author acknowledges the Research Fellowship from Kerala State Council for Science, Technology and Environment (KSCSTE).

References

Alon U (2007) An introduction to systems biology: design principles of biological circuits. Chapman & Hall/CRC, Boca Raton

Barberis A, Petrascheck, M (2003) Transcription activation in eukaryotic cells. Encyclopedia of life sciences. doi:10.1038/npg.els.0003303

Barcroft J, Hill AV (1910) The possible effects of the aggregation of the molecules of hemoglobin on its dissociation curves. J Physiol 40:iv–vii

Griffth JS (1968) Mathematics of cellular control processes I. Negative feedback to one gene. J Theor Biol 20:209–216

Guldberg CM, Waage P (1864) Studies concerning affinity. J Chem Educ 63:1044–1047 (Forhandlinger: Videnskabs-Selskabeti Christiana, 35. English edition: Guldberg CM, Waage P (1986) Studies Concerning Affinity (trans: Abrash HI))

Hofmeyr JHS, Cornish-Bowden H (1997) The reversible Hill equation: how to incorporate cooperative enzymes into metabolic models. Computer Appl Biosci 13(4):377–385

Klipp E, Liebermeister W, Wierling C et al (2009) Systems biology-a textbook. Wiley, Weinheim

Michaelis L, Menten ML (1913) Die kinetikder invertinwirkung. Biochem Z 49:333–369 Biochemistry 50(39):8264–8269 (English edition: Michaelis L, Menten ML (2011) The original Michaelis constant: translation of the 1913 Michaelis–Menten paper (trans: Goody RS, Johnson KA))

Santillan M (2008) On the use of Hill functions in mathematical models of gene regulatory networks. Math Model Nat Phenom 3(2):85–97

Sauro HM (2011) Enzyme kinetics for systems biology. Future Skill Software and Ambrosius Publishing, Lexington

Slonczewski J, Foster JW (2009) Microbiology: an evolving science. W. W Norton & Co., New York

Weiss JN (1997) The Hill equation revisited: uses and misuses. FASEB J 11(11):835–841

Westermark PO, Hellgren-Kotaleski J, Lansner A (2004) Derivation of a reversible Hill equation with modifiers affecting catalytic properties. WSEAS Trans Biol Med 1:91–98

Wikibooks (2013). Structural Biochemistry/Protein function/Binding Sites/Cooperativity. Wikibooks, The free textbook project. http://en.wikibooks.org/w/index.php?title=Structural_Biochemistry/Protein_function/Binding_Sites/Cooperativity&oldid=2559676. Accessed 18 Oct 2013

Chapter 6
Molecular Modeling

Dr. Preethi Badrinarayan, Chinmayee Choudhury and Prof. G. Narahari Sastry

Abstract Recent advances in theoretical methods based on quantum mechanics and classical mechanics with visualization tools have played a very important role in chemistry and biology. Further, computers have a profound influence on the way we do science in the last few decades. The current chapter provides a preliminary exposure to a range of molecular modeling approaches applicable to small to medium sized molecules to proteins. Recent advances in the theoretical and computational methodologies which are aimed to treat large molecules are described. Emphasis is given on methods of computer aided drug design. These are preceded by a simple introduction to quantum mechanics, classical mechanics and molecular dynamics. A major area in modelling biomolecules is a proper quantitative treatment of non-bonded interactions. The importance of understanding various non-bonded interactions is highlighted. The role of these non-bonded interactions which determines the biological structure and functions is described. Thus, the current chapter provides a brief overview of computational methods applied to biomolecules.

Keywords Quantum mechanics · Molecular dynamics · QM/MM · Coarse grained simulations · Computer aided molecular design · Dokcing · Quantitative structure activity relationships · Virtual screening · Multi-scale modeling

List of Abbreviations

ANN	Partial least square, artificial neural network
CADD	Computer aided drug design
DFT	Density function theory
FBDD	Fragment based drug design
FEP	Free energy perturbation
F-value	Fisher statistic
FTA	Fragment tailoring approach

Prof. G. N. Sastry (✉) · Dr. P. Badrinarayan · C. Choudhury
Centre for Molecular Modeling, CSIR-Indian Institute of Chemical Technology, Tarnaka, Hyderabad 500 607, India
e-mail: gnsastry@gmail.com

V. Singh, P. K. Dhar (eds.), *Systems and Synthetic Biology*,
DOI 10.1007/978-94-017-9514-2_6

GFA	Genetic function approximation
HTS	High-throughput screening
Ki	Inhibitory constant
LBVS	Ligand-based virtual screening
MCSCF	Multi-configurational self-consistent field
MD	Molecular dynamics
MLR	Multi-linear regression
MM	Molecular mechanics
PCR	Principal component regression
PDE	Phosphodiesterase
QM	Quantum mechanics
QM/MM	Quantum mechanics/molecular mechanics
QSAR	Quantity structure activity relationship
SBVS	Structure-based virtual screening
SCF	Self-consistent field
TI	Thermodynamic integration
VS	Virtual screening

6.1 Introduction

Models are essential to comprehend the incomprehensible reality. While biology and chemistry are empirical sciences, which are largely based on observations and observing, it has become necessary to develop the understanding at atomistic level. Therefore, modeling molecules has become an indispensable tool to not only complement the structural and spectroscopic attempts to characterise molecules, but also to provide the basic tool for imaging molecular action. Thus, tools for molecular visualization on one hand and computational approaches based on rigorous theory have occupied central stage in the computer age. When dealing with small molecules, the objective has been to obtain highly reliable properties often comparable with those of experiments. Quantum mechanical treatment of atoms and molecules thus provide the most fundamental dimensions to treat small molecules (Cramer 2004; Leach 2001; Hinchliffe 2003). In natural sciences there are different kinds of theories. While theory has been a strong and integral part of physics, the science of chemistry has grown with experimentation and biology with observations. In addition to theory and experiment, the entry of computations has provided another dimension to pursue science. While the *ab initio* computational approaches have the ability to model and obtain every possible experimental property, their applicability becomes very limited as the size of the system increases. Consequently, for macro-molecules, the application of quantum mechanical approaches is severely limited. When the time dependency in solving the Schrödinger wave equation is considered, its application is restricted still further. Also the time scales that can be probed using *ab initio* molecular dynamics (MD) are very small. Thus application of quantum chemical methods based on *ab initio* theory is practically limited to systems with very limited length and

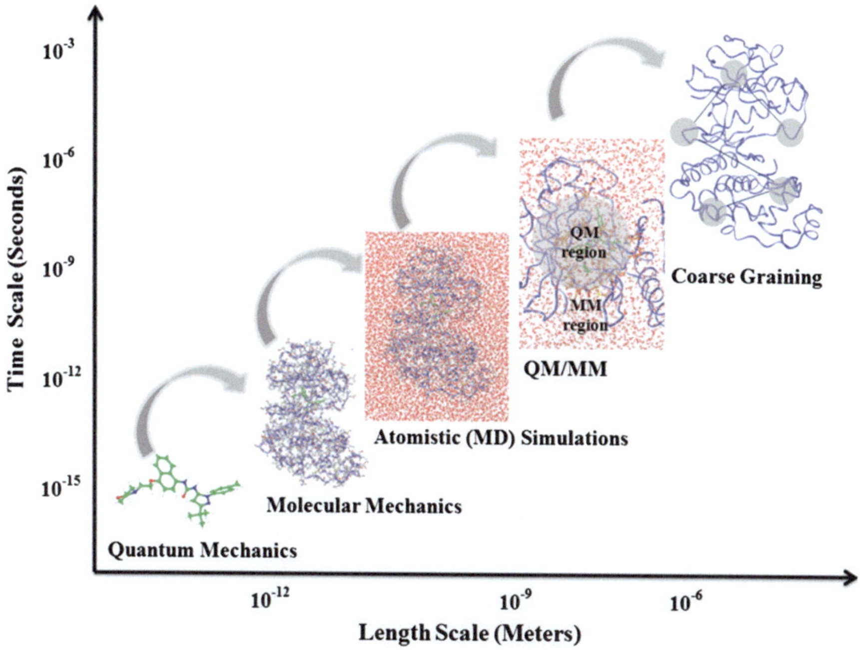

Fig. 6.1 Hierarchical order of molecular modeling approaches at different time and length scales. The figure depicts typical systems and methods which can be useful for varying time and length scales

time scales. Therefore, one needs to resort to methods based on classical mechanics to employ computational methods with larger length and time scales (Fig. 6.1).

In addition to the quantitative theories, many qualitative theories have emerged out of a large body of data obtained from experiments and observations. These methods based on informatics have been extremely successful in analysing the massive data of protein and nucleic acid sequences and thus ushering the area of bioinformatics. Most of the data in the area of bioinformatics is experimental. However, in the area of small molecules, computations have played a very important role and produced a large body of data of high reliability obtained by employing computational tools. Therefore, molecular modeling and computation have been complementary to the experimental efforts as far as the small molecules are concerned.

In this chapter, we introduce various computational methods which may be applied to molecular systems with varying length and time scales. Further, we discuss the basic principles of structure and analogue based approaches briefly. Finally, a cursory outlook on the application of these methods on biomolecules and materials has been given.

6.2 Quantum Mechanics

Computational chemistry is aimed at theoretically determining the properties of the molecules based on quantum chemical or classical mechanical equations of motion. Quantum chemical approaches are needed to accurately model the systems at atomistic scale and more importantly for obtaining electronic structure information. According to quantum mechanics (QM), all possible information on a molecular system can be obtained from a wavefunction, ψ which is obtained by solving the Schrödinger wave equation. However, the Schrödinger wave equation can be solved only for one electron systems thus rendering it unsolvable for many electron systems. The Schrödinger equation is the fundamental equation in QM and provides the basis for providing a complete description of the electronic structure of a molecule. Due to the difficulty associated with solving the Schrödinger equation for many electronic systems a large number of approximations were provided. There are excellent treatises available in the literature on computational QM, which deals with electronic structure calculations based on either *ab initio* molecular orbital theory or density functional theory (Cramer 2004; Jensen 2007; Levine 2013). Thus, we refrain from providing any further details on this section for two reasons. The first being the limitation of the space and the second and most important one is the limitation of the applicability of quantum chemical approaches to large biomolecules in general. However, in the following sections we discuss about the hybrid methods and multiscale modelling approaches, which employ and also largely based on the principle of quantum mechanics.

Herein we provide a cursory look at the various approximations that are being employed to solve the Schrödinger wave equation for medium sized molecules. The primary approximation for most of the electronic structure calculations is the Born-Oppenheimer approximation which essentially decouples the nuclear and electronic part of the kinetic energy operator. Following this variation theorem has become extremely effective for setting a lower bound for the energy. Such a condition has played a very important role in getting a better wavefunction through iterative procedures. The second important approximation is perturbation theory truncated to second, third or higher orders. However, the most effective theory based on electronic structure methods is *ab initio* self-consistent field (SCF) theory, where the fundamental level of reference wavefunction for the single determinant wavefunction is obtained by using the Hartree-Fock method. Electron correlation, which in principle may be divided into static and dynamic, is one of the most important parameter which needs to be included for the accurate description of the wavefunction. Thus, methods which go beyond Hartree-Fock level were warranted to obtain reliable properties of the molecular systems. The most popular variants of these methods, where a single Slater determinant can reasonably describe the system, are based on Moller-Plesset perturbation theory, configuration interaction and couple cluster methods. However, for open shell systems, the non-dynamic electron correlation becomes important, and one needs to have more than one Slater determinant for the reference wavefunctions. In such conditions the multi-configurational self-consistent field (MCSCF) procedures become imminent.

Methods based on these approximations have become very popular and have contributed greatly to the understanding of molecular structure, function and property relationships. The most rigorous method among these is based on the *ab initio* molecular orbital theory. One of the main bottlenecks in the application of the rigorous *ab intio* calculations to large molecules is computational capacity. In order to overcome that, several economical semiempirical SCF methods have emerged. However, the recent advances in the density functional theory have become very effective in dealing with the medium to large biomolecules.

6.3 Molecular Mechanics

Molecular mechanics (MM) applies Newtonian mechanics on a molecule or a molecular system to model its detailed structure and physical properties by calculating the energy of a molecule in terms of the bonded and non bonded interactions. MM is useful to study a broad range of molecular systems starting from small molecules to large biological systems or material assemblies with many thousands to millions of atoms (Field et al. 2007). The atomistic MM methods treat molecules as balls joined by springs wherein each atom is a single particle with an assigned radius (typically the van der Waals radius), polarizability, constant net charge (generally derived from quantum calculations and experiment). The bonded interactions are treated as springs with an equilibrium distance equal to the experimental or calculated bond length. All these bonded and non bonded terms all together are represented as a functional abstraction or force field to calculate the potential energy of a molecular system in a given conformation.

6.3.1 Energy of a Molecule

The steric energy of a molecule is the energy due to the geometry of a molecule. By nature, a molecule always tends to be in its lowest energy conformation to attain stability. As stated earlier, MM assumes the steric energy of a molecule to arise from a few, specific interactions within a molecule. These interactions include the stretching or compressing of bonds beyond their equilibrium lengths and angles, torsional effects of twisting about single bonds, the van der Waals attractions or repulsions of atoms that come close together, and the electrostatic interactions between partial charges in a molecule due to polar bonds (Hirschfelder 1954). To quantify the contribution of each, these interactions can be modeled by a potential function that gives the energy of the interaction as a function of distance, angle, or charge. The total steric energy of a molecule can be written as a sum of the energies of the interactions:

$$E = E_{bonded} + E_{non-bonded} \tag{6.1}$$

$$E_{bonded} = E_{stretch} + E_{bend} + E_{stretch-bend} + E_{dihedral} + E_{improper} \tag{6.2}$$

Table 6.1 Bonded and non-bonded energy components

S. No.	Energy component	Energy function	Constants and variables
	Bonded		
1	Stretch	$E_{str}(r) = \sum_{bonds} \frac{1}{2}k_r(r - r_0)^2$	k_r = force constant for the bond, r = actual bond length between the two atoms defining the bond and r_0 = equilibrium distance for the bond
2	Bend	$E_{bend}(\Theta) = \sum_{angles} \frac{1}{2}k_\Theta(\Theta - \Theta_0)^2$	Θ = bond angle, Θ_0 = equilibrium angle, k_Θ = force constant
3	Strech-bend	$E_{str-bend} \sum_{bonds,angles} = \frac{1}{2}k_{sb}\,(r - r_0)\,(\Theta - -\Theta_0)$	k_{sb}= stretch-bend force constant
4	Dihedral	$E_{dihedral} = \sum_{dihedrals} \frac{1}{2}V_n[1 + \cos(n\varphi - \delta)]$	n = periodicity of the angle, δ = phase of the angle, V_n = force constant
5	Improper	$E_{improper}(\omega) = \sum_{improper} \frac{1}{2}k_\omega(\omega - \omega_0)^2$	k_ω and ω_0 = force constant
	Non-bonded		
6	Electrostatic	$E_{elec} = \frac{q_i q_j}{r_{ij}}$	q_i and q_j = partial atomic charges for atoms i and j separated by a distance r_{ij}
7	van der Waals	$E_{LJ} = \sum \frac{A_{ij}}{r_{ij}^{12}} - \frac{B_{ij}}{r_{ij}^{6}}$	Aij and Bij are positive constants whose values depend on the depth of the Lennard-Jones well Ïµij

$$E_{non-bonded} = E_{electrostatic} + E_{van\ der\ Waals} \tag{6.3}$$

The bond stretching, bending, torsion and improper interactions are called bonded interactions because the atoms involved must be directly bonded or bonded to a common atom. The van der Waals and electrostatic interactions are between non-bonded atoms (Table 6.1).

6.3.2 The Force Fields

Force field refer to a mathematical function with a set of parameters (obtained experimentally as well as theoretically from computer intensive quantum calculations)

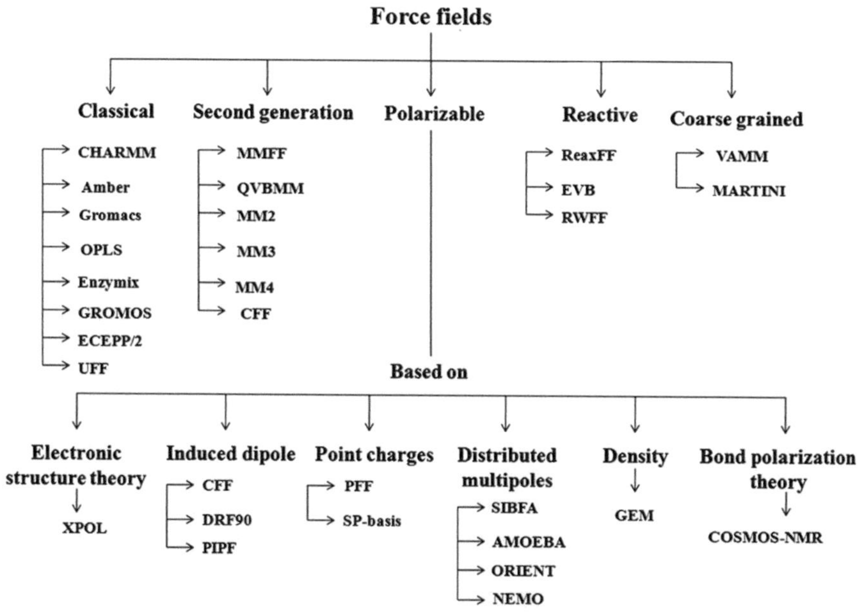

Fig. 6.2 Examples of various types of fields

to represent the potential energy of a molecular system. There are various types of force fields depending upon the level of accuracy (Fig. 6.2). For example, the "coarse grained" force fields which are used to simulate large proteins provide a crude representation to save computational time, while the "all atom" force fields, although computationally expensive, can accurately treat even the terminal hydrogen atoms (Ponder 2003). The basic functional form of a typical force field is given by Eqs. 1–3 which has already been discussed. In this section we discuss the various classes of force fields.

Apart from a representative function for the potential energy, each force field has a set of parameters for each bonded and non-bonded terms. Also, each force field has a particular atom typing. For example, the parameters for an oxygen atom in a carbonyl group and in a hydroxyl group are given distinct parameters. The typical parameter set includes values for atomic mass, van der Waals radii, and partial charge for various atom types, and equilibrium values of bond lengths, bond angles, dihedral angles, impropers and also the spring constants associated with them. The parameters for given atom types are generally derived from observations on small organic molecules that are more tractable for experimental studies and quantum calculations and extrapolated for larger molecules like proteins and DNA.

6.4 Molecular Dynamics

Biological processes are complex and involve a repertoire of atomic interactions. Although experiments help to deduce the molecular level understanding of the biological processes, the atomic interactions need to be modelled computationally. Thus MD simulations are used to estimate the microscopic properties and dynamic motions of assemblies of a biomolecular structure. MD simulations provide an access to the thermally-accessible states and help to correlate them with the functions of biomolecular systems (Frenkel and Smit 2002). It is thus a method, which integrates the Newtonian equations of motion for '*N*'particles of a system over a period of time resulting in a trajectory which is used for the calculation of the micro and macroscopic properties. The calculation of the MD trajectories is based on the principles of statistical mechanics (Allen and Tildesley 1987). MD simulations calculate the microscopic properties of the system such as position and velocities of each individual atom of the system. However, the properties that are of higher practical value are the macroscopic properties such as number of particles (N), volume (V), energy (E), temperature (T), pressure (P), chemical potential of particles μ (Rapaport 2004). These bulk properties are used to gauge the thermodynamic modulation with time.

The positions and momenta of all the particles of a system define a microscopic state. The positions and momenta of all the particles in the system are adjudicated as coordinates of a 6N dimensional space also called as phase space. Thus at any given time, the system corresponds to a point of the multidimensional space. The evolution of a system with time therefore corresponds to a trajectory in the phase space and can be determined by solving the equations of motion based on the potential energy (PE). There are different ways to distribute the total energy among the N particles of the system. An ensemble (EN) constitutes a collection of systems with similar macroscopic properties wherein each system corresponds to a point in the phase space. There are different types of ensembles based on the set of constant macroscopic properties such as canonical (NVT), the grand canonical (μVT), microcanonical (NVE) and the isothermal-isobaric (NPT) ensemble.

The partition function as given in Table 6.2 defines the microscopic state of a system explicitly. However due to the existence of large number of microscopic states in a biomolecular system and their sampling according to Boltzmann distribution in canonical ensemble, direct calculation of Z_{NVT} is not feasible. A MD simulation is initialized with the assignment of initial positions and velocities of all particles in the system. The initial velocities are assigned to particles in such a way that the total momentum is ensured to be zero, whereby the Maxwellian velocity distribution law is obeyed (Zeigler et al. 2000).

$$v^2{}_\alpha = \frac{\kappa_\mathrm{B} T}{m} \tag{6.4}$$

With the initial velocities assigned, the next step is the calculation of potential energy of the system using Eqs. (1–3). This is followed by deducing the force acting on each particle of the system by differentiating the calculated energy with respect to

Table 6.2 Types of statistical ensembles used in MD simulations

S. No.	Ensemble	Features	Partition function	Remarks
1	Microcanonical	Constant number of particles (N), volume (V) and energy (E). Entropy of the system increases continuously	$\Omega(E)$	$\Omega(E)$ is the number of micro-states corresponding to the system's energy E
2	Canonical	Number of particles (N), volume (V) and temperature (T) are constant	$Z_{NVT} = \sum_{j=1}^{j_{max}} e^{-\beta PE(x)}$	$\beta = \frac{1}{K_B\ T}$, $PE(x)$ = energy of the$x$thmicrostate of the system
3	Grand canonical	Chemical potential or fugacity (μ), volume (V) and temperature (T) are constant	$Z_{\mu VT} = \sum_i e^{(N_i \mu - PE(x))/TK_B}$	$\beta = \frac{1}{K_B\ T}$, $PE(x)$ = energy of the $x$thmicrostate of the system

Table 6.3 Mathematical functions used to generate velocity and position at each time step (Δt) with Verlet and Verlet-like integrators. (Frenkel and Smit 2002; Fermann and Valeev 1997)

S. No.	Integrators	Equation to update new position (x)	Equation to update new velocity (v)
1	Verlet	$x(t + \Delta t) = 2x(t) - x(t - \Delta t) + \frac{f(t)}{m}(\Delta t)^2$	$v(t) = \frac{x(t+\Delta t)-x(t-\Delta t)}{2x(t)} + \mathcal{O}(\Delta t)^2$
2	Velocity Verlet	$x(t+\Delta t) = x(t)+v(t)\Delta t+\frac{f(t)}{m}(\Delta t)^2$	$v(t + \Delta t) = v(t) + \frac{f(t+\Delta t)+f(t)}{2m}\Delta t$
3	Leap-Frog	$x(t + \Delta t) = x(t) - \Delta t v\left(t + \frac{\Delta t}{2}\right)$	$v\left(t + \frac{\Delta t}{2}\right) = v\left(t - \frac{\Delta t}{2}\right) + \Delta t \frac{f(t)}{m}$
4	Velocity corrected Verlet	$x(t + \Delta t) = x(t) + v(t)\Delta t + \frac{4f(t)-f(t-\Delta t)}{6m}\Delta t^2$	$v(t + \Delta t) = v(t) + \frac{2f(t+\Delta t)+5f(t)-f(t-\Delta t)}{6m}\Delta t$

the atomic positions (Fig. 6.3). After calculating the force on each particle, Newton's laws of motions are integrated to generate new positions and velocities for specified time-steps (Table 6.3).

The behaviour of a system is determined by its thermodynamic properties such as free energy, entropy and enthalpy. In a biomolecular system, the formation of a protein-ligand complex involves a change in free energy (Reddy and Erion 2001). The free energy determines the equilibrium properties of a system and is usually considered as the Gibbs or Helmholtz free energy. The configurational Helmholtz free energy (A) and can be represented as follows for a canonical ensemble:

$$A = -\beta^{-1} ln Z_{NVT} \tag{6.5}$$

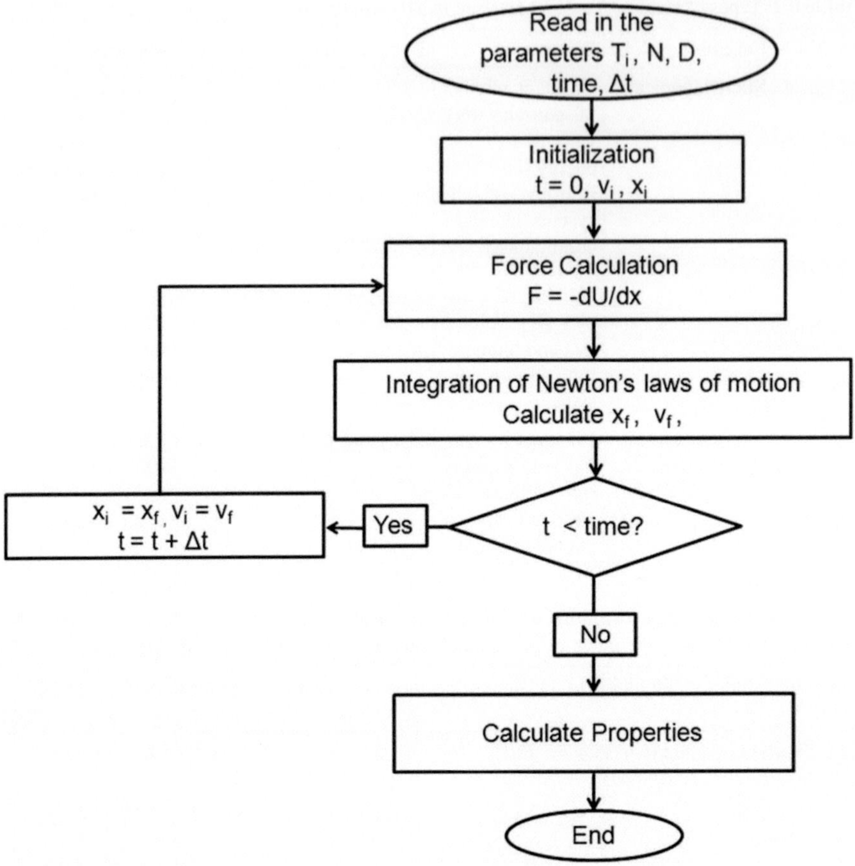

Fig. 6.3 Flow-chart depicting the general steps of a typical MD simulation

Calculation of absolute free energy is difficult due to inadequate sampling. There is therefore a need for different methods to calculate free energy. The free energy simulation techniques aim at computing ratios of partition functions using various techniques. The most common methods include free energy perturbation, thermodynamic integration, umbrella sampling and potential of mean force. The free energy calculation methods thus calculate the ratio between the two partition functions Z_{a1} and Z_{a2} to obtain the difference in free energy (ΔA). The free energy is calculated between two states. Thus the free energy differences between states '*a1*' and '*a2*' with partition functions Z_{a1} and Z_{a2} respectively can be calculated as:

$$\Delta A = -\beta^{-1} ln \frac{Z_{a2}}{Z_{a1}} \tag{6.6}$$

The ΔA thus obtained, is used to calculate the binding affinity of a protein-ligand complex. The solvation-free energy or binding-free energy is usually calculated through alchemical transformations through addition or removal of ligand related energy terms from the total Hamiltonian. The calculation of free energy by two most widely used methods namely free energy perturbation (FEP) and thermodynamic integration (TI) have been described below.

Free energy is a state function therefore the difference in free energy in the two states can be represented as:

$$\Delta A = -\beta^{-1} \ln \langle \exp[-\beta(V_{a2} - V_{a1})] \rangle_{a1} \tag{6.7}$$

In the TI approach, the difference in free energy between two states is calculated by integrating over enthalpy changes between the transition states. These states can be described with a parameter λ wherein λ_0 and λ_1 represent states 0 and 1 respectively (Wang and McCammon 2012). Thus the difference in free energy between the two states 0 and 1 is (i.e. from λ_0 to λ_1) obtained by as

$$\Delta A = \int_{\lambda_0}^{\lambda_1} \frac{\partial V}{\partial \lambda}_{\lambda} \tag{6.8}$$

6.5 Computer Aided Molecular Design

Computational approaches have become an integral and indispensable part of both academia and industry. Deciphering of the human genome is one of the first definitive accomplishments towards the molecular level understanding of biology. This has provided a quantitative understanding of the structural and functional aspects of biology unraveling a multitude of disease targets for drug discovery (Hopkins and Groom 2002). New drugs are constantly required for improving the treatment of existing and the newly identified diseases, in addition to the production of safer drugs by the reduction or removal of adverse side effects. Consequently huge investments are being channeled from pharmaceutical industries in research and development activities. The interdisciplinary nature of drug discovery warrants a fruitful collaboration among chemists, biologists, pharmacologists, physicians, computational and informatics scientists etc. New lead design is now more a strategic than a serendipity driven process. Thus, in the last couple of decades, *in silico* approaches have become an integral part of essentially all rational drug discovery programs. The rational approaches in drug discovery are traditionally classified as structure and analogue based (Fig. 6.4).

Medicinal chemistry driven approaches for several decades have relayed on the analogue based approaches wherein finding the quantity activity relationship was the key. However the recent advances in structure and molecular biology have provided more fundamental insights at molecular level. These approaches have been applied to obtain insights on the binding characteristics of the drug with the target. A drug

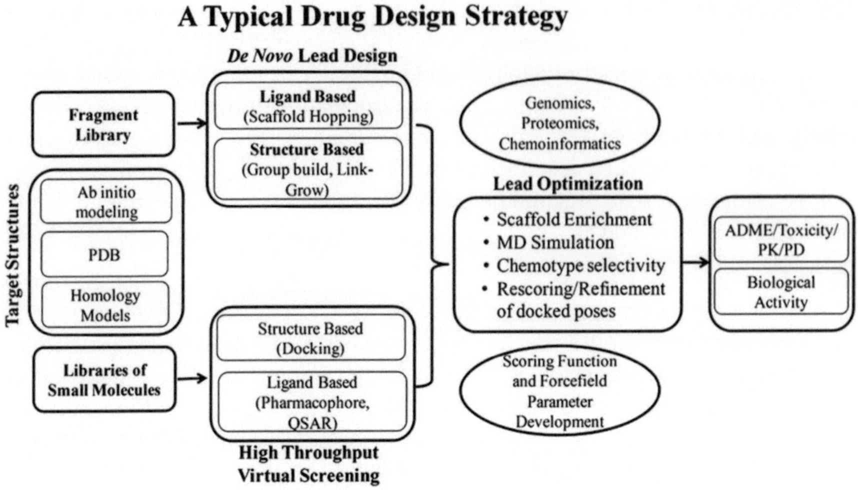

Fig. 6.4 A general work-flow of computer aided molecular design

contains various sub-units which contribute to the druglikeliness parameters such as the ADMET, pK/pD, blood brain barrier, drug metabolism, human intestinal absorption and permeability (Lipinski 1997). The long term use of a drug is however restricted by a multitude of inter-related factors such as development of resistance due to mutations, drug-drug interactions and most importantly target specificity. Rational design of a target specific drug is capable of overriding the other restricting factors. Specificity is of prime importance in the design of leads (Badrinarayan and Sastry 2013). There are different types of specificities such as target specificity, chemotype specificity and sub-type specificity. The different kinds of enzyme specificity that can be obtained by targeting different binding sites in a protein like the active site, allosteric sites. An inhibitor scaffold constitutes several fragments or chemotypes which individually can contribute to selectivity which can be used in the design of inhibitor for the active site to obtain high efficacy. This can possibly be achieved through molecule design or fragment based drug design (FBDD) using selectivity rendering fragments (Ringe and Reynolds 2010). The difference in the shape and constituency of the additional binding site called allosteric site can be exploited to design selective inhibitors for targets which share the same active site. Most proteins or enzymes exist in several isoforms which share high structural similarity among themselves. In such cases, small binding pockets or sub-pockets can be detected which are non-conserved and these can be used to obtain selectivity among the various subtypes.

Quantum chemical calculations based on *ab initio* have been proved to be highly reliable. However, the computational time rises exponentially as the number of electrons in the system increases thereby precluding their practical application to molecules with more than few dozens of atoms. Although it is possible to treat small ligand sized molecules quantum mechanically, it is expensive to apply them for larger bimolecular systems like proteins and nucleic acids. Thus MM has become a definitive choice of application for biomolecular targets. Computer aided drug design (CADD) focuses on understanding three essential factors for the design of drugs namely the features which render a macromolecule druggable, the properties which distinguish a drug from a small molecule and the interactions which facilitate an optimal fit of a drug-like molecule into a druggable target. A disease target is one which plays a pivotal role in the cause and expression of a disease phenotype and can be modulated by a drug. The therapeutic targets are thus both disease modifying and druggable. At present the currently approved drugs interact with only 2 % of human proteins hence there still exists a repertoire of undiscovered targets (Hopkins and Groom 2002). CADD uses amalgamation of structure and analogue based approaches. The structure based approaches include homology modeling, docking, virtual screening (VS), MD, MM-PBSA/GBSA, free energy calculations (FEP, TI) while the analogue based approaches include quantity structure activity relationship (QSAR), pharmacophore mapping, toxicity prediction and chemoinformatics methods.

6.5.1 Structure Based Drug Design

Advances in sophisticated large-scale automation were expected to generate an unprecedented number of novel leads resulting in a substantial increase in novel drug entities to be launched in market every year. This could not materialize as the discovered hits failed to optimize into actual leads. Thus, the initial euphoria associated with these approaches has subsided owing to the significantly high costs and disappointingly low hit rates involved in high-throughput screening (HTS). This calls for the rational application of drug design approaches such as virtual screening or docking to obtain lead compounds, which can be optimized further as drugs.

6.5.1.1 Docking

Docking is carried out using an automated computer algorithm that determines the binding of a compound to the active site of a protein (Stahl and Rarey 2001). This includes determining the orientation of the compound, its conformational geometry, and the scoring. There are two key components of a docking program namely the search algorithm and the scoring algorithm (Table 6.4). The search algorithm positions molecules in a multitude of locations, orientations, and conformation within the active site (Young 2009). The identified orientations are sampled further through downhill minimization to obtain bioactive conformations. The choice of the search algorithm determines the thoroughness of the program in checking the possible positions of the molecule and time taken.

Table 6.4 Scoring functions based on different algorithms implemented in some of the popular docking softwares. (Friesner et al. 2004; Morris et al. 1998; Rarey et al. 1996; Jones et al. 1997)

Software	Search algorithm	Scoring function
GOLD	Genetic Algorithm	$GOLD_{Fitness\ Score} = HB_{ext} + 1.3750 * vdW_{ext} + HB_{int} + 1.0000 * Lig_{int}$; $Li{{g}_{int}}=vd{{W}_{int}}+{torsion}$
GLIDE	Systematic Conformational Search	SP : $\Delta G_{bind} = C_{lipo-lipo}\sum f(r_{lr}) + C_{hbond-neut-neut}\sum g(\Delta r)h(\Delta\alpha) + C_{hbond-neut-charged}\sum g(\Delta r)h(\Delta\alpha)$ $+ C_{hbond-charged-charged}\sum g(\Delta r)h(\Delta\alpha) + C_{max-metal-ion}\sum f(r_{lr}) + C_{rotb}H_{rotb}$ $+ C_{polar-phob}V_{polar-phob} + C_{coul}E_{coul} + C_{vdW}E_{vdW} + solvation\ terms$
	XP : $XP_{Glide\ Score} = E_{coul} + E_{vdW} + E_{bind} + E_{penalty}$	
FlexX	Incremental Construction	$\Delta G = \Delta G_0 + \Delta G_{rot}\ N_{rot} + \Delta G_{hb}\sum_{neu\ Hbonds} f(\Delta R,\ \Delta\propto) + \Delta G_{ion}\sum_{ionic\ int} f(\Delta R,\ \Delta\propto)$ $+ \Delta G_{aro}\sum_{aro\ int} f(\Delta R,\ \Delta\propto) + \Delta G_{lipo}\sum_{lipo\ int} f^{*}(\Delta R,\ \Delta\propto)$
AutoDock	Lamarckian Genetic Algorithm	$\Delta G = \Delta G_{vdW}\sum_{ij}\left(\frac{A_{ij}}{r_{ij}^{12}} - \frac{B_{ij}}{r_{ij}^{6}}\right) + \Delta G_{hbond}\sum_{ij} E(t)\left(\frac{C_{ij}}{r_{ij}^{12}} - \frac{D_{ij}}{r_{ij}^{10}}\right)$ $+ \Delta G_{elec}\sum_{ij}\frac{q_i q_j}{\varepsilon(r_{ij})r_{ij}} + \Delta G_{tor}N_{tor} + \Delta G_{sol}\sum_{ij}(S_i V_j + S_j V_j)e^{(-r_{ij}^2/2\sigma^2)}$

Orientations with closely placed atoms are scored and others discarded. The energetically favorable modes of binding of a ligand are stored as different poses. These poses representing the protein-ligand interactions are scored in terms of binding enthalpies or Gibbs free energies, or a qualitative numerical measure or from a potential of mean force equation. This is concomitant to the inhibitory constant (K_i) calculated experimentally. Most of the scoring functions correlate well with the K_i values whereas others provide a qualitative ranking of the compounds tested. Some programs retain all the poses generated whereas some provide a scrutinized list based on the scores. Evaluation of closely placed atom pairs using full force field equations consumes a lot of computer time (Table 6.5).

This is overcome by using a grid based algorithm wherein potential fields are created which can be numerically evaluated over a grid generated over the active site. At a given point, the value of the potential on the grid is equivalent to the energy required for placing a unit charge at that point. Thus different types of scoring functions such as knowledge based, empirical, force field and many more have been developed to gauge the strength of interactions between the receptor and the small molecules. Most of the components of these scoring functions predict the non-covalent interactions and they are hardly any accounting for the covalent interactions (Cross et al. 2009). Considering the lacuna in individual scoring functions, consensus scoring is in vogue. Although all the small molecules undergo a conformational change during docking, the protein is held rigid in a fixed geometry in majority of the cases. Some programs facilitate alteration in the conformation of the active site as in flexible docking. This takes a longer time therefore options such as side chain repositioning and scaling are introduced which results in an induced fit approach so as to mimic the physiological conditions as far as possible.

In addition to search algorithms, scoring functions and flexibility, solvation is a major issue in defining the accuracy of results obtained as it has a direct impact on the binding energies of dissimilar molecules. An algorithm deficient of a solvation term results in identification of charged ligands which are large in size. The free energy of interaction relative to the free energies in solution of two molecules determines the binding affinity of the ligand for a particular receptor. FEP techniques accurately calculate the relative free energy however these are time consuming and are biased for similar set of molecules and are impractical for application for screening huge datasets of relatively diverse molecules. Therefore, energy correction for the solvent surrounding the protein needs to be included rather than considering only those occupying the active site (Gohlke and Klebe 2002). This has made this method as a protocol of choice in both academia and industry prediction of binding mode of a ligand during lead optimization as well in the identification of the potent lead itself through virtual screening.

6.5.1.2 Virtual Screening

Drug here is treated as a chemical substance that is used to prevent or cure diseases. In ancient times, a wide range of natural products obtained from animal, vegetable and

Table 6.5 Force field parameters used in docking algorithms. (Leach 2001; Ponder 2003; Friesner et al. 2004; Morris et al. 1998; Rarey et al. 1996)

S. No.	Force Field	Energetics
1	CVFF	$E_{MMFF} = \sum_{i}^{3N} \frac{p_i^2}{2m_i} + \sum_{ij}^{nbonds} V_b(r_{ij}) + \sum_{ijk}^{nangles} V_a(\Theta_{ijk}) + \sum_{ijkl}^{ntorsions} V_t(\varphi_{ijk}) + \sum_{ij}^{npairs} V_{np}(r_{ij})$
2	Tripos	$E = \sum E_{str} + \sum E_{bend} + \sum E_{oop} + \sum E_{tors} + \sum E_{vdW} + \sum E_{elec} + \sum E_{multifit}$
3	MMFF94	$E_{MMFF} = \sum EB_{ij} + \sum EA_{ijk} + \sum EAB_{ijk} + \sum EOOP_{ijk;l} + \sum ET_{ijkl} + \sum EvdW_{ij} + \sum EQ_{ij}$
4	OPLS	$E(r^N) = E_{bond} + E_{angle} + E_{dih} + E_{ab}$
5	CHARMM	$V = \sum_{bonds} k_b(b - b_0)^2 + \sum_{angles} k_\Theta(\Theta - \Theta_0)^2 + \sum_{dihedrals} k_\phi[1 + cos(n\phi - \delta)] + \sum_{impropers} k_\omega(\omega - \omega_0)^2 + \sum_{Urey-Bradley} k_u(u - u_0)^2$ $+ \sum_{nonbonded} \epsilon \left[\left(\frac{R_{min_{ij}}}{r_{ij}} \right)^{12} \left(\frac{R_{min_{ij}}}{r_{ij}} \right)^{6} \right] + \frac{q_i q_j}{\epsilon r_{ij}}$
6	AMBER	$V(r^n) = \sum_{bonds} \frac{1}{2} k_b(l - l_0)^2 + \sum_{angles} \frac{1}{2} k_a(\Theta - \Theta_0)^2 + \sum_{torsions} \frac{1}{2} V_n[1 + \cos(n\omega - \gamma)]$ $+ \sum_{j=1}^{N-1} \sum_{i=j+1}^{N} \left\{ \epsilon_{i,j} \left[\left(\frac{\sigma_{ij}}{r_{ij}} \right)^{12} - 2 \left(\frac{\sigma_{ij}}{r_{ij}} \right)^{6} \right] + \frac{q_i q_j}{4\pi\epsilon_0 r_{ij}} \right\}$

mineral sources were used for medicinal purposes but with the increase in knowledge the focus is centered on the use of pharmaceutically active compounds as starting point for the development of drugs. However, the increase in chemical space has made the identification of lead a tedious and erudite process. Therefore computational based screening methods such as VS are used for the identification of these pharmaceutically active compounds from a core set of molecules (Badrinarayan and Sastry 2011; Reddy et al. 2007a). The approaches used can be classified as ligand-based and structure-based methods. The availability of the physicochemical information dictates what strategy is more likely to be applied. Existence of structural data of the target protein calls for the application of structure-based virtual screening (SBVS) strategies. In the absence of structural information, ligand-based virtual screening (LBVS) protocols are usually applied. Virtual screening is carried out in concord with a number of different tools such as informatics (chemo and bio), docking, QSAR, pharmacophore mapping, machine learning tools (MLT), fingerprints, quantum mechanics/molecular mechanics (QM/MM), QM etc.

LBVS functions on the similarity principle which considers that the molecules which are structurally similar have similar biological activities. In the absence of structural information, LBVS use similarity, QSAR and pharmacophore based methods to correlate the physiochemical properties of known ligands with their structural characteristics and generate a query. LBVS methods look out for desired patterns in molecules such as fragments, pharmacophore, and core scaffold through graph theory like approaches or use molecular descriptors. Molecular fingerprints are however emerging as the most sought after options due to the ease of handling and speed. Fingerprints can be formulated with both 2D and 3D features (Sastry et al. 2010). The fingerprints are defined in the form of vectors which constitute bit strings of ones and zeros or position vectors indicating the presence or absence of a particular feature. A large part of the LBVS work being done is driven by informatics wherein the similarity indices are used to scale the nature if identity between the query and the database molecules.

With the increase in the repertoire of crystal structure data and the efficiency of docking in deducing the binding modes of ligands, SBVS is still employed fervently. Given a 3D structure, the SBVS approaches employ docking to generate the binding modes of the database compounds which are then shortlisted based on their scores. VS is a multi-step protocol and with the advent of multi-drug resistant strains and cross-target reactions, each step of the screening process is embedded with filters for druglikeness, Lipinski, target-selectivity, toxicity-ADME etc. in order to garner the best of the lot at every step and curb the percentage of false positives (Klebe 2006). These filters vary in complexity and dimension (1D-3D). Time and precision are the two endearing factors of VS. The time required depends on the type of query used and the complexity of the databases (molecule or fragment) being screened. The query used for screening can be simple 2D co-ordinates in the form of SMILES, fingerprints, bit vectors or a complex 3D representation of the active site, ligand template, pharmacophore or surface maps. The precision of the endeavor relies greatly on the stringency of filters used. The query can be complex 3D active site, molecule, pharmacophore, surface-volume or simple 2D co-ordinates, and feature-trees. SMILES, correlation vectors, bit strings and fingerprints are the simplest. The discrepancies in scoring functions are being encountered through incorporation of parameters extracted from MM-PBSA or QSAR calculations (Stahl and Bohm 1998). Although such procedures increase the accuracy of the hits obtained they however increase the overall time to get the desired outcome. The numbers of molecules that can be screened with virtual screening are several orders of magnitude higher than that of HTS. This number crunching ability of virtual screening and its inexpensive execution makes it endearing. However the crux of the process lies in the development of a screening strategy with efficient filters to obtain target specific leads.

We have developed a three step filtering strategy to identify target specific allosteric fragments for the inflammatory target p38 MAP kinase (Badrinarayan and Sastry 2012; Badrinarayan and Sastry 2010). The study entails the design of two target specific virtual screening filters based on docking score components and substructure interaction fingerprints. The components of the scoring function of two well-known docking protocols were evaluated to gauge their individual contribution in identification of lead. Eight thresholds were identified for the active and inactive conformations of kinase and were used in the identification of lead for the inactive conformation of the target kinase. The fragments or chemotypes demonstrating specific interactions with the study target were garnered from the known set of inhibitors. These interacting chemotypes were converted into substructure interaction fingerprints. The filters were used to screen a database of 10 million compounds and extract the interacting chemotypes from the identified leading hits. The extracted allosteric fragments itself constitute a new library of target specific allosteric fragments and are a good starting point for many lead design endeavors. Such protocols can easily be extended for different druggable targets to ensure the retrieval of target specific hits positively.

6.5.1.3 Fragment Based Methods

A drug or an inhibitor molecule constitutes a number of sub-parts called fragments whose presence either enhances their efficacy or renders them synthetically feasible. The action of drugs emancipate either from their physicochemical properties or from their chemical structure. The former are non-specific, act in large doses by forming a monomolecular layer over the entire cellular surface of an organism as in case of general anaesthetics, hypnotics such as aliphatic alcohols, antiseptics and antifungals (Lemke and Williams 2008). Those which are structure driven are specific and act in small doses on specific protein molecules which are usually located in the cell membrane to trigger a series of physiological and biochemical response. The specific recognition of receptors is driven by a fragment called chemotype which specifically endears the small molecule to that particular receptor (Badrinarayan and Sastry 2010). Identification of specific low molecular weight fragments in a molecule sets a stage for the stepwise design of new leads incorporating the identified chemotype. FBDD samples the chemical space to a greater extent than virtual screening of

Table 6.6 Postulates for the design of drugs, leads, scaffolds and fragments

S. No.	Properties	Drug-like (RO5)	Lead-like	Scaffold-like	Fragment-like (RO3)
1	Hydrogen bond donors (sum of -OH and -NH)	≤ 5	≤ 5	≤ 3	≤ 3
2	Hydrogen bond acceptors (sum of O and N)	≤ 10	≤ 8	≤ 8	≤ 3
3	Molecular weight (daltons)	< 500	< 450	< 350	< 300
4	Lipophilicity (clog *P*)	≤ 5.0	−3.0 to 4.5	≤ 2.2	≤ 3.0
5	Number of rotatable bonds (NROT)	–	≤ 9	≤ 6	≤ 3
6	Polar surface area (PSA) Å^2	–	–	–	60

molecules (Hajduk and Greer 2007). The fragments adhering to a set of properties as specified by the 'rule of three' are used embellished, linked and then grown into new leads. Identification of the right fragment through virtual screening or optimizing a prioritized one is computationally complicated since the existing scoring functions have been formulated for molecules and the cut offs prescribed do not suit the rule of three (Table 6.6).

However, fragments are more rigid with lesser numbers of degrees of freedom as compared to small molecule and are therefore can be easily docked. Majority of the initial work in FBDD has been carried out for the kinase targets. The fragment libraries are designed using reduced topological graph to compare the modes of the feature tree as in LoFT (Fischer et al. 2010) or use a set of bond rules to model ring substitution and cleavage sulphur groups as in BRICS (Degen et al. 2008). Certain protocols such as BROOD identify the fragment similar to the query template and design leads through bioisosteric replacements (Chen and Wang 2003). SeeDs on the other hand use pharmacophore fingerprints to screen for fragments (Baurin et al. 2004). The design of lead from fragments is usually carried out by linking fragments that bind to different parts of the target active site through a linker. We have developed a new fragment based lead design called 'Fragment Tailoring Approach (FTA)' based on similar principle wherein the existing set of kinase inhibitors binding to its highly conserved ATP site is reengineered into a target specific inhibitor by linking it with a chemotype which binds to its non-conserved allosteric site in the inactive conformation. The newly designed leads thus acquire efficacy from the ATP site fragment and specificity from the allosteric fragment (Badrinarayan and Sastry 2010). Self binding fragments as in click-chemistry on the other hand need no linker for connecting to each other and form a lead. The leads can also be derived by embellishing an individual fragment with function groups complementing the target active site. This has led to the development of different FBDD protocols such as BREED, LUDI, RECAP, ADAPT, LEA3D, LigBuilder based on genetic algorithm and SKELGEN, SMoG, SPROUT based on Monte Carlo simulations. FBDD has thus popularized the concept of 'prioritized sub-structures'. FBDD has resulted in the successful design of leads for several important diseases such as BACE-1, Phosphodiesterase (PDE) 4, Bcl-XL, Urokinase, Thrombin and Aurora kinase (Loving et al. 2010) to name a few.

6.5.2 *Analogue Based Drug Design*

Analogue based approaches for rational drug design have also emerged in parallel to the structure based approaches. These approaches complement the structure based approaches where the structure of the target is unknown, but the active inhibitors for the target are known. The main concept of analogue based drug design is based on a belief that chemical structure and biological activity of the analogues of a drug are often similar to the lead drug. In the last few decades, computational methods have significantly contributed to model new analogues for an existing drug as well as to predict the activities of new analogues. These predictive models rapidly screen large databases to identify new hit and lead molecules with improved biological activity profile and greater potency, thus opening up the way to new types of structures for drug research. QSAR modeling, pharmacophore modeling is some of the most important methods in analogue based drug design.

6.5.2.1 QSAR

The QSAR modeling is one of the analogue based computational tools, which establishes a quantitative correlation between biological activity/toxicity/property of a molecule and its structural features. In QSAR study, the variations of biological activity/toxicity/property within a series of compounds are correlated with changes in a group of computed features of the molecules referred to as descriptors.

QSAR method to predict a certain property of a molecule from its structure as a mathematical expression in the form of

$$y = m_1x_1 + m_2x_2 + \ldots + C \quad (6.9)$$

Where, y is the predicted property (the dependent variable) and x_1, x_2, ... are the known molecular properties called descriptors. QSAR uses descriptors that are a single number describing some aspect of the molecule, such as molecular weight, number of atoms, topological indices etc. The coefficients $m_1, m_2 \ldots$ in the QSAR equation are weights of the descriptors obtained by using various curve fitting methods.

The activities and properties being modeled by QSAR/QSPR are known as dependent variables (y) of the QSAR model. A dependent variable can be a biological

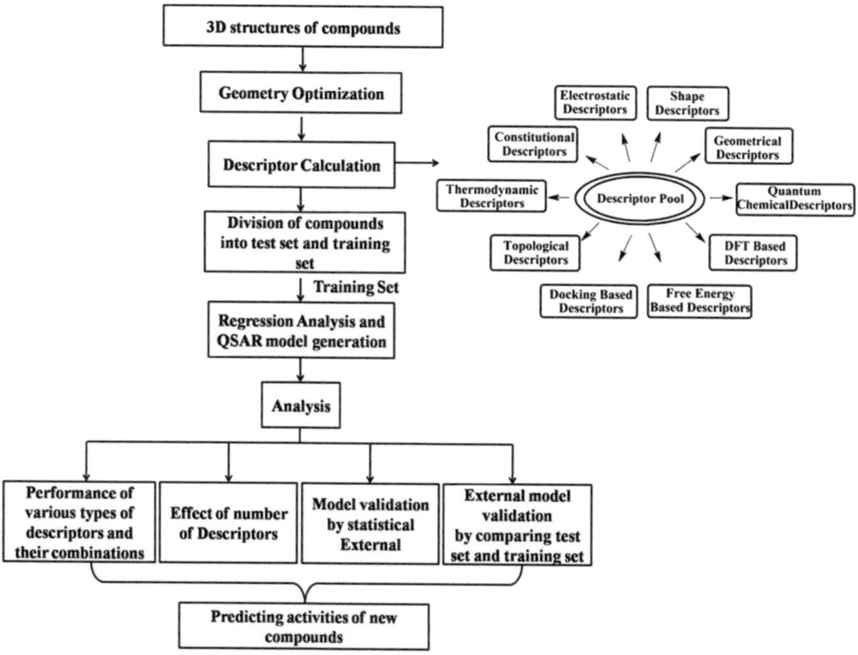

Fig. 6.5 Steps of QSAR modeling

property such as receptor binding, inhibition constant, permeability, pharmacokinetics, biodegradation, carcinogenicity, drug metabolism and clearance, mutagenicity, toxicity etc. or a chemical property such as boiling point, chromatographic retention time, dielectric constant, diffusion coefficient, dissociation constant, melting point, reactivity, solubility, stability, thermodynamic properties, viscosity etc. (Young 2009).

QSAR modeling typically describes molecular structures in terms of the descriptors and then correlates these molecular descriptors with observed activities using various statistical methods. The first step of QSAR modeling is preparation of a dataset of molecules with their activities, which follow a uniform distribution and calculation of descriptors. Molecular descriptors are chemical information that is encoded within the molecular structures and are collectively responsible for a particular activity of the molecule (Todeschini and Consonni 2000). The descriptors serve as the independent variables of a QSAR model. Various categories of descriptors employed in QSAR (Katritzky et al. 1994; Karelson et al. 1996). Constitutional descriptors are simple descriptors that represent only the molecular composition of the compound independent of the geometry and electronic structure (Fig. 6.5).

Examples are number of atoms, number of bonds, molecular weight etc. Topological descriptors/topological indices describe the atomic connectivity in the molecule. Examples are Wiener index, Randic and Kier & Hall indices, Kier flexibility index,

Information content index and its derivatives etc. (Katritzky et al. 1994). Geometrical descriptors are dependent upon 3D-coordinates of the atoms in the given molecule. For example, moments of inertia, shadow indices, molecular volume, molecular surface area, gravitation indexes etc. Electrostatic descriptors are calculated based on the charge distribution of the molecule. Examples are topological electronic index and charged partial surface area descriptors. Quantum-chemical descriptors are calculated from quantum chemical data at various levels of theory. For example Extreme (maximum and minimum) values of the atomic nucleophilic (N_A), electrophilic (E_A) and one-electron (R_A) Fukui reactivity indices, ε_{LUMO} and ε_{HOMO} etc. (Karelson et al. 1996). Hydrophobicity descriptors such as log P, aqueous solubility and chromatographic parameters are also very useful for QSAR studies (Helguera et al. 2008). However, development of simple and new descriptors is still a topic of high interest (Badrinarayan et al. 2011; Srivani et al. 2007). Among the new descriptors the density function theory (DFT) based ones are extensively studied. In many studies DFT based descriptors show good performance in predicting the biological activities (Parr 1983; Singh et al. 2004; Wadehra and Gosh 2005; Srivastava and Sastry 2012). Employment of docking scores as QSAR descriptors is one of the new approaches. The free energies of binding calculated by MMPBSA/GBSA methods are also tested in several studies and they show excellent correlation with the bioactivities (Srivastava et al. 2012). Once descriptors are computed, it is very crucial to choose the descriptors that should be included in the QSAR model. Preprocessing of the dataset should also be performed carefully as anomalies, errors, missing/incomplete data may lead to severe erroneous/misleading predictions. The data should also be normalized or standardized where there is a large range of variability in the dataset. Inter-correlated descriptors should be removed from the dataset before the model construction. (Nantasenamat et al. 2009).

Various techniques based on the multi-linear regression (MLR) analysis are employed in order to achieve the QSAR equation. This equation essentially correlates the variation of activities of the molecules as a function of the variations of the molecular structures present in the molecular data set (Kubinyi 1993). MLR analysis is usually used to correlate a given bioactivity with molecular descriptors. Different statistical methods come into play for building a QSAR model. Depending on the type of dataset and other parameters, however, it is possible to generate nonlinear equations that contain exponents of best fit, logarithms of descriptors, etc. MLR, principal component regression (PCR), partial least square, artificial neural network (ANN), genetic function approximation (GFA), factor analysis, discriminant analysis, cluster analysis are a few of the statistical methods that can be employed in the QSAR modeling (Dehmer et al. 2012).

For the linear QSAR equations the correlation coefficient r^2 gives a quantitative measure of how well the descriptor describes the activity (Wold 1991). r^2 is calculated as follows

$$r^2 = \frac{1 - \sum (y_{obs} - y_{calc})^2}{\sum (y_{obs} - y_{mean})^2} \tag{6.10}$$

where, y_{calc}, y_{obs} and y_{mean} are predicted, actual, and mean values of the target property respectively. Thus, the descriptors with the highest correlation coefficient can be selected. The predictive power of a QSAR model can be verified through statistical measures such as the correlation coefficient between actual and predicted values. Various statistical parameters such as cross validated correlation coefficient, Fisher statistic (F-value) values etc.

Crossvalidated r^2, lso called as q^2 signifies how best the model predicts. It is calculated by omitting each compound once from the training set, then predicting its activity using the model constructed from the remaining compounds. The model thus built with the remaining molecules is used to predict the response of the deleted compound/compounds. This cycle is repeated till all the molecules of the dataset have been deleted once. The cross-validated squared correlation coefficient q^2 is calculated as follows

$$\frac{1-\sqrt{\sum (y_{obs}-y_{calc})^2}}{\sum (y_{obs}-y_{mean})^2} \tag{6.11}$$

where, y_{calc}, y_{obs} and y_{mean} are predicted, actual, and mean values of the target property respectively. F-value is also an important measure of the statistical significance of the regression model, which is given by the following equation (Wold 1991).

$$F=\frac{r^2}{1-r^2} \tag{6.12}$$

where r^2 is the correlation coefficient. Also as an external validation, some of the compounds with known results are left out of the training set to be used as a test of the predictive ability of the QSAR model.

QSAR is a valuable tool for predicting molecular properties that cannot be computed any other way. It is very useful or the prediction of a wide range of biological properties, essential to identify potential leads (Nantasenamat et al. 2009). Although it may not be a reliable tool to predict drug activity, pharmacokinetic properties, such as blood–brain barrier permeability and passive intestinal absorption etc. can be fairly predicted by QSAR method. Hence, QSAR models are of immense help to predict the properties of new and untested compounds possessing analogous molecular structures as compounds used in the development of the models.

6.5.2.2 Pharmacophore Modeling

Pharmacophore modeling has gained immense importance as an analogue based approach in past few days because of its simplicity. According to IUPAC definition (Wermuth et al. 1998), "*A pharmacophore is the ensemble of steric and electronic features that is necessary to ensure the optimal supra-molecular interactions with a specific biological target structure and to trigger (or to block) its biological response.*" However, different researchers define it through their own view glasses

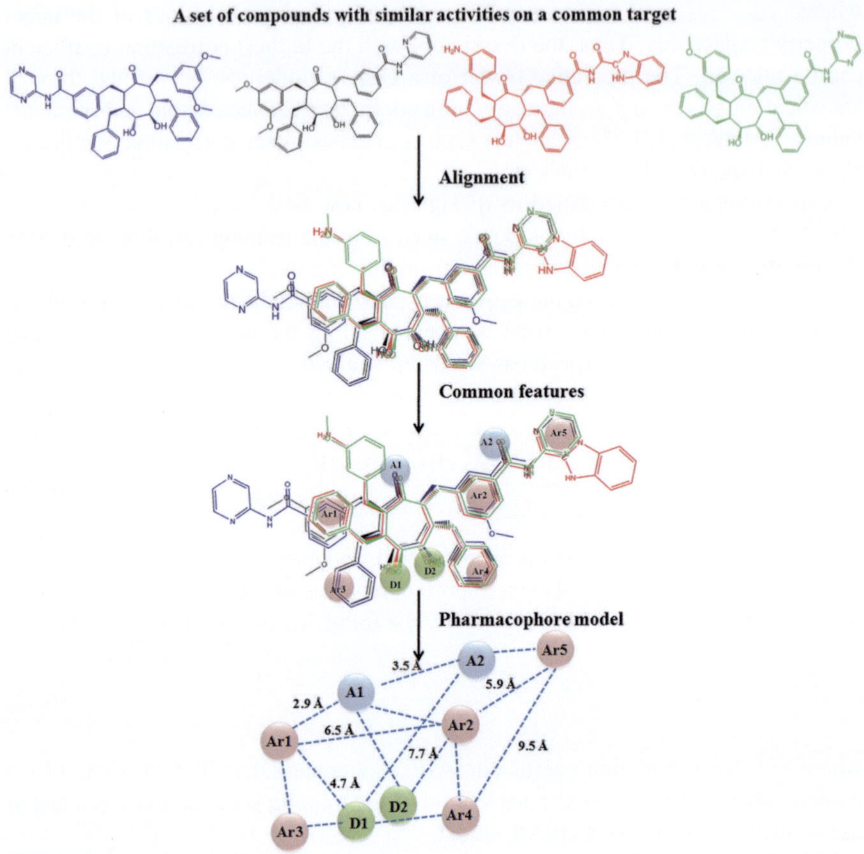

Fig. 6.6 Steps of pharmacophore modeling

depending on the suitability. A pharmacophore can be considered as the maximal set of common features extracted from a group of molecules exhibiting a similar pharmacological profile on a common target protein (Guner 2000).

A pharmacophore does not represent any real molecule, but represents the common interaction pattern of a group of compounds with their target (Wermuth 2006). The chemical signatures identified in a molecule which are actually responsible for making a certain type of non-covalent interaction with the receptor are called as pharmacophore features (Fig. 6.6).

A few examples of such functional features are hydrogen bond donors, hydrogen bond acceptors, aromatic rings (may be ring atoms, ring center, or normal to the ring), hydrophobic centers (also called neutral centers), positive charge centers, negative charge centers, acidic groups, basic groups, bulky groups engaged in steric interactions, planar atoms, CO_2 centroid (i.e., ester or carboxylic acid), metal (also called a metal ligator) and excluded volumes—forbidden regions, where the protein is and the ligand cannot have functional groups (Dror et al. 2006). Pharmacophore models provide a reasonable qualitative prediction of binding by modeling the spatial arrangement of a small number of atoms or functional groups (Yang 2010; Ekins et al. 2001). A detailed quantitative prediction of active molecules based upon the binding pattern requires sophisticated computational techniques as well as lots of computer time. Pharmacophore models are of immense use in analogue based virtual screening. Its usefulness covers three major domains. The generation of a relevant pharmacophore model, consistent with structure property relationship in a series of molecules helps in design of optimal ligands. Scaffold hopping may be an important implication of pharmacophore modeling, which consists in the design of functional analogues by searching within large virtual compound libraries of structures with similar activity profiles, but based on a different scaffold. New active compounds can also be designed by combining the key pharmacophore features of two different pharmacophore models (Langer and Hoffmann 2006; Wolber et al. 2008).

6.6 Modeling Large Molecular Systems

Large-scale biomolecular simulations are significantly important to study the functionality of large biomolecular systems. MD simulations have substantially contributed to the advancement of knowledge in biology, chemistry and material science. Although the MD simulations are being conducted for systems with millions of atoms and for millisecond timescale, the atomistic MD simulations fare to be too long and large when studying a biological phenomenon. The functioning of the living cell is a complex process, characterised by multiple interactions between macromolecules that act across multiple levels of structural and functional organisation—from molecular reactions to target-drug binding to protein-protein interactions. Since biological systems are multiscale in nature, there should be efficient model building and biological knowledge integration and prior data at all biological scales. Hence to explore such multiscale systems quantitatively, one has to integrate several different simulation techniques at different time and length scales. This calls for a paradigm shift in the simulation techniques wherein the atomistic treatment of the large biomolecular system as in MD simulations is replaced by the partitioning of the system. Such approaches either partition the large systems based on the level of precision required for its various components as in case of QM/MM wherein the protein is divided into the active site region which is treated quantum mechanically and the non-active site region which is treated with molecular mechanics. The other approaches include the multiscale approach which constitutes a framework comprising of different levels accuracy and couples them to enable a hierarchical handshake which leads to effective transfer of information across the different scales. These approaches divergent from the basic atomistic MD are useful in predicting the structure activity relationships and provide a fundamental mechanistic understanding of biological process. This facilitates efforts in predictive modeling and molecule design efforts.

6.6.1 QM/MM

The QM methods are too complex to be applied on the large biomolecular systems whereas MM methods fail to model the enzyme mediated reaction mechanisms. Therefore considering the individual shortcomings of each of these methods, a hybrid method such QM/MM employing their individual strengths is warranted (Ayton et al. 2007). The QM/MM partitions the biomolecular system into two regions. The active site comprises the smaller region and is treated quantum mechanically while the rest of the system is treated with the classical molecular mechanics force fields. There are two schemes to calculate the total energy of the system namely the additive and subtractive (Sherwood et al. 2008; Senn and thiel 2009; Sherwood et al. 2003). The subtractive scheme consists of four components namely the total energy of the system $E_{QM/MM}$ (system), E_{MM} (system) the MM energy of the entire system, E_{QM} (QM) the QM energy of the QM region andE_{MM} (QM) the MM energy of the QM region. The equation used to calculate the energy of the system though the subtractive scheme is represented as follows:

$$E_{QM/MM}\,(system) = E_{MM}\,(system) + E_{QM}(QM) - E_{MM}\,(QM) \tag{6.13}$$

The scheme encounters shortcomings due to the treatment of interactions between the QM and MM region only at MM level which is inaccurate. The scheme requires MM parameters for the QM region. Parameters are not usually available for those systems which are present in excited electronic states or contain transition metals. The additive scheme has therefore gained popularly. In this scheme, the total energy of the system $E_{QM/MM}$ (system) comprises of only three components viz., E_{MM} (MM) the MM energy of the MM region only, E_{QM} (QM) the QM energy of the QM region and the E_{QM-MM} (QM, MM) a term which interfaces between the QM/MM through the inclusion of bonded and non-bonded interactions. The bonded interactions account for bond stretching, bending and torsion while the non-bonded account for the van der Waals and electrostatic interactions.

$$E_{QM/MM}(system) = E_{MM}(MM) + \; E_{QM}\,(QM) - E_{QM-MM}(QM,\, MM) \tag{6.14}$$

The key to such QM/MM methods is the coupling between the electric field from the surrounding and the QM Hamiltonian in the active-site region. This requires careful treatment of the boundary between the QM and MM regions, either by using hybrid orbitals for the connection or a linked atom approach. The calculation of free energies from QM/MM simulations can be performed by averaging over the system's configurations via perturbations from a reference surface; however, such sampling for accurate free energy evaluations as well as calculations of pKa values remain challenging and form an active area of research.

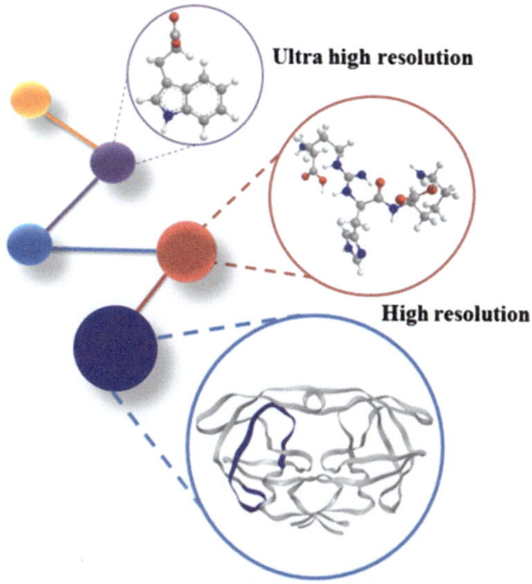

Fig. 6.7 Different types of coarse grained models: the low resolution model constituting the major functional domains as beads and used to study molecular level interactions, the high resolution mesoscale model clustering groups of amino acid residues of the protein as a bead. The third is the ultra high resolution model considering each amino acid residue as a bead and used to study atomistic level interactions

6.6.2 Coarse Graining

MD simulations, scaling over longer timescales, work well with biomolecular systems such as proteins, lipids and nucleic acids however they fall short in investigating complex phenomenon such as protein-protein assembly, vesicle diffusion, membrane deformation, DNA super coiling, DNA packaging in bacteriophage, folding of RNA in ribosome etc. Therefore, approaches such as coarse grained simulations are used wherein the single complex system is divided in to a couple of systems by grouping several atoms (Saunders and Voth 2013). Coarse graining clusters groups of atoms into beads or sites. Based on the accuracy desired either one amino acid is defined as a bead or a group of amino acids form a bead (Fig. 6.7).

These beads which are a kind of quasi-particles interact with each other. The combination of these interactions and the reduce degrees of freedom help to span the spatiotemporal scales. The accuracy and utility of a coarse grained model is largely dependent on the force field parameterization which implicitly account for the enthalpic and entropic contributions of free energy. The key steps in coarse graining include development of primary models based on experimental results followed by large scale simulation and identification of interactions influencing the energetic of the model system. Coarse graining retains the primary physical features of the system thus distills the atomistic scale information into simplistic but low resolution models. The final step therefore is to link with the molecular scale through all atom MD or Monte Carlo simulations based on the previous coarse grained simulation results so as to bridge the atomistic and mesoscopic scales. The precision in such cases can be obtained with multiple iterations of the entire protocol. Coarse grain simulations provide information on complex phenomenon such as biomolecular self-assembly at the mesoscopic scale and with iterative information transfer guides leverage of the atomistic details of the studied phenomenon through MD and Monte Carlo simulations. Thus the property or responses which are inaccessible at the atomistic or continuum levels of theory can be effectively simulated through coarse grained approach.

There are two main approaches to coarse graining called the inversion approach and the multiscale approach. The inversion approach employs thermodynamics, structural and experimental properties to developed coarse grained models. There a gamut of inverse coarse grained methods such as the Monte Carlo inverse Newton method (Lyubartsev and Laaksonen 1995), direct Boltzmann inversion approach (Tschop et al. 1998), iterative Boltzmann inversion method by Muller-Plathe (Muller-Plathe 2002). Most of these methods use reduced statistical distributions such as radial distribution instead of calculating the many body coarse grained potential mean force functions and detect the most appropriate coarse grained potential by inversing the data. The multiscale approach builds a hierarchical ladder to bridge atomic interactions to the mesoscale coarse grained model. In this the basic functions depicting the many body coarse grained potential mean force is mapped by atomistic scale forces. One of the earliest contributions has been made by Levitt & Warshel who identified the essential components contributing to the problem of protein folding and constructed a coarse grained model (Levitt and Warshel 1975). Gholke et al. have developed a three step multiscale coarse grained approach to model the conformational changes in proteins (Kruger et al. 2012). A high resolution coarse grained model with multiple beads per amino acid residue in protein is can effectively delineate the atomistic level interactions. However, to decipher the molecular scale motions occurring at the cell level necessitates simulation of large protein assemblies using the multiscale modeling approaches.

6.6.3 Multiscale Modeling

Biological systems are made up of several individual components or strata organised in a hierarchical manner (Schnell et al. 2007). The transfer of information among them leads to the functioning of the system as a whole. There are two different ways to scale the biological systems namely the spatial and temporal scales. The spatial scales hierarchically classify the biological processes based on the organization of biological systems. These scales are called 'levels of organization' and range from quantum, molecular, cellular, tissue, organ, organism to its ecosystem (Southern et al. 2008). Associated with the spatial levels of organization are the temporal scales of biological processes which range from microsecond for molecular interactions to years for an average lifespan of human being (Walker and Southgate 2009). According to this theory, a cell is made up of millions of molecules, while a tissue is made up of billions of cells and the number game thus augments. These key components of the biological system have intra- and inter-connections (Twycross 2010).

The diversity and connectivity among these scales increase the complexity of the biological systems. It is therefore necessary to model the individual components at multiple scales and integrate them to understand the impact of intra- and inter-scale interactions on the system and its surrounding ecosystem (Noble 2002). The multiscale modeling is an integrated and iterative approach which couples information obtained from various scales.

Mathematical representation of a complex system is termed as a 'model'. Models representing complex systems span a wide range of time and length scales and such models are termed as 'multiscale models'. The use of such multiscale models in addition with experimental data to understand the functioning of a biological system is termed as 'systems biology' while the engineering of these multiscale models to construct an artificial biological system to study its functioning comes under the preview of 'synthetic biology'. The modeling of biological systems is associated with different levels of complexity and therefore it requires a laddered approach instituting simulations at various time and length scales using methods offering varied degrees of precision and speed. Multiscale approaches thus encompass the combined use of computations and mathematics to obtain a simulated representation of a physiological system at different scales of time and biomolecular organization. This is an ingrained concept in the areas of engineering, aerodynamics and fluids associated with physics and material science however it is still comparatively raw to chemical and biological science.

The multiscale approach in modeling biological systems integrates the well-established disciplines like quantum chemistry, classical MD, systems biology, pathway modeling, and bioinformatics. They lie at the crossroads of frontier research areas in physics, biology, chemistry, and medicine. The multiscale models integrate (QM), molecular mechanics (MM), hybrid QM/MM, MD (MD), coarse-grained (CG), linear scaling and heuristic approaches. Multiscale modeling of biological systems is thus a measure to understand various scales of life at different resolutions. Each scale offers different features and therefore it is up to the discretion of the modeller to choose the appropriate strategy for the maximum abstraction of data and to bridge the gap between the various scales. There are two main approaches in multiscale modeling namely the 'top-down approach' and 'bottom-up approach' (Qu et al. 2011). The top-down approach treats the system as an individual entity and studies the macroscopic properties of the system. Hodgkin et al. created an action potential model of the giant axion using this approach (Hodgkin and Huxley 1952). To do so the individual ion-channels were overlooked and the voltage dependence of whole currents was modelled based on experimental data. This simplified the process to a great extent but they fail to account for the impact of individual components which participate in the expression of studied phenomenon. The bottom-up approach on the contrary simulates each individual component of a system and models their interactions to understand the nature of the system as a whole. This approach is useful in studying the behaviour of the interactive elements of a system and is therefore used in the study of cell-transport, protein folding and working of ion-channels (Kamerlin and Warshel 2011). The main aim of multiscale modeling is not only to model a particular system at different scales but also to conserve the data accurately during its transit from a lower scale to a higher scale or vice-versa. It has been employed to understand the functioning of important biological processes such as protein folding, membrane remodeling, drug metabolism and nucleic acid packaging.

The main objectives of a multiscale protocol are the identification of the individual processes constituting a complex system, the scales for modeling them and development of a link to couple these individual processes. In a multiscale strategy, the system is first decomposed into several sub-units. The temporal and spatial scales are allocated to model each of these sub-units. For example if the diffusion process is modelled then the temporal scale would define the rate of diffusion (Dada and Mendes 2011). The coupling of the micro-, meso- and macro-level processes leads to the development of multiscale model. Establishing coupling between scales is an intricate process (Martins et al. 2010). Solutions like multiscale Simulation Library and Environment (MUSCLE), Model Coupling Toolkit, XML based multiscale model management in systems biology have been developed to ensure smooth coupling of the scaled models.

6.7 Non-covalent Interactions

The drug once taken, travels through the body and elicits a pharmacological response. The site of drug action is the receptor while the pharmacodynamics is controlled by the different forces of interaction which bind a drug to a specific receptor (Holtje et al. 2008). The drugs and receptors exist as an ensemble of conformers in solvent. Thus to form a solvated complex with the receptor, the drug molecule needs to displace the solvent molecules occupying the binding site of the receptor. This is possible only when the interactions between the drugs and receptor are stronger than their individual interactions with the solvent molecules (Bissantz et al. 2010). The complex formation is entropically unfavorable and induces a loss in the conformational, rotational and translational degrees of freedom of both the drug and the receptor. The entropic loss is therefore expected to be compensated by favorable enthalpic contacts i.e. interactions. The bonds are spontaneously formed between atoms with a decrease in free energy (ΔG) i.e. when ΔG is negative. The activity of a small molecule (drug) is initiated by its atomic level interaction with the macromolecule (receptor or target). This association is stabilized by a plethora of intermolecular drug-receptor interactions which are either covalent or non-covalent in nature. The interaction of a drug with the binding site of a receptor depends on the complementarity of fit between the two molecules as stated in the Lock and Key Hypothesis by Emil Fischer (Silverman 2004). The interactions comply with the law of mass action. The binding of a drug to its receptor is therefore usually orchestrated though a gamut of non-covalent interactions rather than the covalent ones.

One of the most important and exhaustively studied drug-receptor interactions is the H-bond. Strong H-bonds like N-H. . . O, N-H. . . N and O-H. . . O are formed by the Glu, Leu and His residues which interact with the donor atom of inhibitor whereas the Leu and Gly residues interact with the acceptor atom (Sarkhel and

Desiraju 2004). The linked heterocyclic systems of the inhibitors are stabilized by the weak H-bonds such as C-H. . . N, C-H. . . O. Of the 20 amino acids comprising the protein, Gly and Glu play a substantial role as H-bond donor and acceptor. The propensity and strength to form H-bond varies with different functional groups. Thus the constitution of protein's active site has a substantial influence on the desolvation effects and SAR (Foloppe et al. 2005). The ammonium groups found in drug-receptor complexes are usually not permanently charged quaternary ions. This leaves at least one proton on the nitrogen atom which can be used in binding. The strength of the H-bond shows a dramatic increase when augmented with additional H-bonds (Neela et al. 2010). The hydrophobic nature of the active site can be attributed to a large extent to the side-chains of several aromatic residues which open into it. The aromatic rings of Phe, Trp, Tyr, and His form cation-π interactions with the cationic side-chains of Lys and Arg (Reddy and Sastry 2005). The cation-π interactions in biological systems stem from the interaction of nitrogen, phosphorous, oxygen and sulphur based onium ions (Mahadevi and Sastry 2013). The shape and electronic properties of the aryl rings of the aromatic amino acids give rise to large polarizabilities and a considerable quadrupole moment. The π-motifs engage in a T-shaped edge-to-face and the parallel-displaced stacking arrangement and interact with the heterocyclic rings of inhibitor. In proteins, the π-systems cluster into networks of various sizes. A database study by our group has shown that the CH-π and π-π stacking interactions formed by the side-chains of the aromatic residues provide stability to the protein hydrophobic pockets (Reddy et al. 2007b). The correlation between π-π stacking and hydrogen bonding is a very well studied example, owing to its relevance in nucleic acids (Vijay et al. 2008). The π-motifs form networks and their influence manifest strongly on the nature of inhibitor binding (Chourasia et al. 2011). The binding and stabilization is also contributed by the alkyl-aryl interactions as well. The aromatic π-motif forms one of the strongest non-covalent interactions on interacting with a metal ion. Such an interaction is a key player in enzyme regulation, stabilization, and functioning of nucleic acids. A subtle competition is also seen to exist between the π and σ- (in plane) approach of metal ion with the aromatic motifs (Reddy et al. 2006). The non-covalent interactions either complement or compete with each other in a cooperative or non-cooperative manner (Mahadevi and Sastry 2013; Vijay and Sastry 2010). The cooperativity of non-covalent interactions is an interesting phenomenon which is known to influence stability, conformational transitions and allosteric interactions in addition to inhibitor binding. The array of non-covalent interactions and their role in bio-macromolecules contribute to in the drug-receptor interactions, stabilization, and functional reorganization necessitating their consideration in design of leads (Fig. 6.8).

The non-covalent interactions engage in reversible binding are therefore preferred in CNS drugs, depressants etc. where the pharmacological effect needs to be terminated after some time. The role and relevance of non-covalent interactions in biological systems and the ability of computational methods to model them makes them a topic of high contemporary interest.

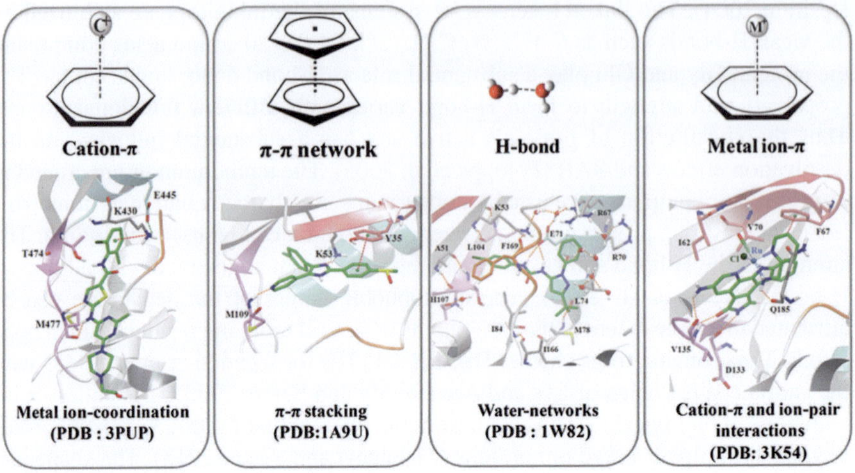

Fig. 6.8 Non-covalent interactions at the interface of chemistry and biology

6.8 Outlook

Molecular modeling has occupied the central space in basic, applied and industrial research. At the interface of chemistry, biology and material science, computational modeling has played a pivotal role in understanding the structure function relationships at atomistic levels. Although reliable and rigorous approaches have strong limitations in their applicability as the size of the system increases, several practical alternatives have been steadily emerged. This chapter provides a brief overview of the computational methods which can be applied to small and large molecules particularly bio-molecules.

Acknowledgement Department of Science and Technology, New Delhi is thanked for the Swarnajayanti Fellowship to GNS, Women Scientist Fellowship to PB and INSPIRE Fellowship to CC. Department of Biotechnology and Council of Scientific and Industrial Research, New Delhi are also thanked for financial assistance.

References

Allen MP, Tildesley DJ (1987) Computer simulations of liquids. Clarendon Press, Oxford

Ayton GS, Noid WG, Voth GA (2007) Multiscale modeling of biomolecular systems: in serial and in parallel. Curr Opin Struct Biol 17:192–198

Badrinarayan P, Sastry GN (2010) Sequence, structure, and active site analyses of p38 MAP kinase: exploiting DFG-out conformation as a strategy to design new type II leads. J Chem Inf Model 51:115–129

Badrinarayan P, Sastry GN (2011) Virtual high-throughput screening in new lead identification. Comb Chem High T Scr 14:840–860

Badrinarayan P, Sastry GN (2012) Virtual screening filters for the design of type II p38 MAP kinase inhibitors: a fragment based library generation approach. J Mol Graph Modell 34:89–100

Badrinarayan P, Sastry GN (2013) Rational approaches towards lead optimization of kinase inhibitors: the issue of specificity. Curr Pharm Des 19:4714–4738

Badrinarayan P, Srivani P, Sastry GN (2011) Design of 1-arylsulfamido-2-alkylpiperazine derivatives as secreted PLA2 inhibitors. J Mol Model 17:817–831

Baurin N, Aboul-Ela F, Barril X, Davis B, Drysdale M, Dymock B, Finch H, Fromont C, Richardson C, Simmonite H, Hubbard RE (2004) Design and characterization of libraries of molecular fragments for use in NMR screening against protein targets. J Chem Inf Comput Sci 44:2157–2166

Bissantz C, Kuhn B, Stahl MA (2010) Medicinal chemist's guide to molecular interactions. J Med Chem 53:5061–5084

Chen X, Wang W (2003) The use of bioisosteric groups in lead optimization. Annu Rep Med Chem 38:333–346

Chourasia M, Sastry GM, Sastry GN (2011) Aromatic-aromatic database, A2ID: an analysis of aromatic π-networks in proteins. Int J Biol Macromol 48:540–552

Cramer CJ (2004) Essentials of computational chemistry: theories and models, 2nd edn. Wiley, Chichester

Cross JB, Thompson DC, Rai BK, Baber JC, Fan KY, Hu Y, Humblet C (2009) Comparison of several molecular docking programs: pose prediction and virtual screening accuracy. J Chem Inf Model 49:1455–1474

Dada JO, Mendes P (2011) Multi-scale modelling and simulation in systems biology. Integr Biol 3:86–96

Degen J, Wegscheid-Gerlach C, Zaliani A, Rarey M (2008) On the art of compiling and using 'drug-like' chemical fragment spaces. Chem Med Chem 10:1503–1507

Dehmer M, Varmuza K, Bonchev D (eds) (2012) Statistical modeling of descriptors in QSAR and 881 QSPR. Wiley-Blackwell, Weinheim

Dror O et al (2006) Predicting molecular interactions *in silico*. I. An updated guide to pharmacophore identification and its applications to drug design. Front Med Chem 3:551–584

Ekins S, De Groot MJ, Jones JP (2001) Pharmacophore and three-dimensional quantitative structure activity relationship methods for modeling cytochrome P450 Active Sites. Drug Metab Dispos 29:936–944

Fermann JT, Valeev EF (1997) Fundamentals of molecular integrals evaluation. Tech. rep

Field MJ, Moleculaire LD, Grenoble (2007) Practical introduction to the simulation of molecular systems, 2nd edn. Cambridge University Press, Cambridge

Fischer JR, Lessel U, Rarey MJ (2010) LoFT: similarity-driven multiobjective focused library design. Chem Inf Model 50:1–21

Foloppe N, Fisher LM, Howes R, Kierstan P, Potter A, Robertson AG, Surgenor AE (2005) Structure-based design of novel Chk1 inhibitors: insights into hydrogen bonding and protein-ligand affinity. J Med Chem 48:4332–4345

Frenkel D, Smit B (2002) Understanding molecular simulations: from algorithms to applications, 2nd edn. vol 1, Computational Science Series Academic Press, San Diego

Friesner RA, Banks JL, Murphy RB, Halgren TA, Klicic JJ, Mainz DT, Repasky MP, Knoll EH, Shelley M, Perry JK, Shaw DE, Francis P, Shenkin PS (2004) Glide: a new approach for rapid, accurate docking and scoring. 1. Method and assessment of docking accuracy. J Med Chem 47:1739–1749

Gohlke H, Klebe G (2002) Approaches to the description and prediction of the binding affinity of small-molecule ligands. Angew Chem Int Ed Engl 41:2644–2676

Guner OF (ed) (2000) Pharmacophore perception, development, and use in drug design. International University Line, La Jolla

Hajduk PJ, Greer J (2007) A decade of fragment-based drug design: strategic advances and lessons learned. Nat Reviews Drug Discov 6:211–219

Helguera AM, Combes RD, Gonzalez MP, Cordeiro MN (2008) Applications of 2D descriptors in drug design: a DRAGON tale. Curr Top Med Chem 8:1628–1655
Hinchliffe A (2003) Molecular modelling for beginners, vol xviii. Wiley, Chichester
Hirschfelder JO, Curtiss L, Bird RB (1954) Molecular theory of gases and liquids. Wiley, New York
Hodgkin AL, Huxley, AF (1952) A quantitative description of membrane current and its application to conduction and excitation in nerve. J Physiol 117:500–544
Holtje HD, Sippl W, Rognan D, Folkers G (2008) Molecular modeling: Basic principles and explanations, 3rd edn. Wiley-VCH, Weinheim
Hopkins AL, Groom CR (2002) The druggable genome. Nat Rev Drug Discov 1:727–730
Jensen F (2007) Introduction to computational chemistry, 2nd edn. Wiley, UK
Jones G, Willett P, Glen RC, Leach AR, Taylor R (1997) Development and Validation of a genetic algorithm for flexible docking. J Mol Biol 267:727–748
Kamerlin SC, Warshel A (2011) Multiscale modeling of biological functions. Phys Chem Chem Phys 13:10401–10411
Karelson M, Lobanov VS, Katritzky AR (1996) Quantum-chemical descriptors in QSAR/QSPR Stud. Chem Rev 96:1027–1043
Katritzky AR, Lobanov VS, Karelson M (1994) CODESSA 2.0 comprehensive descriptors for structural and statistical analysis. University of Florida, U.S.A.
Klebe G (2006) Virtual ligand screening: strategies, perspectives and limitations. Drug Discov Today 11:580–594
Kruger DM, Ahmed A, Gohlke H (2012) NMSim web server: integrated approach for normal mode-based geometric simulations of biologically relevant conformational transitions in proteins. Nucleic Acids Res 40:W310–316
Kubinyi H (1993) QSAR, Hansch analysis and related approaches. In: Timmerman H, Mannhold R (eds) Methods and principles in medicinal chemistry. Wiley-VCH, Weinheim
Langer T, Hoffmann RD (eds) (2006) Pharmacophore and pharmacophore searches, vol 32. Wiley-VCH, Weinheim
Leach AR (2001) Molecular modelling—principles and applications, 2nd edn. Prentice Hall, Essex
Lemke TL, Williams D, Roche VF, Zito SW (eds) (2008) Foye's principles of medicinal chemistry, 6th edn. Lippincott Williams & Wilkins, Philadelphia
Levine IN (2013) Quantum chemistry, 7th edn. Prentice Hall, UK
Levitt M, Warshel A (1975) Computer simulation of protein folding. Nature 253:694–698
Lipinski CA, Lombardo F, Dominy BW, Feeney PJ (1997) Experimental and computational approaches to estimate solubility and permeability in drug discovery and development settings. Adv Drug Deliv Rev 23:3–25
Loving K, Alberts I, Sherman W (2010) Computational Approaches for Fragment-Based and *De Novo* Design. Curr Top Med Chem 10:14–32
Lyubartsev AP, Laaksonen A (1995) Calculation of effective interaction potentials from radial distribution functions: a reverse Monte Carlo approach. Phys Rev E 52:3730–3737
Mahadevi AS, Sastry GN (2013) Cation-π interaction: its role and relevance in chemistry, biology, and material science. Chem Rev 113:2100–2138
Martins ML, Ferreira SC Jr, Vilela MJ (2010) Multiscale models for biological systems. Curr Opin Colloid Interface Sci 15:18–23
Morris GM, Goodsell DS, Halliday RS, Huey R, Hart WE, Belew RK, Olson AJ (1998) Automated docking using a lamarckian genetic algorithm and an empirical binding free energy function. J Comput Chem 19:1639–1662
Muller-Plathe F (2002) Coarse-graining in polymer simulation: from the atomistic to the mesoscopic scale and back. Chemphyschem 3:754–769
Nantasenamat C, Ayudhya CIN, Naenna T, Prachayasittikul V (2009) A practical overview of quantitative structure-activity relationship. EXCLI J 8:74–88
Neela YI, Mahadevi AS, Sastry GN (2010) Hydrogen bonding in water clusters and their ionized counterparts. J Phys Chem B 114:17162–17171
Noble D (2002) Modeling the heart from genes to cells to the whole organ. Science 295:1678–1682

Parr RG (1983) Density functional theory. Annu Rev Phys Chem 34:631–656
Ponder JW, Case DA (2003) Force fields for protein simulation. Adv Prot Chem 66:27–85
Qu Z, Garfinkel A, Weiss JN, Nivala M (2011) Multi-scale modeling in biology: how to bridge the gaps between scales? Prog Biophys Mol Biol 107:21–31
Rapaport DC (2004) The art of MD simulation, 2nd edn. Cambridge University Press, New York
Rarey M, Kramer B, Lengauer T, Klebe G (1996) A fast flexible docking method using an incremental construction algorithm. J Mol Biol 261:470–489
Reddy MR, Erion MD (eds) (2001) Free energy calculations in rational drug design. Kluwer/Plenum Press, New York
Reddy AS, Sastry GN (2005) Cation [$M = H^+$, Li^+, Na^+, K^+, Ca^+, Mg^{2+}, NH_4^+, and NMe_4^+] interactions with the aromatic motifs of naturally occurring amino acids: a theoretical study. J Phys Chem A 109:8893–8903
Reddy AS, Vijay D, Sastry GM, Sastry GN (2006) From subtle to substantial: role of metal ions on pi-pi interactions. J Phys Chem B 110:247924-81
Reddy AS, Pati SP, Kumar PP, Pradeep HN, Sastry GN (2007a) Virtual screening in drug discovery—a computational perspective. Curr Prot Peptide Sci 8:329–351
Reddy AS, Sastry GM, Sastry GN (2007b) Cation-aromatic database. Proteins: Struc Func Bioinfo 67:1179–1184
Ringe D Jr, Reynolds CH, Merz KM (eds) (2010) Drug design: structure- and ligand-based approaches. Cambridge University Press, UK
Sarkhel S, Desiraju GR (2004) N–H. . . O, O–H. . . O, and C–H. . . O hydrogen bonds in protein-ligand complexes: strong and weak interactions in molecular recognition. Proteins 54:247–259
Sastry M, Lowrie JF, Dixon SL, Sherman W (2010) Large-scale systematic analysis of 2D fingerprint methods and parameters to improve virtual screening enrichments. J Chem Inf Model 50:771–784
Saunders MG, Voth GA (2013) 1. Coarse-graining methods for computational biology. Annu Rev Biophys 42:73–93
Schnell S, Grima R, Maini PK (2007) Multiscale modeling in biology. Am Sci 95:134–142
Senn HM, Thiel W (2009) QM/MM methods for biomolecular systems. Angew Chem Int Ed Engl 48:1198–229
Sherwood P, de Vries AH, Guest MF et al (2003) QUASI: a general purpose implementation of the QM/MM approach and its application to problems in catalysis. J Mol Struct Theochem 632:1–28
Sherwood P, Brooks BR, Sansom MS (2008) Multiscale methods for macromolecular simulations. Curr Opin Struct Biol 18:630–640
Silverman RB (2004) The organic chemistry of drug design and drug action, 2nd edn. Elsevier Academic Press, San Diego
Singh PP, Srivastava HK, Pasha FA (2004) DFT-based QSAR study of testosterone and its derivatives. Bioorg Med Chem 12:171–177
Southern J, Francis JP, Whiteley J, Stokeley D, Kobashi H, Nobes R, Kadooka Y, Gavaghan D (2008) Multi-scale computational modelling in biology and physiology. Prog Biophys Mol Biol 96:60–89
Srivani P, Srinivas E, Raghu R, Sastry GN (2007) Molecular modeling studies of pyridopurinone derivatives—potential phosphodiesterase 5 inhibitors. J Mol Graph Model 26:378–390
Srivastava HK, Sastry GN (2012) MD investigation on a series of HIV protease inhibitors: assessing the performance of MM-PBSA and MM-GBSA approaches. J Chem Inf Model 52:3088–3098
Srivastava HK, Choudhury C, Sastry GN (2012) The efficacy of conceptual DFT descriptors and docking scores on the QSAR models of HIV protease inhibitors. Med Chem 8:811–825
Stahl M, Bohm HJ (1998) Development of filter functions for protein-ligand docking. J Mol Graph Model 16:121–132
Stahl M, Rarey M (2001) Detailed analysis of scoring functions for virtual screening. J Med Chem 44:1035–1042

Todeschini R, Consonni V (2000) Handbook of molecular descriptors. In: Mannhold R, Kubinyi H, Timmermann H (eds) Methods and principles in medicinal chemistry. Wiley-VCH, Weinheim

Tschop W, Kremer K, Batoulis J, Burger T, Hahn O (1998) Simulation of polymer melts. I. Coarse graining procedure for polycarbonates. Acta Polym 49:61–74

Twycross J, Band LR, Bennett MJ, King JR, Krasnogor N (2010) Stochastic and deterministic multiscale models for systems biology: an auxin-transport case study. BMC Syst Biol 4:34–45

Vijay D, Sastry GN (2010) The cooperativity of cation-π and π-π interactions. Chem Phys Lett 485:235–242

Vijay D, Zipse H, Sastry GN (2008) On the cooperativity of cation-π and hydrogen bonding interactions. J Phys Chem B 112:8863–8867

Wadehra A, Ghosh SK (2005) A density functional theory-based chemical potential equalization approach to molecular polarizability. J Chem Sci 117:401–409

Walker DC, Southgate J (2009) The virtual cell-a candidate co-ordinator for 'middle-out modelling of biological systems. Briefings Bioinf 10:450–461

Wang Yi, McCammon JA (2012) Introduction to MD: theory and applications. In: Dokholyan NV (ed) Biomolecular modeling computational modeling of biological systems. From molecules to pathways. Springer, USA, pp 3–30

Wermuth CG (2006) Pharmacophores: historical perspective and viewpoint from a medicinal chemist. In: Langer T, Hoffmann RD (eds) Wiley-VCH Verlag GmbH & Co. KGaA

Wermuth CG, Ganellin CR, Lindberg P, Mitscher LA (1998) Glossary of terms used in medicinal chemistry (IUPAC Recommendations 1998). Pure Appl Chem 70:1129–1143

Wolber G, Seidel T, Bendix F, Langer T (2008) Molecule-pharmacophore superpositioning and pattern matching in computational drug design. Drug Discov Today 13:23–29

Wold S. (1991) Validation of QSARs. Quant Struct Act Relat 10:191–193

Yang SY (2010) Pharmacophore modeling and applications in drug discovery: challenges and recent advances. Drug Discov Today 15:444–450

Young DC (2009) Computational drug design: a guide for computational and medicinal chemists. Wiley, Hoboken

Zeigler B, Praehofer H, Kim T (eds) (2000) Theory of modeling and simulation: integrating discrete event and continuous complex dynamic systems, 2nd edn. Academic Press, New York

Chapter 7
Complex Networks and Systems Biology

Ushasi Roy, Rajdeep Kaur Grewal and Soumen Roy

Abstract Modern biology has decisively moved in a direction where we scrutinise systems holistically rather than looking at entities in different levels discretely or in isolation. Unlike previous reductionist approaches; in this new approach called Systems Biology, networks play a crucial role in arriving at and summing up the holistic picture and in understanding the emergent properties of the system. In this chapter, we give an overview of how network approaches are useful at various levels in biology. After a conceptual introduction to networks and various network metrics used to quantify networks; we discuss various concepts like network motifs and random networks. We then examine at length about how networks shed insight at virtually every layer of life like gene regulatory networks, networks involving proteins and metabolic networks. We end the chapter with a discussion of the application of networks to epidemiology.

Keywords Network · Directed networks · Weighted networks · Degree · Degree distribution · Assortavity · Shortest path length · Connectedness · Eccentricity · Diameter · Closeness centrality · Betweenness centrality · Clustering coefficient · Cliques · Community structure · Modularity · k-core decomposition · Erdos-Renyi graphs · Small-world · Scale-free · Motifs · Feed forward loops · Gene Regulatory Network (GRN) · Protein Structure Network (PSN) · Protein Energy Network (PEN) · Allostery · Protein Protein Interaction network (PPI networks) · Protein folding network · Metabolic networks · Epidemiology · Susceptible Infectious Recovered (SIR) · Susceptible Infectious Susceptible (SIS)

7.1 Introduction

7.1.1 Systems Biology

The study of biological systems has historically been a largely phenomenological or observational science. However, in the last quarter of the twentieth century; in-depth

S. Roy (✉) · R. K. Grewal · U. Roy
Bose Institute, 93/1 Acharya Prafulla Chandra Roy Road,
Kolkata 700 009, India
e-mail: soumen@jcbose.ac.in

V. Singh, P. K. Dhar (eds.), *Systems and Synthetic Biology*,
DOI 10.1007/978-94-017-9514-2_7

quantitative studies of various biological phenomena started gaining momentum. Over the course of the last decade and half, the advent of high-throughput technologies have only made the application of quantitative techniques imperative to biology. They also inculcated the realisation that biological systems are far too complex to be solved by classic reductionist approaches. It was becoming increasingly apparent that the study of biological systems need an integrated, multidisciplinary approach whose essence is underscored by an effective cycle of modelling and experimentation. "Systems" approaches are definitely poised to occupy mainstream biology over the course of the next decade or so. These approaches examine the structure and dynamics of cellular and organismal function, contrary to the study of isolated parts of cell or organism (Kitano 2002). Thus, "Systems Biology" is a new branch of science which integrates techniques from Mathematics, Physics, Chemistry, Computer Science, Engineering and Information theory to model various biological phenomena from a holistic point of view.

Intrinsic to this development, is the concept of "emergent properties" which refer to holistic properties at the system level, since the behaviour of the system as a whole will not merely be an agglomeration of the properties of its segregated constituents. For studying this composite system, consolidation of the diverse interactions among various components of the system is required. The theory of networks which is based on a well established graph-theoretic approach; enables us to do so efficiently (Albert et al. 2002; Newman 2010).

7.1.2 Networks

From the perspective of Graph Theory, a network can be represented by a graph. A graph is defined as $G = \{V, E\}$ where V is the set of nodes (or vertices or simply points) and E denotes the set of edges (or links or arcs or simply lines), which establishes an interconnection among the nodes. A real complex system can be mapped onto a network structure where one needs to identify the major components of the system as the nodes and the interactions among them as the edges. This concept has been illustrated below by two simple graphs. In Fig. 7.1a, the set of nodes $V = \{a, b, c, d\}$ and the set of edges is given by $E = \{(e_1 = (a, b)), (e_2 = (b, c)), (e_3 = (a, c)), (e_4 = (c, d))\}$. Similarly, $V = \{v_1, v_2, v_3, v_4, v_5\}$ and $E = \{(e_1 = (v_1, v_5)), (e_2 = (v_2, v_5)), (e_3 = (v_2, v_3)), (e_4 = (v_3, v_5)), (e_5 = (v_4, v_5))\}$ correspond to the set of nodes and edges in Fig. 7.1b.

7.1.2.1 Subgraph

A subgraph $G' = \{v', e'\}$, having v' vertices and e' edges is defined to be a subgraph of $G = \{V, E\}$ if v' is a subset of V and e' is a subset of E.

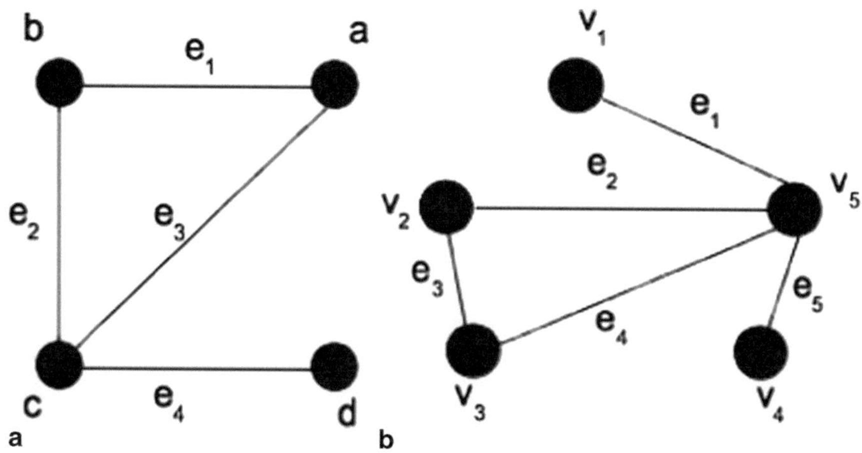

Fig. 7.1 (**a**) and (**b**): Simple graphs

7.1.2.2 Directed Networks

In a directed network, the edges have a direction, i.e., identification of the "source" and "sink" nodes for a particular connection is important. Thus, a particular node will have both incoming and outgoing edges and will have different in and out degree distributions. Many important networks, viz., World Wide Web (WWW) and metabolic networks are directed in nature.

7.1.2.3 Weighted Networks

Generally we construct binary networks with the edge weights having two possible values, 0 and 1; representing absence and presence of connections respectively. In contrast, many real networks are weighted in nature. In these networks, in addition to the binary values, edge weights can have any fractional values in between 0 and 1, depending on the strength of interactions. Here all the edges are not equally important and the edge with higher edge weight will have a higher significance in the network. Examples are social networks, internet and cellular networks as they are characterized by the level of acquaintance between individuals, band widths and reaction rates which may have different values (Fig. 7.3).

7.2 Network Metrics

Network metrics help in the characterisation of a given network—both quantitatively and qualitatively. Their significance lies in analysing both the local property, i.e., the individual behaviour of nodes or edges, as well as the global property of the whole

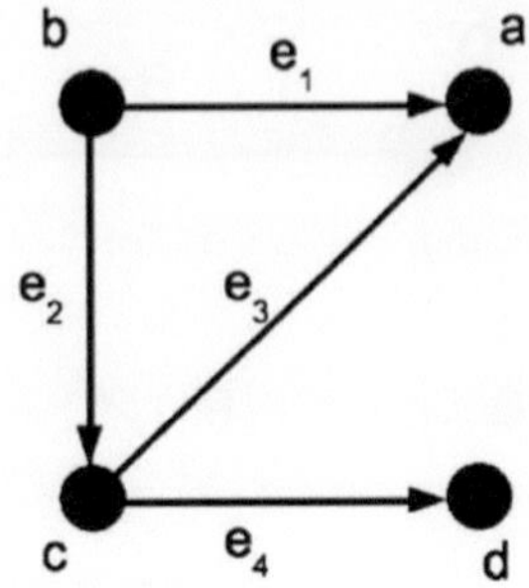

Fig. 7.2 A simple directed graph with node set $V = \{a, b, c, d\}$ and the set of directed edges $E = \{(e_1 = (b, a)), (e_2 = (b, c)), (e_3 = (c, a)), (e_4 = (c, d))\}$ where the first node in the edge set denotes origin while the second one represents the end of an edge

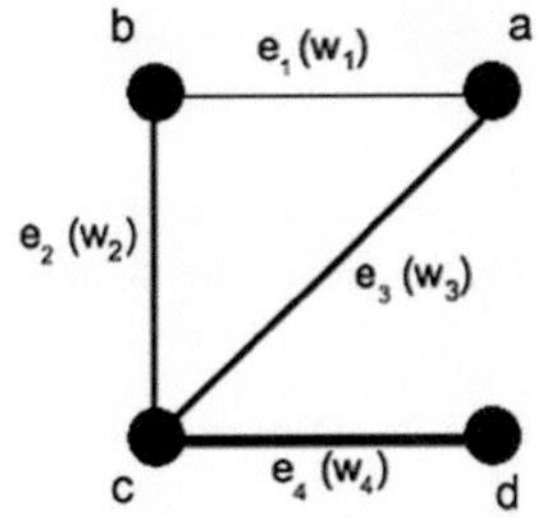

Fig. 7.3 A simple undirected weighted graph with the set of nodes defined as $V = \{a, b, c, d\}$ and the set of edges $E = \{e_1, e_2, e_3, e_4\}$ having edge weight $W = \{w_1, w_2, w_3, w_4\}$

network. These structural network metrics may also serve as a great tool for exploring the unified behaviour of the network.

7.2.1 *Degree*

A degree of a node is defined as the number of edges incident on that node. It signifies the number of connections made by a node i with the remaining nodes in the network, termed as neighbours of node i. The nodes which have comparatively much higher degree than that of the other nodes in a network correspond to the *hub*.

For directed networks, degree of a node is specified using two distinct centrality measures *in-degree* and *out-degree*. In a directed network, the number of edges directing outward from the particular node is its out-degree and the number of nodes directing towards it correspond to the in-degree of that node in a network. In Fig. 7.1a, the degree of each of the nodes {a,b,c,d}in the graph G are {2,2,3,1} respectively. For the directed graph H in Fig. 7.2 the in-degree and out-degree of the nodes {a,b,c,d} are {2, 0, 1, 1} and {0, 2, 2, 0} respectively.

7.2.2 *Degree Distribution*

The degree distribution P(k), the probability that a randomly chosen node has degree k or fraction of nodes in the network having degree k, of a network provides one of the basic topological characterisation of a network. Various types of networks can sometimes be distinguished by their degree distribution. For instance, scale free networks have a power law degree distribution,

$$P(k) \approx k^{-\gamma} \tag{7.1}$$

and it has been claimed that when $2 \leq \gamma \leq 3$; the hubs play a significant role in the network (Barabasi and Oltvai 2004). In contrast, small random networks follow Binomial distribution which in the limit of large N approaches the Poisson distribution

$$P(k) \approx e^{-\langle k \rangle} \frac{\langle k \rangle^k}{k!} \tag{7.2}$$

where $\langle k \rangle$ denotes the average degree of the graph. For directed networks, there might be different distributions of in-degree, out-degree and total degree of the nodes in the network.

7.2.3 *Assortativity*

Assortativity refers to the affinity of nodes in a network to become linked to other nodes having similar degree distribution. This tendency of correlation among nodes of similar degree is also sometimes called as assortative mixing. In contrast, sometimes high degree nodes are somewhat inclined towards low degree nodes. This kind of dissimilar preferential attachment gives rise to a disassortative network. Most biological and technological networks exhibit disassortative mixing while social networks belongs to the former class, i.e., they are assortative in nature. Mathematically, *assortativity* of a complex network can be expressed as

$$r = \frac{\langle k_1 k_2 \rangle - \langle k_1 \rangle \langle k_2 \rangle}{\sigma_k^2} \tag{7.3}$$

where the averages are taken over all edges and σ_k^2 is the variance of the node-degree k. For all practical purposes, calculating assortativity of real world networks, the above equation can be modified as (Newman 2002)

$$r = \frac{E^{-1} \sum_e j_e k_e - \left[E^{-1} \sum_e \frac{1}{2}(j_e + k_e)\right]^2}{E^{-1} \sum_e \frac{1}{2}(j_e^2 + k_e^2) - \left[E^{-1} \sum_e \frac{1}{2}(j_e + k_e)\right]^2} \tag{7.4}$$

where j_e, k_e are the degrees of the nodes at the ends of the eth edge, with $e = 1, 2, \ldots .., E$.

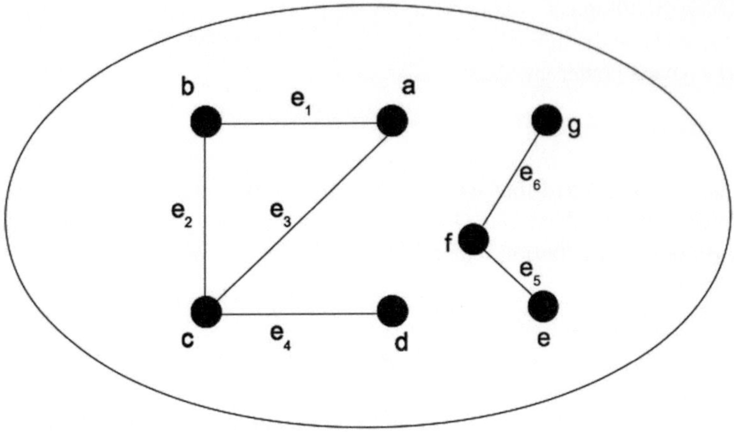

Fig. 7.4 A graph G with two disconnected components

7.2.4 Shortest Path Length

A *path* is an alternate sequence of nodes and edges, starting and ending with a node, such that each edge in the sequence is incident on the node preceding and following it. There is no repetitions of nodes and edges in a path. In Fig. 7.1a {a,e1,b,e2,c,e4,d} represents a path connecting the nodes a and d. *Shortest path* between a pair of vertices (i, j), where $i, j \in V$,in a graph is the geodesic distance (d_{ij}) between them i.e the minimum number of edges traversed while moving from node i to node j.

7.2.5 Connectedness

A graph is said to be connected if there exists at least a path between any pair of nodes constituting the graph. It may so happen that there exits a pair of nodes in a graph having no path connecting them. Such graphs are known as disconnected graphs. For a disconnected graphs, each connected component is termed as a *cluster*. *Giant cluster* in a network refers to the largest connected component of the network (Fig. 7.4).

Directed graphs, in terms of connectedness, are defined to be strongly or weakly connected graphs. If each pair of nodes in the directed graph has at least one directed path (each edge in the sequence is incident out- and in- on the node preceding and following it, respectively) between them, the graph is said to be strongly connected. If the underlying undirected graph (graph obtained from the directed graph by removing the directions of edges from it) of the directed graph is connected, we call it as weakly connected graph. It is quite obvious that a strongly connected graph will definitely be a weakly connected graph (Fig. 7.5).

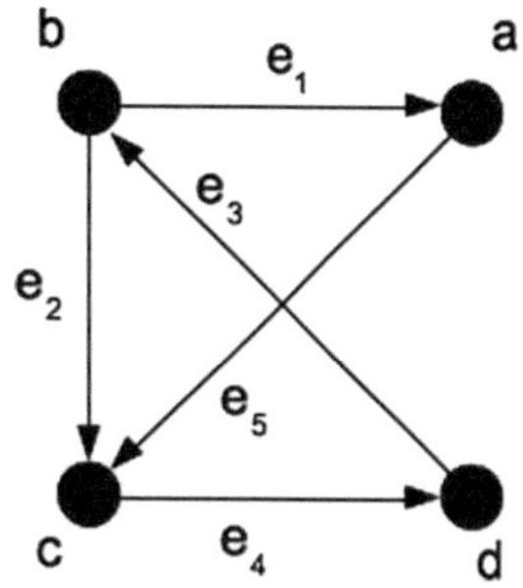

Fig. 7.5 An example of a strongly connected graph: Say, for example, the set of directed paths P from node a to the other three nodes is given by $P = \{(a, e_5, c, e_4, d, e_3, b), (a, e_5, c), (a, e_5, c, e_4, d)\}$

The connection between a pair of nodes in a network is often represented by *adjacency matrix* or connection matrix. The adjacency matrix of the graph in Fig. 7.1a of N nodes and no parallel edges is an N by N symmetric binary matrix $A = [a_{ij}]$, where

$$x_{ij} = 1, \text{ if there is an edge between node i and j}$$
$$= 0, \text{ if there is no edge between them}$$

$$\mathcal{A} = \begin{array}{c} \\ a \\ b \\ c \\ d \end{array} \begin{array}{c} \begin{array}{cccc} a & b & c & d \end{array} \\ \begin{pmatrix} 0 & 1 & 1 & 0 \\ 1 & 0 & 1 & 0 \\ 1 & 1 & 0 & 1 \\ 0 & 0 & 1 & 0 \end{pmatrix} \end{array} \tag{7.5}$$

7.2.6 Average Shortest Path Length

Average Shortest Path Length (l) or the characteristic path length of a network is the sum of all the shortest path lengths between each pair of nodes in a graph averaged over all possible edges in a network.

$$L = \frac{1}{N(N-1)} \sum_{i,j \in V, i \neq j} d_{ij} \tag{7.6}$$

The above definition, however, fails in case the network has more than one connected component. One way of dealing with it is to restrict the sum over the nodes belonging to the largest connected component of the network. Another approach is to assign infinite distance between the pair of disconnected nodes or the pair of nodes having no connected path, and then take the harmonic mean of the shortest path between the pair of nodes in the network. The latter gives a quantitative measure, called the Efficiency of the network, which is defined as follows

$$E = \frac{1}{N(N-1)} \sum_{i,j \in V, i \neq j} \frac{1}{d_{ij}} \tag{7.7}$$

7.2.7 *Eccentricity*

Eccentricity $E(i)$ of a node i in a graph G is the maximum value of all the geodesic distances calculated from that particular node i to all other nodes j in the network.

$$E(i) = \max_{i \in V} d(i, j) \tag{7.8}$$

The eccentricity of a node i represents how close or distant is i from the farthest node of the network. The node with minimum eccentricity in graph G is called the *centre* of G.

7.2.8 *Diameter*

The diameter of a graph refers to the maximal distance between any pair of its nodes. The diameter of a disconnected network, composed of more than one isolated components or clusters, is infinite. So, for practical purposes, in such cases, it may be defined as the maximum diameter of its components.

7.2.9 *Closeness Centrality*

The closeness centrality C of a node n_i is the inverse of the sum of its distances to all other nodes, n_j. Mathematically, it is defined as

$$C(n_i) = \frac{N-1}{\sum_{j=1}^{g} d(n_i, n_j)} \tag{7.9}$$

Closeness of a node signifies the efficiency of a node to convey information within the network. For example, consider a star graph as shown in Fig. 7.6. In this graph the node i is the most centrally located node in the graph. Thus, it spreads information much faster than any other node in the network can.

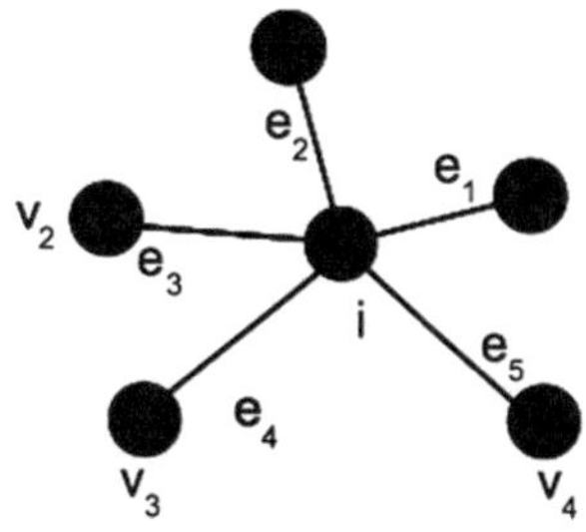

Fig. 7.6 Star graph

Fig. 7.7 In this figure, the node having highest betweenness is *i*

7.2.10 Betweenness Centrality

The betweenness centrality of a node measures the node's involvement in the communication paths of other nodes in the network.

$$B(v) = \sum_{s \neq v \neq t} \frac{\sigma_{st}(v)}{\sigma_{st}} \tag{7.10}$$

where σ_{st} is the total number of shortest paths from node s to node t and $\sigma_{st}(v)$ is the number of those paths that pass through v (Freeman 1977).

For better understanding of this centrality, consider the graph shown in Fig. 7.7. Here, nodes in the graph can be divided into two groups. These two group of nodes are connected by a single node i. Hence the betweenness centrality value of node i is the highest among others. If one wants to travel from one node lying in one cluster to another in the other cluster, then the path passing through node i is the only way. Another important realisation of this centrality can be gained while analysing this graph. If the node i from the network is removed (along with the edges incident on it), the graph becomes disconnected, with two connected components. Removal of the high betweenness nodes will result in either of the two following consequences. In one case, the communication among different clusters may get completely lost, as in the above mentioned example. In the other one, the cost of traveling may get enhanced since the path will comprise of more edges than before. These high betweenness nodes are often called as bottlenecks of the network.

7.2.11 Clustering Coefficient

It is a measure which accounts for the tendency of a node in a network to cluster together. This behaviour is commonly observed in most real world networks, in particular social networks. Clustering can either be global or local, depending on the overall clustering of nodes in the whole network or the property of the single node. The definition of *Global Clustering Coefficient* (GCC) is based on the concept of triples of nodes. A triple consists of three nodes which remain connected by either three (closed triple) or two (open triple) undirected edges. GCC is the ratio of the number of triangles to the number of connected triples.

$$C' = \frac{3 \times \text{Number of triangles}}{\text{Number of connected triples}} \tag{7.11}$$

The *Local Clustering Coefficient* (LCC) of a node in a graph gives a quantification of the proximity of its neighbours from becoming a completely connected graph. It can be defined in the following way. A node, i with k_i neighbours, can have, at most, ${}^{k_i}C_2 = \frac{k_i(k_i-1)}{2}$ number of possible edges in its neighbourhood. Suppose, the neighbours of node i are connected by e_i edges, then the LCC of that node is defined as

$$c_i = \frac{2e_i}{k_i(k_i - 1)} \tag{7.12}$$

Therefore the Clustering Coefficient of the whole graph can be obtained by taking average of c_i over all the nodes in G:

$$C = \langle c \rangle = \frac{1}{N} \sum_{i \in N} c_i \tag{7.13}$$

7.2.12 *Cliques and Community Structure*

In a complex network having large number of nodes and edges, a *k-Clique* is defined as a completely connected subgraph having a set of k nodes in which each node is connected to every other node by an edge in that subgraph. Two k-cliques will belong to the same *community* when they share $k - 1$ nodes.

7.2.13 *Modularity*

A relatively independent unit, called modules (also called groups, clusters or communities), is often present in a complex network. Modularity is a quantitative measure which describes the extent to which a system is divided into modules. A network with high modularity value will be endowed with intense connections among nodes within a module but sparse or minimal links to other modules in the network. Mathematically, modularity is defined as

$$M = \sum_{i=1}^{m} = \left[\frac{e_i}{E} - \left(\frac{d_i}{2E} \right)^2 \right] \tag{7.14}$$

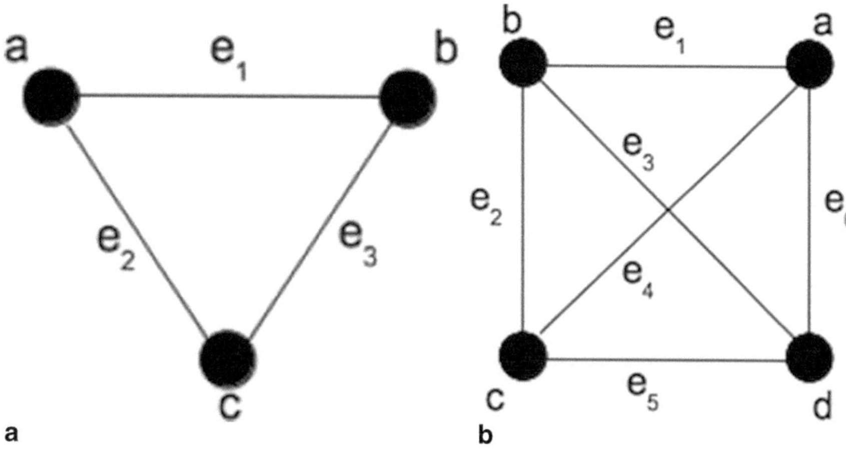

Fig. 7.8 **a** 3-clique and **b** 4-clique

where E is the total number of edges in the network, e_i is the number of edges within module i, d_i is the sum of degrees of all the nodes of module i, and the summation runs over total number of modules m in the network (Fig. 7.8).

7.2.14 k-Core (or k-Shell) Decomposition

K-core decomposition method provides us a hierarchical representation of the network. A k-core of a graph G is a maximal subgraph of G in which each node is connected to at least k other nodes in the subgraph. A node i belongs to a *k-shell* if and only if it belongs to the kth-core but not to the $k+1^{\text{th}}$-core.

The k-core decomposition is based on sequential removal of nodes along with its edges. Let us consider a connected graph G. At first, all nodes with degree d = 1 are removed from the graph G. After their removal, new nodes with degree d = 1 may appear in G. The pruning process is continued until all the nodes with degree d = 1 are removed. These nodes together with their incident edges forms the $k_s = 1$ shell. In a similar fashion the higher degree nodes are removed to obtain the $k_s = 2$ shell and so on. The process is repeated until all the nodes from the graph G have been removed.

The network topology plays a significant role in portraying the interactions within the nodes. Such decomposition have been used by many researchers to analyse the real world networks (Wuellner et al. 2010). The k-core decomposition of PPI network of yeast has revealed that the proteins belonging to the innermost core have higher probability of being both essential and evolutionary conserved (Wuchty et al. 2005). Judicious introduction of new parameters like synthetic accessibility have demonstrated sufficient promise in predicting the viability of knockout strains with accuracy comparable to approaches using biochemical parameters (like FBA etc.)

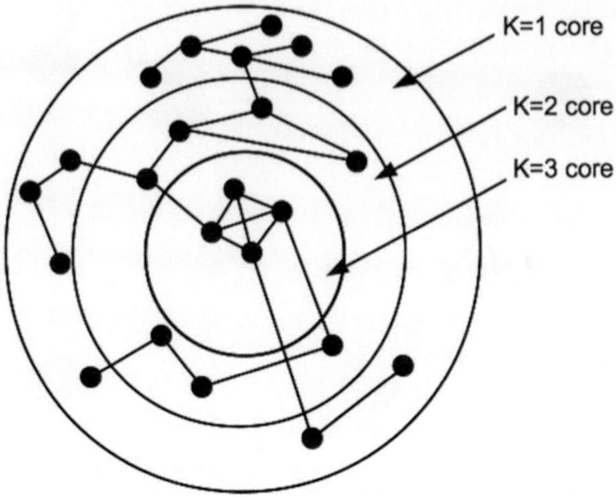

Fig. 7.9 *k-core* decomposition of a simple graph

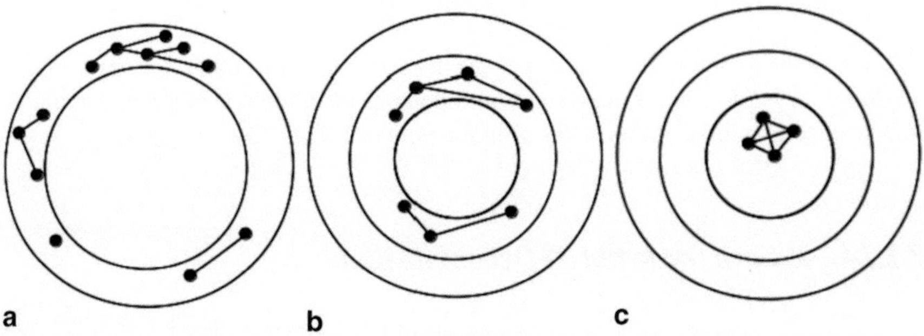

Fig. 7.10 *k*-shells (k_s) of the graph *G* in Fig. 7.9. **a** $k_s = 1$, **b** $k_s = 2$ and **c** $k_s = 3$

on large, unbiased mutant data sets (Wunderlich et al. 2006). Another recent topic where network metrics are thought to play a significant role is the controllability of biological networks (Banerjee et al. 2012; Fig. 7.10).

In this section we have hopefully presented an elaborate introduction to network metrics. Recent research has however conclusively shown that instead of looking at just one or two metrics, it is imperative that we look at multiple metrics in parallel to get the most informative picture (Filkov et al. 2009; Roy 2012, 2014).

7.3 Random Graph Theory

7.3.1 Erdos-Renyi Graphs (ER Graphs)

Erdos-Renyi Graphs are random graphs where edges are constructed between all pairs of nodes with some equal probability (say *p*), independent of one another. The

Fig. 7.11 **a** Positive autoregulation: activation of gene *A* by its own product, **b** Negative autoregulation: deactivation/inhibition of gene *B* by its own product

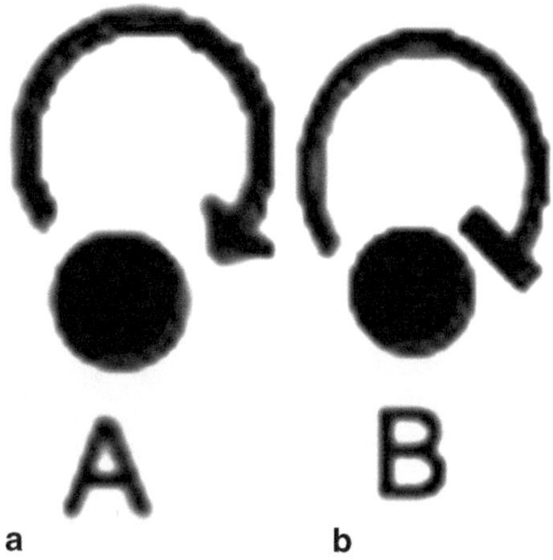

degree distribution profile of ER graphs shows Poisson distribution. The ER Graphs have low clustering coefficients and the average path length are found to be smaller compared to the real world networks.

7.3.2 *Small World Networks*

Networks having smaller average path length comparable to the ER graphs of similar size and order but larger clustering coefficient than ER graphs are termed as small world networks. The average shortest path length of the small world networks scale as logarithm of the number of nodes in the network i.e.

$$L \propto logN \tag{7.15}$$

Most of the real networks exhibit small-wold property. The small world feature is thus common to most biological networks such as neural network of *C. elegans* and Food web.

7.4 Motifs in Network

Motifs in a network refer to a particular pattern of subgraphs that appear more commonly than what is expected to occur in a random graph. Motifs are much more abundantly present in biological networks than other type of networks. Self loops, i.e., the edges which originate and terminate in the same node, can be thought of as the simplest network motif. This will refer to autoregulation, or autogeneous control, e.g., regulation of a gene by its own gene product, in a transcription network (Fig. 7.11).

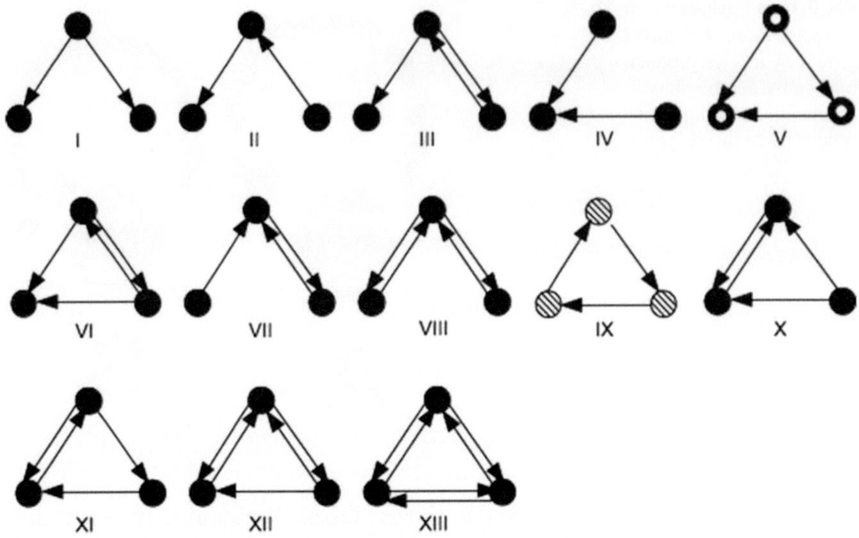

Fig. 7.12 The 13 possible three-node directed subgraphs. Subgraph *V*, having annular nodes, is the Feed Forward Loop (FFL), while subgraph *IX*, with striped nodes, is the Feed Back Loop (FBL)

Autoregulatory network may be positive or negative. For instance, in the former case, the genes activate their own transcription, while in the latter, the genes act as repressors. Negative autoregulation has many advantages. It speeds up the response time of gene circuits. Also, it promotes robustness of the steady-state expression level to fluctuations in production rate. In contrast, positive autoregulation slows down responses. In addition, the system exhibits bistability when the rate of positive autoregulation is strong compared to the degradation/dilution rate. The next interesting step will be to look at three-node patterns. There are 13 such patterns, as shown in Fig. 7.12. Out of these thirteen patterns, the only significant one is the Feed Forward Loop (FFL), Fig. 7.12 (V), as found in the sensory transcription network of *E. coli* and yeast (Lee et al. 2002; Milo et al. 2002). It is a strong network motif which appears more often than its randomised version. A straight forward description of a FFL would be as follows. It is composed of a transcription factor, say X, which regulates a second transcription factor, Y, and both X and Y regulate gene Z. It has two parallel paths of regulation, a direct path that goes from X to Z, consisting of a single edge, and another indirect one via Y, having a cascade of two edges. A plus sign or a minus sign is assigned to each of the edges corresponding to activation and repression respectively. So there are $2^3 = 8$ possibilities, out of which four are coherent FFL and the rest four are incoherent. This grouping is based on the comparison between the signs of the direct and the indirect paths. If both comes out to be the same, then we get coherent FFLs, and incoherent ones have opposite signs. Incoherent FFLs have an odd number of minus signs and the two paths possess an antagonistic effect. Among all the eight different types, Coherent Type-I, followed by the Incoherent Type-I, are the two most abundant FFLs present across various biological networks. Feedback Loops (FBL) (Fig. 7.13).

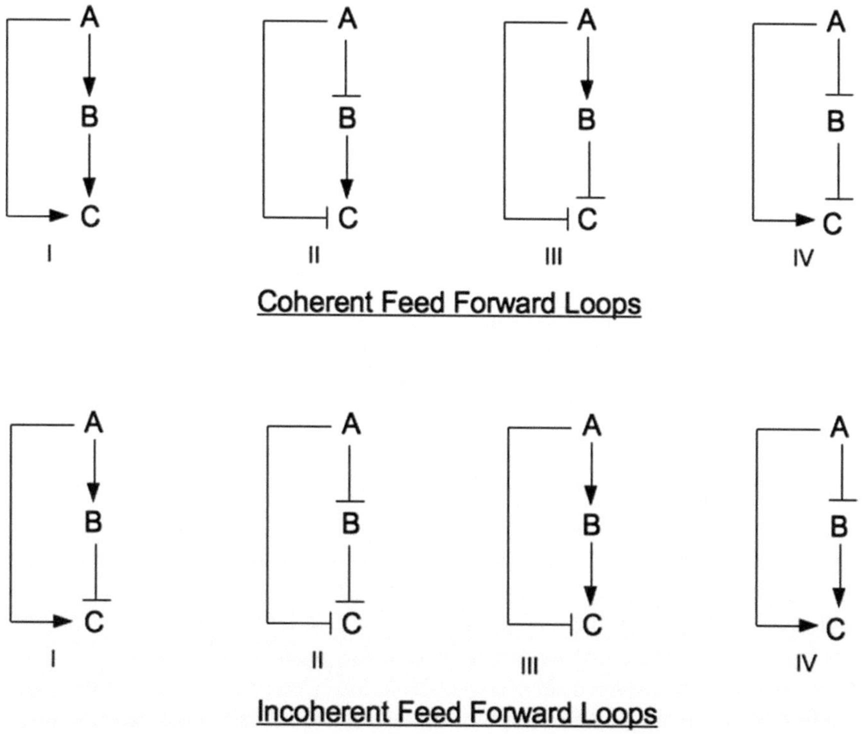

Fig. 7.13 The eight possible Feed Forward Loops (FFLs). The *upper four* are the coherent FFLs, while *lower four* are incoherent FFLs. $\downarrow$ denotes the activation (+ sign) and $\perp$ denotes inhibition (− sign)

7.5 Gene Regulatory Network (GRN)

Genes are fragments of DNA molecules which carry the genetic code in the form of a sequence constituting four nucleotides, viz., adenine (A), thymine (T), guanine (G) and cytosine (C). Each individual gene has its own characteristic genetic code and genes are collectively responsible for various functions in a living organism. The two step process in which at first the information encoded in the nucleotide sequence of a DNA gets decoded to messenger RNA (mRNA) and then proteins are synthesised to perform all the essential biochemical functions is called gene expression. The former step is called *Transcription* while the latter is the *Translation*. A number of genes act together to perform a definite biological function. To depict this, we can think of an interactive network of fragments of DNA or mRNA (nodes) which governs the rate of gene expression, i.e., the rate of protein synthesis, which is known as a Gene Regulatory Network or GRN.

7.6 Networks of Proteins

Protein, the most important biological macromolecule, which performs almost all the essential functions in a living organism; is a polypeptide chain formed from 20 possible amino acids. To accomplish various biological functions, the protein folds to attain a well defined three dimensional spatial conformation (often called as the native state). This native state correspond to the global minima of the energy landscape. The protein folding is driven by a number of non covalent interactions, viz., hydrogen bonding, van der Waals force, ionic and hydrophobic interactions, among its constituent amino acids. To visualise this interaction, one may take recourse to networks. Proteins can be modelled into a network containing amino acid residues as nodes and two of the residues are linked together if they interact.

7.6.1 Protein Structure Network (PSN)

Protein Structure Networks (PSN) are based on the geometrical distance between different amino acids. Geometrical considerations provide deep insights to protein folding. PSN's identify the $C\alpha$ atoms of the amino acid residues as nodes. Two residues are said to interact with each other if the geodesic distance between their $C\alpha$ atoms is less than a fixed cut-off value like 8.5 A^o (Vendruscolo et al. 2002). Such a representation mainly emphasises the backbone chain interactions of the proteins. A few selected nodes (often called *key residues*), from these networks which have high betweenness centrality; correspond to the previously known nucleation centres for protein folding. The residues identified by such graphical properties are sometimes investigated further for their role in providing unique structure to the protein native structure. However such a formalism of PSN disregards the side chain interactions of the amino acids within the polypeptide chain. Side chain interactions are essential for maintaining the 3D structure of the protein. To encapsulate these interactions, a different mechanism for designing PSN has been proposed. Instead of considering the $C\alpha$ atoms only, connections were established for any two atoms of the amino acid residues whose distance falls within the fixed cut-off. Many such PSNs with varying cut-off distances to probe the long-range and short-range interactions within a protein have been explored (Greene et al. 2003). The short-range interactions networks show small world property while single-scale behaviour in degree distribution was observed for long-range interactions networks. The latter was thought to confer robustness in the overall topology of the protein structure against random mutations. An alternative study incorporated only the non-covalent side chain interactions of the amino acid residues (Kannan et al. 1999). The interactions were defined on the basis of specific minimum *interaction strength*. The cluster profile and hubs in these networks were identified to play a significant role in secondary structural integration in a tertiary structure of proteins. The hubs also play a crucial role in enhancing the thermal stability of the thermophilic proteins when compared to their mesophilic counterparts (Brinda et al. 2005).

7.6.2 *Protein Energy Network (PEN)*

Thus, we have seen that PSNs can capture the atomic interactions of proteins at geometric level very well. Though they overlook the the basic chemistry of bonded and non-bonded interactions. The energies of these interactions result from various types of interactions, e.g., hydrogen bonding, hydrophobic interactions, cation-pi interactions etc. taking place within a protein. The networks, which account only non-bonded interaction energies, viz., van der Waals interaction (vdW) and electrostatic interaction energy of the side chain atoms of the amino acid residues, are termed as Protein Energy Networks (PEN) (Vijayabaskar et al. 2010).

The various amino acid residues are the nodes of the network. Edges are defined between the residues i and j, if the non-bonded interaction energy, E_{ij}; is less than a cut-off energy e. Since interaction energies between different pairs vary, the resulting PEN is an undirected weighted network. Vijayabaskar et al. had explored PEN for six different proteins. The interaction energies were calculated from equilibrium ensembles obtained by performing Molecular Dynamics (MD) simulations. They observed that the networks are densely connected i.e they have more number of interactions for small energy cut-off e (less negative, $\sim -5\ kJ/mol$). As the cut-off interaction energy is increased to high negative values ($\sim -25\ kJ/mol$) the network becomes more sparsely connected i.e it has low number of interactions or edges connecting the nodes. The fractional contribution of vdW and electrostatic energy to the total energy was also analysed. The vdW interaction energy dominates the region of low interaction energy (less negative values) and its value falls off to zero for $e \sim -35\ kJ/mol$ while reverse is the case for electrostatic interaction energy which dominates high interaction energy region (high negative values). Another important observation was that the PEN breaks down into small independent clusters within a small window of e. For less negative values of e, a large cluster percolates within the network which can be quantified by the tethering together of small independent clusters within the PEN by weak vdW interactions; as the value of e is made to have less negative values. This provides an evidence for weak interactions (rather than strong interactions) holding together the 3D structure of a protein. The cluster profile of the network helps in understanding the structural integrity of the proteins.

7.6.3 *Allostery and Protein Energy Network*

Recently allosteric mechanism has drawn much attention in the field of research. Allostery can be defined as the control of protein structure, function and/or flexibility induced by the binding of a ligand or another protein, which is called an effector, at a site away from the active site (allosteric site) (Goodey et al. 2008)

Loosely speaking, allostery is a regulation between two distant sites of a protein caused by binding of ligands. PEN serves as a useful tool to explore this mechanism of communication within the proteins. The communication paths between the two functional sites of a protein can be elucidated by tracking the shortest path in the weighted PEN (Bhattacharya et al. 2011). The shortest paths between a pair of residues in these networks, from energy point of view, will be the ones which are less costly or energetically more favourable. To achieve this weights assigned to the edges have values proportional to the reciprocal of the interaction energy among the pair of residues. The suboptimal paths of the network with reduced efficiency were also explored by deleting all the edges incident on any one of the residues belonging to the optimal paths (the shortest path). An interesting observation was the presence of these suboptimal paths as the optimal paths in less frequently accessed conformations during MD simulations and thus effectively act as alternate paths of communication adapted due to mutation/ligand induced perturbations. Such insights gained by analysing PENs support theoretical as well as experimental observations of the concept of transmission of allosteric signals through multiple, preexisting pathways (de sol et al. 2009).

7.6.4 Protein Protein Interaction Network (PPI Networks)

Most fundamental biological processes are carried out by proteins and their interactions. Proteins usually execute their functions through interactions with other biomolecular units, rather than acting in isolation. In this type of networks, proteins are nodes and if there is an experimental verification regarding binding between two proteins, then an edge is drawn between the two. Previous studies have discussed whether PPI networks are scale-free in nature. Such a study of a PPI network for yeast shows that its degree distribution follows a power law with an exponential cut-off (Jeong et al. 2001). In scale-free protein networks, most proteins participate in very few interactions, while few hubs are involved in most of the interactions. Another characteristic property is that small-world effect is also present in PPI networks which indicates that any two proteins are connected by a short path of very few links. These networks are disassortative in nature, i.e., highly connected nodes are seldom connected among themselves. The elimination of a protein often causes functional disruption of a module in a PPI network. Such proteins are termed as *lethal*. Thus lethality of a protein is the decisive factor characterising the biological indispensability of a protein.

7.6.5 Protein Folding Network

During folding, a protein takes up consecutive conformations. Distinct conformational states are represented by nodes in the network and two of them are linked by an edge if one can be obtained from another by an elementary move. It has been studied that the network formed by the various conformations of a 2D lattice polymer has small world properties (Scala et al. 2001). The degree distribution has been found to be consistent with a Gaussian (Amaral et al. 2000)

7.7 Metabolic Networks

Metabolism, a set of biochemical reactions essential for sustaining life, is one of the various life processes taking place within an organism. The metabolism of a compound involves a sequence of reactions, termed metabolic pathway, in which the initial compound is transformed into various other intermediary compounds to get the product by the action of enzymes. The intermediaries and the products of such chain reactions are termed as metabolites. It may happen that the product of one pathway is served to initiate some other pathway.

In metabolic networks, the nodes correspond to the substrates (ADP, ATP, H2O) and the edges represent the predominantly directed chemical reactions among these substrates. For 43 organisms, these networks have been studied (Jeong et al. 2001) and for all of them; the degree distribution of the incoming and outgoing links have been claimed to follow a power law, with the exponent value in the range 2.0–2.4. There have also been alternate representation of these networks: ATP, ADP, NADH are included as nodes only if they directly take part in the reaction (Ma et al. 2003). Such metabolites are called current metabolites and are ignored while measuring the average path length of the network during their indirect participation in the reaction.

It was found that the path lengths of the metabolic networks in eukaryotes are longer than that of bacteria. Small world property was found in *E. coli* by representing metabolic networks as two complementary networks—substrate graph and reaction graph. It was hypothesised that since metabolic networks respond to perturbations (like changes in concentration of the metabolite or the enzyme), their function could be optimised by the small-world behaviour of the network (Wagner et al. 2001).

7.8 Networks and Epidemiology

We can get deep insights into the dynamics of disease spreading in an interacting population of species by applying network theory. Here, we briefly describe two well known spreading models on networks and recent developments about influential spreaders in networks.

7.8.1 *Susceptible Infectious Recovered (SIR)*

In a network of N nodes, initially we assume one node is in the infectious state (I) and the rest in the susceptible state (S). This node, denoted by I, is the origin of Infection. The infection gets propagated in successive time steps. In each time step, nodes of type I infects neighbours, which are susceptible to infection, with some probability β. They then enter the recovered state (R), where they cannot be infected again, i.e., they achieve immunity against infection.

7.8.2 Susceptible Infectious Susceptible (SIS)

Here the immunised or recovered state of the origin, just after infecting the neighbours, is absent. Infected individuals still possess the capability of infecting their neighbours with probability β. However, they may subsequently return to the susceptible state with probability λ; thus remaining infectious with probability $(1-\lambda)$.

7.8.3 Influential Spreaders in Networks

A common belief related to infection or disease spreading is that the best (efficient) spreaders will correspond to a highly connected nodes (high degree) or to the most central nodes (having high betweenness value). It has been argued that the network topology should naturally play an important role in infection spreading or information spread. The position of a node in the network serves as a deciding factor for it to be the most influential spreader. The k-shell decomposition method was performed on a set of eight real social networks and both SIS and SIR model were studied (Kitsak et. al. 2010). The nodes in the innermost k-shell were claimed to be the most efficient spreaders.

7.9 Conclusion

In this chapter, we have hopefully given an overview of how complex networks are important at every level in biology. In Sect. 7.1, we mention how biology has shifted from a reductionist approach to holistic approach. Hence deriving a network picture is of immeasurable value because complex networks understandably play an integral part in this new approach. We went on to introduce the very basics of a network or graphical representation; namely nodes, edges, weighted networks etc. In Sect. 7.2, we dwell in-depth on common network metrics like degree, shortest path length, connectedness, giant clusters, cliques and community structure, eccentricity, diameter, closeness and betweenness centralities, clustering coefficient, assortativity, k-core and modularity. In the next section, we briefly discuss about small-world properties and random networks which serve as a good reference points in networks. We then discuss the concepts regarding motifs and their importance in biological networks. In Sect. 7.5, we discuss about interactive Gene Regulatory Networks of fragments of DNA or mRNA (nodes) which governs the rate of gene expression, i.e., the rate of protein synthesis. In Sect. 7.6, we discuss about networks of proteins: protein structure networks, protein energy networks and protein-protein interaction networks and protein folding networks. In Sect. 7.7, we discuss about metabolic networks. Finally, in Sect. 7.8 we end this chapter with a discussion of concepts and models which deal with spread of infection on networks. Thus, we have hopefully been able to portray the importance of complex networks to understand processes at virtually every level of life.

References

Albert R, Barabasi A-L (2002) Statistical mechanics of complex networks. Rev Mod Phys 74:47–97

Albert R, Jeong H, Barabasi A-L (2000) Error and attack tolerance of complex networks. Nature 406:378–382

Amaral LAN, Scala A, Barthelemy M, Stanley HE (2000) Classes of small-world networks. Proc Natl Acad Sci U S A 97:11149–11152

Banerjee SJ, Roy S (2012) Key to network controllability arxiv:1209.3737

Barabasi AL, Oltvai ZN (2004) Network biology: understanding the cells functional organization. Nat Rev: Genet 5:101–113

Bhattacharyya M, Vishveshwara S (2011) Probing the allosteric mechanism in Pyrrolysyl-tRNA synthetase using energy-weighted network formalism. Biochem 50:6225–6236

Brinda KV, Vishveshwara S (2005) A network representation of protein structures: implications for protein stability. Biophys J 89:4159–4170

Filkov V, Saul ZM, Roy S, D'Souza RM, Devanbu PT (2009) Modeling and verifying a broad array of network properties. EPL (Europhys Lett) 86:28003

Freeman LC (1977) A set of measures of centrality based on betweenness. Sociometry 40:35–41

Goodey NM, Benkovic SJ (2008) Allosteric regulation and catalysis emerge via a common route. Nat Chem Biol 4:474–482

Greene LH, Higman VA (2003) Uncovering network systems within protein structures. J Mol Biol 334:781–791

Hongwu M, An-Ping Z (2003) Reconstruction of metabolic networks from genome data and analysis of their global structure for various organisms. Bioinformatics 19:270–277

Jeong H, Tombor B, Albert R, Oltvai ZN, Barabasi AL (2000) The large-scale organization of metabolic networks. Nature 407:651–654

Jeong H, Mason SP, Barabasi A-L, Oltvai ZN (2001) Lethality and centrality in protein networks. Nature 411:41–42

Kannan N, Vishveshwara S (1999) Identification of side-chain clusters in protein structures by a graph spectral method. J Mol Biol 292:441–464

Kitano H (2002) Systems biology: a brief overview. Science 295(5560):1662–1664

Kitsak M, Gallos L, Havlin S, Liljeros F, Muchnik L, Stanley HE, Makse HA (2010) Identification of influential spreaders in complex networks. Nat Phys 6:888–893

Lee TI et al (2002) Transcriptional regulatory networks in *Saccharomyces cerevisiae*. Science 298(5594):799–804

Ma H et al (2003) Reconstruction of metabolic networks from genome data and analysis of their global structure for various organisms. Bioinfomatics 19:270–277

Milo R et al (2002) Network motifs: simple building blocks of complex networks. Science 298(5594):824–827

Newman MEJ (2002) Assortative mixing in networks. Phys Rev Lett 89:208701

Newman MEJ (2010) Networks: an introduction. Oxford University Press, Oxford

Roy S (2012) Systems biology beyond degree, hubs and scale-free networks. Syst Synth Biol 6:31–34. doi:10.1007/s11693-012-9094-y

Roy S (2014) Networks, metrics and systems biology. In Kulkarni V, Stan G-B, Raman K (eds) A systems theoretic approach to systems and synthetic biology I: models and system characterizations. Springer, Heidelberg. DOI:http://dx.doi.org/10.1007/978-94-017-9041-3_8

Roy S, Filkov V (2009) Strong associations between microbe phenotypes and their network architecture. Phys Rev E 80:040902 (R)

Scala A et al (2001) Small-world networks and the conformation space of a short lattice polymer chain. Europhys Lett 55:594

Vendruscolo M, Dokholyan NV, Paci E, Karplus M (2002) Small-world view of the amino acids that play a key role in protein folding. Phys Rev E 65:061910

Vijayabaskar MS, Vishveshwara S (2010) Interaction energy based protein structure networks. Biophys J 99:3704–3715

Wagner A, Fell D (2001) The small world inside large metabolic networks. Proc Roy Soc London Series B 268:1803–1810

Wuchty S, Almaas E (2005) Peeling the yeast protein network. Proteomics 5:444–449

Wuellner DR, Roy S, D'Souza RM (2010) Resilience and rewiring of the passenger airline networks in the United States. Phys Rev E 82:056101

Wunderlich Z, Mirny LA (2006) Using the topology of metabolic networks to predict viability of mutant strains. Biophys J 91:2304–2311

Chapter 8
Systems Approaches to Study Infectious Diseases

Priyanka Baloni, Soma Ghosh and Nagasuma Chandra

Abstract Exposure to infectious agents can either lead to active disease or containment or killing of the pathogen. Outcome of an infectious disease is determined by the complex interplay between the host and the pathogen. Therefore, understanding the crosstalk between the host and the pathogen during infection is crucial to identify molecules that are important for the spread or suppression of the disease and for identification of drug targets. Both the host and the pathogen have several mechanisms for countering each other thereby adding layers of complexity to the host-pathogen interplay. Reconstructing mathematical models of complex processes such as cell regulations, signal transductions and host-pathogen interactions provide a detailed understanding of the various interactions and crosstalks occuring in a biological system and thus form a platform to study the system as a whole. Various experimental methods in functional genomics and proteomics as well as computational approaches have been developed over the years that help in building and modeling the biological systems. These approaches have proved quite helpful in identifying drug targets, generating hypotheses rationalizing and finally predicting the cause andfinal outcome of diseases.

Keywords Infectious disease · Host-pathogen interactions · Flux-balance analysis · Stoichiometric matrix · Response networks · Gillespie's algorithm · Boolean modelling · Networks · Tuberculosis · Malaria · Plasmodium falciparum · plasmoDB · Cholera · targetTB

N. Chandra (✉) · S. Ghosh · P. Baloni
Department of Biochemistry, Indian Institute of Science, Bangalore, India
e-mail: nchandra@biochem.iisc.ernet.in

S. Ghosh
e-mail: soma@mbu.iisc.ernet.in

P. Baloni
e-mail: baloni@mbu.iisc.ernet.in

V. Singh, P. K. Dhar (eds.), *Systems and Synthetic Biology*,
DOI 10.1007/978-94-017-9514-2_8

8.1 Introduction

Infectious diseases are directly responsible for about a third of all deaths occurring worldwide. Tuberculosis, pneumonia, malaria, cholera are among the most fatal infectious diseases, responsible for 58 % child mortality in developing nations (WHO 2012). These infectious diseases can be categorized depending upon their frequency of occurrence into sporadic, endemic, epidemic or pandemic diseases. Although several anti-infective drugs are available for these diseases, they continue to be a burden to human health, a problem further compounded by the emergence of drug resistant varieties of the pathogens (Spellberg et al. 2008), (MacPherson et al. 2009). Discovery of newer, safer and robust drugs require the formulation of new strategies that involve innovative ways of tackling the diseases. It has now become increasingly clear that strategies stemming from holistic system approaches may hold the key for effective and sustained management of infectious diseases (Aderem et al. 2011). A wealth of molecular level data has been gathered over the years on several causative microorganisms, which has increased substantially due to the advances in genomics and other high-throughput technologies. The scale and the complexity of each piece of data, is indeed quite high and requires computational analysis to help in comprehending and making useful inferences from it.

Systems biology is the study of large scale systems, reconstructed from many small scale interactions. This approach is based on the premise that the 'whole is greater than the sum of its parts' (Hood and Perlmutter 2004). It provides a holistic understanding of the biological function from molecular and cellular level to an entire organism and serves as a platform to study and correlate the processes occurring in a complex living system at different scales to understand a biological phenomenon. Application of such computational methods is evident in the field of drug discovery. Simulations using reconstructed models further aid in knowledge based drug target identification, discovery of biomarkers as well as for rational design of vaccines. Overall, studying a system as a whole rather than individual molecular characterizations performed in isolation would be required to understand the phenotypic behaviour of a given system.

With advances in techniques such as high-throughput sequencing, microarrays, nuclear magnetic resonance and mass spectrometry, it is now possible to get better insights into the field of transcriptomics, proteomics and metabolomics, and the data generated using these techniques serve as direct inputs into development of systems level models. The large scale omics data are analyzed using computational methods to derive essential molecular interactions. These molecular interactions are used to build a detailed mathematical model to represent the biological system being studied. Once validated, these models are used to simulate a range of scenarios to predict the behaviour of the system under various conditions. The hypotheses generated can be taken back to the bench again and validated using focused experimental studies (Aderem et al. 2011; Vodovotz et al. 2008). Systems biology, thus, along with different '*omics*' studies is being increasingly used to identify pathways involved in specific disease conditions, establish interconnectedness of different pathways and understand cellular responses to variouscertain conditions including physiological stress and exposure to a pathogenicorganism (da Hora Junior et al. 2012; Day et al. 2010; Kitano 2002; Weckwerth 2003; Weston and Hood 2004).

The study of host-pathogen interactions focuses upon the interactions between microbial or viral pathogens and their plant or animal hosts. The interactions are multi-faceted and form a complex network including moves and counter-moves from both species leading to one of two broad outcomes, either clearance or proliferation of bacteria (Forst 2006; Johanns et al. 2010). Using systems biology approaches it has become feasible to study various phenomena such as recognition of the pathogen by the host immune system, mechanism of virulence, pathogenesis, mechanisms of antibiotic resistance, persistence of disease all as aspects of the complex host-pathogen interplay, the knowledge ultimately useful for biomarker and drug target identification (Weston and Hood 2004; Wang et al. 2010a). Systems biology as a discipline, in fact utilizes both experimental and computational approaches to build computationally amenable mathematical models of complex biological processes. This chapter provides an overview of various systems biology approaches available for studying causative organisms that cause infectious diseases and also the interplay between host and pathogen. In particular, the chapter focuses on the various modeling approaches that are available and being utilized for such studies and summarizes various insights obtained for a few important infectious diseases.

8.2 Modeling Methods

Deciphering functions of individual components even at a genome scale is not sufficient to understand the complexity of the organism or the complex interplay between the host and pathogen. Availability of large scale genomics, proteomics and metabolomics data have led to advances in obtaining pair-wise interactions between pairs of molecules. Different pieces of data are required to be pooled together using mathematical formalisms to build up a biological system, which can be used to address various biological questions. This also provides a handle to the experimentalist to prioritize the proteins for functional studies. Various modeling methods that are commonly used in the field of systems biology are described briefly here and are also depicted in Fig. 8.1. The models are ordered according to the level of granularity in the figure.

8.2.1 Networks

The parts lists obtained from individual *omics* level experiments starting from the genome sequencing are assembled based on various molecular interactions obtained experimentally through a number of studies documented in literature. The list of protein-protein interactions are augmented substantially through a variety of

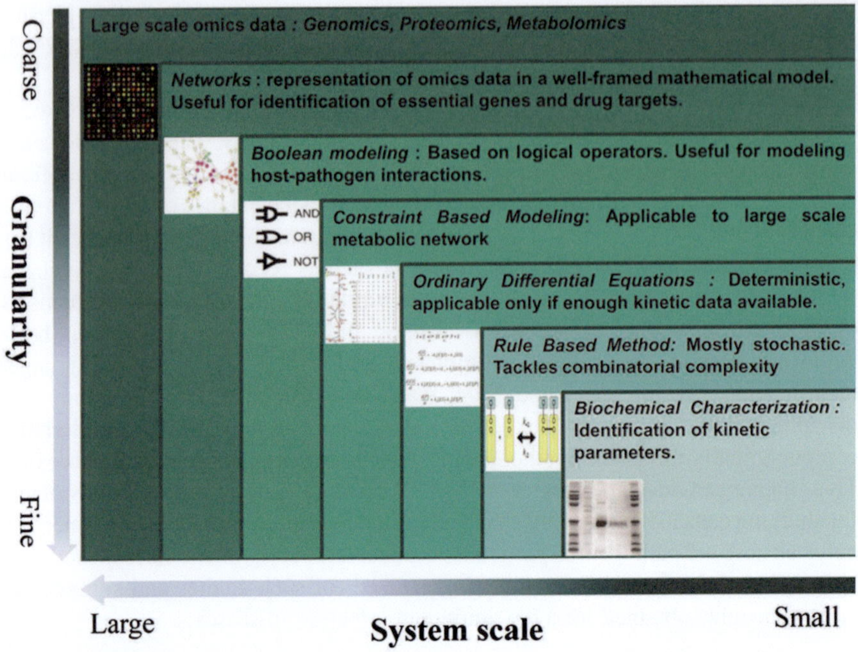

Fig. 8.1 The different modeling methods used in Systems biology. Methods are colour coded based on granularity

knowledge-based predictions using methods based on Rosetta stone concept (Marcotte et al. 1999), phylogenetic profiling (Pellegrini et al. 1999), gene-neighbourhood and its conservation (Dandekar et al. 1998). The set of pair wise interactions and genome-wide functional linkages (Strong et al. 2003) thus identified, ultimately lead to network reconstructions. Databases such as STRING (Szklarczyk et al. 2011)infact make this available to the community in a comprehensive manner.

Individual molecular constituents in the cell form nodes, while interactions between them form edges, put together forming large complex graphs. Graph theory can then be used to understand and explore various aspects of the cell in different conditions (Albert 2007). Depending on the system being reconstructed, directed (eg. signalling networks), undirected (protein-protein interactions) or bipartite networks (metabolite-enzyme) can be generated. The edges can be further weighed if appropriate experimental data is available. Protein-protein interaction networks representing interactomes serve to understand the dynamics of a biological cell. Shortest path analysis has been used to identify criticality of particular nodes in the network (Ravasz et al. 2002). Through systematic knock-outs or node or edge deletions, nodes leading to significant number of broken paths and hence their relative importance in the network is assessed. These networks can be further divided into sub networks based on the intra and inter connectivity and represent the different functional modules present in the system.

Although network analyses helps in identifying important and influential molecules in a system and study the communication between the molecules in detail, it is mostly static in nature and captures a single condition in most cases. Static networks do not provide a complete understanding of the system, but reflect a single snapshot of the numerous possible interactions that can occur as a result of the various adaptive and environmental changes at that instant of time. One approach to overcome this limitation is reported by Ideker et al., who integrated mRNA expression data into a yeast protein-protein and protein-DNA interaction network, to identify subnetworks that were most active under different conditions (Ideker et al. 2002). Active sub networks were identified by calculating the significant fold change of each gene in that subnetwork as a result of changing conditions. The high scoring subnetworks correlated well with known regulatory mechanism. Such active subnetworks that convey a systems response given an experimental condition are termed as response networks (Forst 2006).

Reconstruction of signaling networks, where nodes are signaling components and directional edges are the regulations, helps understand the signaling cascading events taking place inside a cell. Interactions can be tagged as positive or negative or stimulatory or inhibitory (Wang and Albert 2011). Importance of a node is determined by studying the effect of that node's deletion on the propagation of the signal. Minimal set of nodes that can perform signal transduction independently have also been identified using this method.

Organism specific metabolic networks have been constructed and studied using methods such as flux-balance analysis. This requires three basic types of data; (a) enzyme, corresponding substrates and products, (b) stoichiometric matrix of all reactions which gives the ratio in which the substrates and products participate in the reaction and (c) cellular location of the reaction (Feist et al. 2008). Biochemical pathways can be represented using different network types. In a metabolite network, metabolites form nodes and two nodes are connected if they share a substrate-product relationship. In a reaction network nodes represent reactions and two reactions are connected if the product of one forms a substrate for the other. Bipartite networks are useful representations to capture biochemical pathways. A bipartite network contains two types of nodes and an edge can only be drawn between two different types of nodes. In case of biochemical pathways, enzymes form one set of nodes, while metabolites form another set of nodes and a connection can be made only between an enzyme and a metabolite (Raman et al. 2006). Detailed networks can also be built where kinetic information is incorporated as weights in the network. Metabolic networks are analysed using the graph theory tools to identify hubs and cluster the reactions based on their functions. Other tools such as Petri-nets (Pinney et al. 2003) have also been used to study various properties of an organism. Cytoscape (Shannon et al. 2003) is used widely to visualize as well as perform basic network analysis. The Boost Graph Library (Siek et al. 2002) implementation of MATLAB is also frequently used to perform network analysis.

8.2.2 Constraint Based Modeling

Constraint based modeling approaches are being used widely for studying metabolism in a cell. Metabolic reactions are represented using a stoichiometric matrix of size m*n, where rows represent metabolites (m) and columns represent all the reactions (n) present in an organism. Entries in the matrix represent the stoichiometric coefficients of the metabolites in the reaction (Orth et al. 2010; Raman and Chandra 2009). Given the stoichiometric matrix (S), FBA aims to calculate the flux (v) through each reaction at steady state, such that $S.v = 0$. These models are further constrained to mimic biological systems such that a unique flux distribution for the organism is obtained using linear optimization. An interesting feature of FBA is its ability to perform single and multiple gene deletion knockouts. This is done by constraining the bounds of all the reactions coded by that gene to zero. This analysis helps in identifying essential genes and drug targets (Raman et al. 2005). Effect of inhibitors can also be studied by constraining the required reaction to a fraction of the wild type bounds. Segre et al. developed a variant of FBA known as MoMA (Segre et al. 2002), which unlike FBA is not solely based on optimizing the objective function. The idea being that any genetically modified organism may not achieve optimality since the mutant strains are not subjected to long term evolutionary pressures and may perhaps attempt to attain biological function via minimal changes in the flux distribution.

A major advantage of constraint based modeling is that they do not require a detailed understanding of the reaction mechanism or other kinetic parameters to perform *in silico* simulations. Many modifications to the original methods have been reported to incorporate gene expression data (Colijn et al. 2009) and other *omics* data (Schellenberger et al. 2011) to obtain a better mimic of the biological system under investigation. Various tools such as FAME (Boele et al. 2012), FASIMU (Hoppe et al. 2011), COBRA toolbox (Schellenberger et al. 2011), MetaFlux (Latendresse et al. 2012) have been developed over the years to perform FBA and its variants (Lakshmanan et al. 2012).

8.2.3 Kinetic Modeling Using Ordinary Differential Equations

Biochemical reactions have classically been represented as differential equations that define the rate of consumption or production of metabolites. Given the kinetic details of any set of reactions, one can build a mathematical model by forming a system of ordinary differential equations (de Jong 2002). Simulations from ordinary differential equations (ODEs) are much more reliable and precise as they are built and analysed using detailed kinetic parameters. An obvious advantage of this method over FBA is that the time evolution of the model can be studied to obtain a detailed understanding of the system, instead of only analysing the steady state behaviour. However, non-availability of kinetic data limits the broad applicability of this method. MATLAB is widely used to solve the system of ODEs contained in these models. Other software packages such as JDesigner (Sauro 2004), Cell Designer (Funahashi et al. 2003), and Copasi (Hoops et al. 2006)are also commonly used for this purpose.

8.2.4 Boolean Modeling

Boolean modeling also called as logic modeling is being used to model complex biochemical systems and capture the qualitative behaviour of the biological system. Each component in the model can exist in two states, either *on* or *off*. Transition from one state to another is encoded using logical operators. One of the major advantages of logic modeling is the ease with which complex molecular interactions can be represented and therefore these are widely used to model complex biological phenomenon such as apoptosis (Schlatter et al. 2009) or host-pathogen interactions (Raman et al. 2010). New methodologies are being continually developed that transforms Boolean models into a continuous model so as to study the time course evolution of a biological system. State transition rates of each nodes are calculated using mathematical tools such as Markov processes and multivariate polynomial interpolation (Wittmann et al. 2009; Stoll et al. 2012).

8.2.5 Rule Based Modeling

In a rule based model, the biological system is defined using a set of rules. These rules use the notation of a simple chemical reaction and describe the local events taking place inside a cellular system that eventually leads to the emergence of a global property. This method is based on the principle of Gillespie's algorithm (Gillespie 1977), according to which a cell is considered as a well-mixed system and interaction between any two molecules in the cell is dependent on the rate of interaction between the two and the abundances of each molecule interacting. This method is particularly useful when modeling any regulatory system as these systems are inherently complex in nature and have the potential to generate a variety of distinct species as a result of the cascading events that occur in such systems. Formally, due to combinatorial complexity arising from the set of possible interactions in the system, a large number of distinct species are generated, which can all be systematically studied and outcomes of specific scenarios predicted (Hlavacek and Faeder 2009). Rule based methods are also being explored as tools for multi-level modeling of biological systems (Maus et al. 2011). Software tools such as BioNetGen (Blinov et al. 2004), Kappa (Danos et al. 2008), RuleMonkey (Colvin et al. 2010) have been used for rule based modeling. These methods are generally stochastic in nature; however the rules can be rewritten as ODEs to build deterministic models.

8.2.6 Models of Host–Pathogen Interactions

Understanding the outcome of an infectious disease not only requires a detailed study of the host and pathogen system individually, but more importantly, the communication and the crosstalk that occurs between the two systems. Individual models of host and pathogens describing different biological processes are widely available and can be easily manipulated to obtain a host-pathogen model. Such models provide a detailed description of the crosstalk that exists between the two systems as well as the individual processes. This provides a realistic picture of the biological phenomenon being studied and also helps in extrapolating the influence of such crosstalk on host and pathogen.

Host-pathogen interactions have been modeled using several approaches, ranging from simpler models for the prediction of protein—protein interactions between the host and pathogen, to complex models for the metabolic and signal transduction networks. Kirschner and co-workers have developed a virtual model of the host immune response to *M.tb* using agent-based modeling methods (Marino et al. 2011). Numerous insights about critical factors and parameters governing host-pathogen interactions can be obtained through these studies. Integrating the host and pathogen FBA models and further modification of the optimization function have also been used to study host-pathogen interactions(Bordbar et al. 2010).

Different types of approaches can be integrated each of which best describes different aspects of a biological system to obtain overall mechanistic insights. For example, FBA is used for studying metabolic networks while Boolean modeling is used for regulatory networks and the approaches can be clubbed to obtain a metabolic as well as a regulatory model. This is important because the different modules of a biological system interact with each other and influences the functioning of the modules. Covert et al. (2008) have developed a method, iFBA, also known as integrative FBA that integrates FBA with Boolean logic and ODEs to model the dynamics of networks related to the carbohydrate uptake mechanism. They compared the predictions of the integrated model with the individual model and showed that an integrated model is a significant improvement over the individual models. The applications of these methods are described using case studies of different infectious diseases and are presented in the succeeding sections.

8.3 Tuberculosis

According to the sixteenth global report on tuberculosis (TB), published by WHO, an estimated 8.5 – 9.2 million new cases of TB have emerged in the year 2010, while 0.9–1.2 million of the HIV-negative people have succumbed to the disease, and an additional 0.35 million deaths have occurred from the HIV-associated TB cases. Threat from this disease increases drastically with the advent of multidrug resistant (MDR), extremely-drug resistant (XDR) and totally drug resistant (TDR) strains. Unfortunately, no new drugs have come up in the last five decades and the drugs available in the market have their inadequacies. It is thus important to think of newer strategies and develop new classes of drugs to counter the spread of this disease.

The etiological agent of TB, *Mycobacterium tuberculosis* (*M.tb*), enters the host primarily via aerosols containing the bacilli, and on reaching the lungs they are internalized by the alveolar macrophages and undergo phagocytosis. Pathogenesis starts after formation of the phagosome, wherein *M.tb* prevents maturation of the infected macrophage and in this niche the pathogen is able to survive and reproduce. The widespread nature of this disease depends upon its ability to spread easily by aerosol transmission, which is further facilitated by immune-dependent tissue-damaging inflammation (Pieters 2008).

Upon infection, a dynamic interplay occurs between the host and pathogen leading to either of the four outcomes: (a) the initial host response may be completely effective and kill the bacilli; (b) the organisms can grow and multiply immediately after infection resulting in active TB, (c) the bacilli may become dormant and never cause disease at all and (d) the latent bacilli can eventually become active and progress to disease condition (Schluger and Rom 1998). Needless to say, the difference between the outcomes is enormous and results in extreme phenotypes between disease and health. Various experimental as well as computational tools have been used to study the pathogenesis of this disease and its interaction with the host, brieflysummarized here.

Deciphering the whole genome sequence of *M.tb* has been an important landmark in tuberculosis research (Cole et al. 1998). The genome sequence provided a first comprehensive parts-list of the molecular constituents of the cell. This triggered extensive amount of downstream research leading to detailed biochemical and biophysical characterizations of a number proteins (Lew et al. 2011; Galagan et al. 2010). More importantly perhaps, it has provided an impetus for systems level studies. Genome sequence has helped tremendously in completing the gaps in knowledge from decades of biochemical and molecular biology studies of individual molecules in the organism. It has revealed complete lists of proteins belonging to many biochemical pathways, transcription factors, two-component signalling systems (Tyagi and Sharma 2004). It has led to comparative genomics studies through gene and protein sequence comparisons and further to several functional genomics studies (Tucker et al. 2007). Proteins responsible for cellular metabolism are identified comprehensively; indicating that, *M.tb* indeed has most of the standard pathways present in other bacteria such as glycolysis, citric acid cycle, pyruvate, fatty acid, amino acid metabolism to list a few (Cole et al. 1998). There are also interesting differences, for example, presence of mycolic acid and arabinogalacatan pathways, the glyoxylate shunt and beta oxidation pathway for fatty acid metabolism. Identification of such unique features has been useful to obtain direct explanations for phenotypic characteristics of the organism such as the presence of a thick waxy outer cover.

Advances in high-throughput ‘omics’ technologies, that has resulted in a large amount of omics data in the last few years, help significantly in functional characterizations (Kirschner et al. 2010) of both host and pathogen’s genomes. Global gene expression profiles of *M.tb* under different conditions are available. The set of genes in *M.tb* required for optimal growth have been characterized by using the transposon site hybridization (TraSH) method which provides a comprehensive idea about functional significance and essentiality of each gene (Sassetti et al. 2003). The proteome of *M.tb* has also been analyzed by 2D gel electrophoresis and mass spectrometry and also by the isotope-coded affinity tag reagent method coupled with mass spectrometry (Schmidt et al. 2004). Using a guinea pig model of tuberculosis, the bacterial proteome during the early and chronic stages of disease has been examined

(Kruh et al. 2010) by liquid chromatography-mass spectrometry. The study identified numerous *M.tb* proteins, from essential kinases to products involved in metal regulation and cell wall remodeling, present throughout the course of infection. Cell wall processes, intermediary metabolism and respiration were found to be major functional classes of proteins represented in the infected lung. Recently, protein-protein interactions in *M.tb* have been determined experimentally in a high-throughput manner using a bacterial two-hybrid system (Wang et al. 2010a).

Genome scale studies are being carried out for the host systems as well. Several gene expression profiles under different conditions of exposure to *M.tb*, disease and treatment with anti-tuberculars have been obtained, which identify genes that show maximal changes in their expression under different conditions (Boshoff et al. 2004). siRNA screens have been used to systematically knock-out various genes and infer their importance for survival, pathogenesis and stress response (Kumar et al. 2010). Recently many techniques have been developed to visualize spatial features of such interactions inside tissues, which include intravital multiphoton microscopy and four dimensional FRET (Konjufca and Miller 2009; Hoppe et al. 2009). Although these techniques are in their incipient stages of development, they offer promising results and greater understanding of host–pathogen interactions.

The data thus obtained from the above described *omics*-data can be further used to build computational models. One way of incorporating such large scale data is to build a protein-protein interaction network. A comprehensive reconstruction using crowd sourcing based curation from literature and available databases together, capture as many as 71086 interactions in 3967 proteins (Vashisht et al. 2012) adding substantially to the existing resources. Incorporating drug-specific gene-expression fold changes in the network as node weights, Padiadpu et al. (2010) captured the effect of drugs on *M.tb* interactome and the mechanism of triggering resistance. Another study by Kauffman et al. (Rachman et al. 2006) identified genes that are important for the survival and persistence of *M.tb* in a macrophage cell by using a combination of approaches. Using a reconstructed protein—protein interaction network and incorporating genome-wide DNA array into this network, pathways such as iron metabolism, cell wall synthesis, DNA damage repair and fatty acid degradation were identified as important to the pathogen (Rachman et al. 2006).

Yet another method of using experimental data to build computational models is constraint based modeling. Details of this modeling method are provided in the methods section. This method serves as an excellent tool to study genome scale metabolic models. McFadden and co workers (Beste et al. 2007) reconstructed the first genome scale metabolic model for *M.tb*, capturing all known biosynthetic pathways operational in a cell for synthesis of major macromolecular components. This model was calibrated using data from chemostat cultivations of *M.bovis* BCG in continuous culture and measurement of steady state growth parameters. Almost at the same time, an independently reconstructed genome scale network model of *M.tb* H37Rv named iNJ661 was reported by Palsson and coworkers (Jamshidi and Palsson 2007). The authors grew this bacterial model *in silico* on various media, and observed that growth rates were comparable to experimental observations of doubling times in the range of 12–24 h in different media. Using these models, reaction fluxes indicating substrate consumption rates were measured, which correlated well with experimentally determined values. Raman et al. have identified putative drug targets using *in silico* gene deletions for the mycolic acid pathway model in *M.tb* (Raman et al. 2005).

Another classical method to study the dynamics of a cellular system is ordinary differential Equations (ODE), wherein time courses of metabolic reactions are mathematically represented by ODEs. Singh et al. (Singh and Ghosh 2006) built a kinetic model of the tricarboxylic acid cycle and the glycolytic pass of *E.coli* and *M.tb* to compare the two systems and study the effect of enzyme inhibition and thus identify potential drug targets. Kinetic modeling has also been carried out to study the host immune system upon TB infection to reveal the existence of a non-infected steady state and an endemically infected steady state, which can lead to latency or activation of the disease (Ibargüen-Mondragón et al. 2011)

Signalling interactions in a cell can be easily represented by Boolean modeling, also described in the methodology section. Raman et al. built a Boolean model of the host—pathogen interactome (Raman et al. 2010), accounting for several mechanisms of invasion by the pathogen, defense of the host, as well as the defense mechanisms of the pathogen and was simulated under a variety of conditions. The model consisted of 75 nodes that represented the molecules involved in host and pathogen and different states of the molecules and events were governed by logical operators or Boolean rules. This provides a framework to understand the conditions and parameters that favour clearance versus those that favour either active disease or contain the bacteria in a dormant state.

Rule based modeling have also been used to represent signalling processes, especially for those events, wherein the molecule can take up different states depending on its environment. Such models are known to best capture the environmental dependencies. An et al. (An and Faeder 2009) built a rule based model of the Toll-like receptor 4 signal transduction cascade. Simulation of the original model and 'knock-out' were performed to study the behaviour of the system. Ghosh et al., have reported a rule based model to study host-pathogen interaction for TB infection and the role of iron for both host and pathogen during the course of infection has been studied. Regulating the concentration of mycobactin was discussed as one of the strategies to control bacterial infection (Ghosh et al. 2011).

Boolean network models of immunological components of the interplay of various mechanisms of attack and defense in the host and pathogen with respect to M.tb have been developed and provides insights into the immune responses as well as the different outcomes of M.tb infections under different conditions (Raman et al. 2010). Kirschner and co-workers have worked on several mathematical models for the interaction of *M.tb* with the human immune system, some examples of which are a virtual model of the immune response to *M.tb* that characterises the cytokine and cellular network during infection, two compartmental models capturing the important processes of cellular activation and priming capable of reproducing typical disease progression scenarios, agent-based models for simulating granuloma formation (Marino et al. 2011) and a mathematical model describing macrophage biochemical processes based on activation, killing and iron regulation. Host-pathogen FBA models enable studying the metabolic states of the system in an infected condition. Gene essentiality studies were performed and the predictions were shown to be much more accurate in the combined model. The models were further integrated with gene expression data for the different forms of the disease, such as latency, meningeal and pulmonary tuberculosis, to study the subtle metabolic differences amongst the different forms and therefore to have much more accurate perturbation studies for the different forms (Bordbar et al. 2010).

The above methodologies have helped in successfully identifying the different aspects of *M.tb* infection. Protein-protein interactome analyses have helped in identifying highly influential proteins that can form potential drug targets (Padiadpu et al. 2010). Metabolic reconstructions of the host and pathogen as well as the combined models have provided useful insights into genes essential for the survival of the pathogen using FBA (Jamshidi and Palsson 2007). Further, integrating host and pathogen FBA models have provided useful insights into the metabolic changes that occur in the host upon bacterial infection (Bordbar et al. 2010). Host-pathogen interaction studies guide in identifying factors important for virulence, the different immune responses and most importantly understanding the emergence of resistance (Raman et al. 2010). A new concept of *co-targets* was proposed by Raman et al. that inhibited two targets simultaneously to deal with resistance. All these analyses have been integrated into a rational pipeline called targetTB to identify potential drug targets for *M.tb* (Raman et al. 2008), which has yielded a list of about 450 high confidence drug targets.

8.4 Malaria

Malaria caused by *Plasmodium* parasites, is transmitted through the bite of infected Anopheles mosquito. In 2011, an estimated number of 216 million cases of malaria were reported and 655000 deaths were caused by malaria in 2010 (World malaria report 2011), indicating that it is one of the major contributors to global morbidity and mortality rates. Although malaria is curable, it is still a life-threatening disease, and with the emergence of antimalarial resistant strains it has become difficult to tackle this disease efficiently.

Whole genome sequencing of *Plasmodium falciparum* was accomplished in 2002 (Gardner et al. 2002) and it has revealed that approximately 35 % of the proteins encoded have identifiable function and the remaining are uncharacterized. With the availability of genomic sequence of *P.falciparum* it has become easier to identify unique enzymes involved in pathways, which are different from the humans, such that inhibitors can be synthesized against them, thus disrupting the pathway in pathogen. Mass spectrometric studies have been performed in order to understand the mechanism by which the parasite modulates the level of different metabolites taking part in various metabolic processes of the host so as to survive inside the host cell and proliferate (Olszewski et al. 2009). Due to the complex life cycle of the pathogen, it becomes necessary to identify genes expressed at different stages of infection such that they can be used as targets (Winzeler 2005). A combination of genomics and proteomics methods were employed by Hall et al. (2005) in order to identify a conserved set of genes in *Plasmodium spp.* and also emphasize upon genes which have been chosen under selective pressure at different stages of pathogenesis. Flux balance model for *P.falciparum* was constructed in order to study the metabolic state of the pathogen upon perturbation and also predict the essential genes which can also be used as targets (Plata et al. 2010). The model consisted of 1001 reactions and 616 metabolites, of which enzyme-gene associations were reported for 366 genes and 75 % of the total enzymatic reactions known. Models were enriched by incorporating gene-expression data and also the accuracy of the predictions to experimental results was high indicating that *in silico* models can be used for studying the complex pathogen. An open access database called PlasmoDB has been developed which provides information about the transcriptome and protein expression data of *Plasmodium spp.* at different stages of their life cycle, which can be used to investigate the involvement of a gene in a defined process by correlating with gene expression profiles or proteomics or protein-protein interactions data of the species(Aurrecoechea et al. 2009).

Plasmodium spp. is capable of surviving inside the host by synthesizing different chemical compounds during various stages of its life cycle. Although these compounds have been used as targets for vaccine development, not much success has been achieved in eradicating malaria. Due to the complex host-pathogen interaction and prevalence of resistance to antimalarial drugs, efforts have been made to discover newer drugs using a systems biology approach. The immune response of the host plays a complicated role in malaria as it not only helps in evading the pathogen but is also responsible for causing complications in the host (McNicholl et al. 2000). Jomaa et al. reported a non-mevalonate pathway of isoprenoid biosynthesis, located in the apicoplast region of *Plasmodium*, and the drugs effective against the metabolites involved in this pathway as potent antimalarials (Jomaa et al. 1999). Reverse vaccinology approach has been employed to search for antigens in *Plasmodium spp.* which when targeted will appropriately, aid in vaccine development. Systems biology has been used to anticipate the immune response of the host cells upon the interaction with the antigen and also understand the complex life cycle of the parasite (Rappuoli and Aderem 2011). Bioinformatics approaches have been used to annotate the genome of *Plasmodium spp.*, majority of which is still uncharacterized. Fed into systems biology models, simulations help in discovering newer therapies for malaria as the parasite has acquired resistance against known drugs. Number of potent antimalarials (artimesinin and its derivatives) has been synthesized and systems biology based approaches will aid in characterizing the mechanism of action of these newly discovered antimalarial compounds (Dharia et al. 2010).

8.5 Cholera

Reports from WHO indicate that 3.5 million suffer from diarrhoeal infections, the causative agent being *Vibrio cholerae,* capable of secreting the potent cholera toxin (Nelson et al. 2009). This acute intestinal infection is transmitted through contaminated food and water and if left untreated can lead to death of the patients. Although it is curable if treated on time, severe symptoms are observed in immune-compromised patients. The strains of *V.cholerae* have been classified either as classical or El Tor. Two sero groups, *V. cholerae* O1 and *V. cholerae* O139, are mostly responsible for the outbreak of cholera. Multidisciplinary approaches are being used to find new drugs to reduce the number of deaths caused by cholera.

Top down approaches have been used to identify additional genes that are involved in *V.cholerae* virulence and colonization inside host intestine (Kaper et al. 1995). Apart from the enterotoxin produced by *V.cholerae*, Asaduzzaman et al. have also narrowed down on other essential virulence factors present in the bacterium such as toxin-coregulated pilus that functions as a receptor for the bacteriophage and encoding cholera toxin genes (Asaduzzaman et al. 2004). A regulator-centric approach has been used to focus upon LysR-type transcriptional regulators (LTTRs), one of the most diverse families of transcriptional factors in prokaryotes having role in wide range of processes. A few LTTRs were found to be involved in intestinal colonization as well as metabolic regulation *in vivo* (Bogard et al. 2012). Mathematical models have been developed to understand the dynamics of pathogen colonization and indicate the contribution of host and pathogen towards bacterial gut density (Spagnuolo et al. 2011). Such studies are essential to understand pathogenesis of the disease. By performing a high-throughput phenotypic screen of 50,000-compound small molecule library, Hung and coworkers tried to identify inhibitors of *V.cholerae* virulence factor expression (Hung et al. 2005). The authors have reported a compound named virstatin, which is capable of inhibiting virulence expression, ToxT regulation (part of ToxR regulon, responsible for virulence) post-transcriptionally, and also preventing colonization in the intestine of the animal model to an extent.

Although cholera is a re-emerging disease, till date no simple assay has been developed to diagnose this disease efficiently. Oral or IV rehydration are recommended treatment and thus administering immediate oral rehydration therapy, rapid recovery of the patients can be observed. Since the late nineteenth century till 1970s, injections of inactivated whole bacteria were used as a vaccine. However, the limitation of these is that they are effective only for short durations. Oral vaccines against cholera were developed to overcome the shortcomings of parenteral vaccines. Till date two major classes of oral cholera vaccines namely killed WC- based and genetically attenuated live vaccines are used to treat cholera (Shin et al. 2011). Although newer vaccines such as Dukoral and Shanchol have received WHO prequalification, these vaccines also have their own limitations, thus keeping the problem of vaccine discovery as an open challenge (WHO 2012).

Systems biology approaches have been used in order to analyze gene expression of *V.cholerae* to identify virulence genes, which may provide a better insight to the infectious process. Using gene-expression data, comparison of the dynamic transcriptomes was carried out for the pathogen growing in different media at various stages of growth. A set of regulatory interactions for genes involved in virulence were identified (Kanjilal et al. 2010). Using information from different sources regarding the pathogen, gene response network has been constructed which is expected to aid in design of biomarkers and therapeutics. A metabolomics approach has been used to measure the extracellular changes in the flux of certain metabolites upon the administration of cholera toxin in cell lines, and this approach can be extended to study spatial and temporal changes in the metabolites flux, thus providing a clear picture of the metabolic activity in the cell in the presence of toxin (Eklund et al. 2006). Thus, using systems biology approaches it has become possible to identify the genes involved in virulence, interaction of the pathogen with the host, discover new biomarkers for the disease and also develop newer vaccines to overcome the limitations of the already existing vaccines (Hill et al. 2006).

8.6 *Staphylococcus aureus* Infection

Staphylococcus aureus (*S.aureus*), causative agent of nosocomial infection, is a life threatening pathogen to human population due to the wide range of diseases it causes, especially hospital acquired infections. Apart from the number of infections that this microbe is responsible for, it has also been observed that *S.aureus* is acquiring resistance against multiple antibiotics (Kaatz et al. 2005). In some parts of the world, methicillin resistant strains of *S.aureus* (MRSA) have been reported, which is posing a major health problem. Thus, it has become essential to understand the mechanism of pathogenesis of *S.aureus* and also its interaction with the host.

The global transcriptional profile of the pathogen aids in the study of regulatory genes and also gives insight into the expression profile of the genes under different conditions such as exposure to antibiotics (Kuroda et al. 2003) and stress (Anderson et al. 2006). Plikat et al. have constructed a protein expression map to study proteomes of *S. aureus* Mu50 and its mutants. Using GSEA (Gene set enrichment analysis), they have carried out studies to determine the virulence factors and pathways affected in mutants. Capsular polysaccharide of *S.aureus* had been earlier regarded as putative protective antigen and hence as possible vaccine candidate. However, subsequent studies noted that the clinical isolates lack a capsule, hence rendering the vaccine ineffective in the clinical trials. They have also reported that multivalent-antigen vaccine is capable of eliciting both cell-mediated and humoral immunity and in turn induce protection against *S.aureus* thus preventing infections at various anatomical sites (Plikat et al. 2007). Systems biology approaches have been used to identify targets in order to develop multivalent-antigen vaccine and also determine host-microbe interaction which helps in understanding the pathogenesis mechanism and ultimately finding a solution for preventing as well as curing the disease.

8.7 Applications of Systems Biology in 'Anti-Infective' Drug Discovery

With the advent of large scale omics data and the development of various modeling tools, it is possible to build large scale biological models. Although, the reductionist approach provides detailed insights into the molecules responsible for a particular disease, inhibition of a given protein molecule in isolation is insufficient to provide insights into the effect of this inhibition on the system as a whole. Existence of biologically feasible alternate paths may render this inhibition useless. Systems biology provides a mathematical framework to understand the physiological effect of inhibition in a network of interacting components. In the classical drug discovery regime, a major part of it was a black box and a target was selected based on the end result obtained. Mathematical models obtained can be used to study the effect of inhibition of the targets or exposure of the system to the drug, so that a rational behind the working of each drug is understood. TargetTb (Raman et al. 2008) is one such attempt wherein a comprehensive target identification pipeline is developed for *M.tb*. Many known targets were identified, thus validating the model and many more new targets have been suggested. A total of 451 high confidence potential drug targets were listed. The success rates from such pipelines are likely to be high as target selections are knowledge driven. Methods such as FBA have also been successful in identifying set of essential enzymes in *P.falciparum* and form a starting point for antimalarial drug targets (Huthmache 2010). S*ystems vaccinology* is a branch of systems biology that helps in predicting the efficacy of vaccines in a biological system. It is also useful in studying the immunological responses after vaccinations thus helping in vaccine development (Trautmann and Sekaly 2011). Figure 8.2 describes the various applications of Systems biology.

8.8 Conclusion

Understanding a biological phenomenon involves studying the system as a whole rather than as parts. Systems biology provides us with the tools to examine different biological aspects, such as protein-protein interactions, protein-metabolites interactions, regulatory mechanisms, signaling cascades using computational means. This is crucial because a continuous interaction exists between different biological processes and therefore studying these processes individually, as carried out in a reductionist approach, do not provide a holistic view of the system under study. Over the years many computational as well as experimental tools have been developed that help in collation, reconstruction and analysis of large-scale data.

The scale at which various molecular level studies are currently being carried out, is yielding genome-scale and systems level data on many fronts, leading to ready reconstructions of large systems. These can then be integrated with the deep insights already available about individual components. Although a complete systems view of the disease has still not been deciphered, it seems that we have at the least a coarse grained map of the pathogen in many of these cases, helpful for obtaining an aerial view of the disease that can be used for addressing a variety of questions. The map of course is sufficiently fine-grained in parts enabling a more detailed zoomed in version in some pathways especially with respect to intermediary metabolism.

Reconstruction of large scale models encompassing various processes of the bacterium and simulation will be extremely valuable in identifying best strategies for

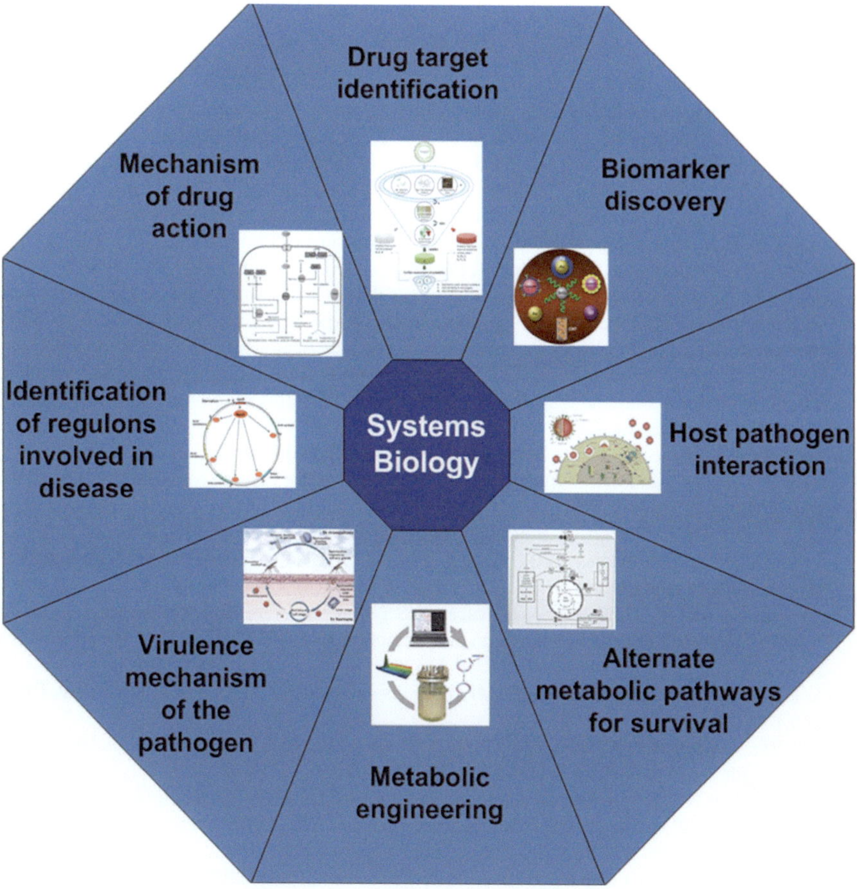

Fig. 8.2 Various applications of systems biology

intervention. Methods to study biological systems at multiple scales and levels and *virtual cells* are not as yet standardized. Nor are the methods required to generate comprehensive omics scale data from multiple perspectives, particularly when it comes to quantitative profiling. Thus, reports in literature of such cellular level models not only for *M.tb*, but in general for any organism are few and far between. Nevertheless, it is quite clear that the virtual cell approach, especially when quantitative aspects are incorporated, holds a lot of promise for picking an efficient or even an optimal strategy for killing the pathogen.

References

Aderem A, Adkins JN, Ansong C, Galagan J, Kaiser S, Korth MJ, Law GL, McDermott JG, Proll SC, Rosenberger C et al (2011) A systems biology approach to infectious disease research: innovating the pathogen-host research paradigm. mBio 2(1):e00325–00310

Albert R (2007) Network inference, analysis, and modeling in systems biology. Plant Cell 19(11):3327–3338

An GC, Faeder JR (2009) Detailed qualitative dynamic knowledge representation using a BioNetGen model of TLR-4 signaling and preconditioning. Math Biosci 217(1):53–63

Anderson KL, Roberts C, Disz T, Vonstein V, Hwang K, Overbeek R, Olson PD, Projan SJ, Dunman PM (2006) Characterization of the *Staphylococcus aureus* heat shock, cold shock, stringent, and SOS responses and their effects on log-phase mRNA turnover. J Bacteriol 188(19):6739–6756

Asaduzzaman M, Ryan ET, John M, Hang L, Khan AI, Faruque A, Taylor RK, Calderwood SB, Qadri F (2004) The major subunit of the toxin-coregulated pilus TcpA induces mucosal and systemic immunoglobulin A immune responses in patients with Cholera caused by *Vibrio cholerae* O1 and O139. Infect Immun 72(8):4448–4454

Aurrecoechea C, Brestelli J, Brunk BP, Dommer J, Fischer S, Gajria B, Gao X, Gingle A, Grant G, Harb OS (2009) PlasmoDB: a functional genomic database for malaria parasites. Nucleic Acids Res 37(Suppl 1):D539–D543

Beste DJV, Hooper T, Stewart G, Bonde B, Avignone-Rossa C, Bushell ME, Wheeler P, Klamt S, Kierzek AM, McFadden J (2007) GSMN-TB: a web-based genome-scale network model of Mycobacterium tuberculosis metabolism. Genome Biol 8(5):R89

Blinov ML, Faeder JR, Goldstein B, Hlavacek WS (2004) BioNetGen: software for rule-based modeling of signal transduction based on the interactions of molecular domains. Bioinformatics 20(17):3289–3291

Boele J, Olivier BG, Teusink B (2012) FAME, the Flux Analysis and Modeling Environment. BMC Syst Biol 6:8

Bogard RW, Davies BW, Mekalanos JJ (2012) MetR-regulated Vibrio cholerae metabolism is required for virulence. MBio 3(5):e00236–12

Bordbar A, Lewis NE, Schellenberger J, Palsson BØ, Jamshidi N (2010) Insight into human alveolar macrophage and M. tuberculosis interactions via metabolic reconstructions. Mol Syst Biol 6(1):422

Boshoff HIM, Myers TG, Copp BR, McNeil MR, Wilson MA, Barry CE 3rd (2004) The transcriptional responses of *Mycobacterium tuberculosis* to inhibitors of metabolism novel insights into drug mechanisms of action. J Biol Chem 279(38):40174–40184

Cole ST, Brosch R, Parkhill J, Garnier T, Churcher C, Harris D, Gordon SV, Eiglmeier K, Gas S, Barry CE 3rd et al (1998) Deciphering the biology of *Mycobacterium tuberculosis* from the complete genome sequence. Nature 393(6685):537–544

Colijn C, Brandes A, Zucker J, Lun DS, Weiner B, Farhat MR, Cheng TY, Moody DB, Murray M, Galagan JE (2009) Interpreting expression data with metabolic flux models: predicting *Mycobacterium tuberculosis* mycolic acid production. PLoS Comput Biol 5(8):e1000489

Colvin J, Monine MI, Gutenkunst RN, Hlavacek WS, Von Hoff DD, Posner RG (2010) RuleMonkey: software for stochastic simulation of rule-based models. BMC Bioinformatics 11(1):404

Covert MW, Xiao N, Chen TJ, Karr JR (2008) Integrating metabolic, transcriptional regulatory and signal transduction models in *Escherichia coli*. Bioinformatics 24(18):2044–2050

da Hora Junior BT Poloni Jde F Lopes MA Dias CV Gramacho KP Schuster I Sabau X Cascardo JC Mauro SM, Gesteira Ada S et al (2012) Transcriptomics and systems biology analysis in identification of specific pathways involved in cacao resistance and susceptibility to witches' broom disease. Mol BioSyst 8(5):1507–1519

Dandekar T, Snel B, Huynen M, Bork P (1998) Conservation of gene order: a fingerprint of proteins that physically interact. Trends Biochem Sci 23(9):324

Danos V, Feret J, Fontana W, Harmer R, Krivine J (2008) Rule-based modelling, symmetries, refinements. Form Methods Syst Biol 5054:103–122

Day J, Schlesinger LS, Friedman A (2010) Tuberculosis research: going forward with a powerful "translational systems biology" approach. Tuberculosis 90(1):7–8

de Jong H (2002) Modeling and simulation of genetic regulatory systems: a literature review. J Comput Biol 9(1):67–103

Dharia N, Chatterjee A, Winzeler E (2010) Genomics and systems biology in malaria drug discovery. Curr Opin Investig Drugs 11(2):131 (London, England: 2000)

Eklund SE, Snider RM, Wikswo J, Baudenbacher F, Prokop A, Cliffel DE (2006) Multianalyte microphysiometry as a tool in metabolomics and systems biology. J Electroanal Chem 587(2):333–339

Feist AM, Herrgård MJ, Thiele I, Reed JL, Palsson BØ (2008) Reconstruction of biochemical networks in microorganisms. Nat Rev Microbiol 7(2):129–143

Forst CV (2006) Host-pathogen systems biology. Drug Discov Today 11(5–6):220–227

Funahashi A, Morohashi M, Kitano H, Tanimura N (2003) CellDesigner: a process diagram editor for gene-regulatory and biochemical networks. Biosilico 1(5):159–162

Galagan JE, Sisk P, Stolte C, Weiner B, Koehrsen M, Wymore F, Reddy TB, Zucker JD, Engels R, Gellesch M et al (2010) TB database 2010: overview and update. Tuberculosis 90(4):225–235

Gardner MJ, Hall N, Fung E, White O, Berriman M, Hyman RW, Carlton JM, Pain A, Nelson KE, Bowman S et al (2002) Genome sequence of the human malaria parasite *Plasmodium falciparum*. Nature 419(6906):498–511

Ghosh S, Prasad KVS, Vishveshwara S, Chandra N (2011) Rule-based modelling of iron homeostasis in tuberculosis. Mol Biosyst 7(10):2750–2768

Gillespie DT (1977) Exact stochastic simulation of coupled chemical reactions. J Phys Chem 81(25):2340–2361

Hall N, Karras M, Raine JD, Carlton JM, Kooij TW, Berriman M, Florens L, Janssen CS, Pain A, Christophides GK et al (2005) A comprehensive survey of the Plasmodium life cycle by genomic, transcriptomic, and proteomic analyses. Science 307(5706):82–86

Hill DR, Ford L, Lalloo DG (2006) Oral cholera vaccines: use in clinical practice. Lancet Infect Dis 6(6):361–373

Hlavacek WS, Faeder JR (2009) The complexity of cell signaling and the need for a new mechanics. Sci Signal 2(81):pe46

Hood L, Perlmutter RM (2004) The impact of systems approaches on biological problems in drug discovery. Nat Biotechnol 22(10):1215–1217

Hoops S, Sahle S, Gauges R, Lee C, Pahle J, Simus N, Singhal M, Xu L, Mendes P, Kummer U (2006) COPASI—a complex pathway simulator. Bioinformatics 22(24):3067–3074

Hoppe AD, Seveau S, Swanson JA (2009) Live cell fluorescence microscopy to study microbial pathogenesis. Cell Microbiol 11(4):540–550

Hoppe A, Hoffmann S, Gerasch A, Gille C, Holzhütter HG (2011) FASIMU: flexible software for flux-balance computation series in large metabolic networks. BMC Bioinformatics 12(1):28

Hung DT, Shakhnovich EA, Pierson E, Mekalanos JJ (2005) Small-molecule inhibitor of *Vibrio cholerae* virulence and intestinal colonization. Science 310(5748):670–674

Huthmacher C, Hoppe A, Bulik S, Holzhutter HG (2010) Antimalarial drug targets in *Plasmodium falciparum* predicted by stage-specific metabolic network analysis. BMC Syst Biol 4:120

Ibargüen-Mondragón E, Esteva L, Chávez-Galán L (2011) A mathematical model for cellular immunology of tuberculosis'. J Math Biosci Eng 8(4):973–986

Ideker T, Ozier O, Schwikowski B, Siegel AF (2002) Discovering regulatory and signalling circuits in molecular interaction networks. Bioinformatics 18(suppl 1):S233–S240

Jamshidi N, Palsson B (2007) Investigating the metabolic capabilities of *Mycobacterium tuberculosis* H37Rv using the in silico strain iNJ661 and proposing alternative drug targets. BMC Syst Biol 1(1):26

Johanns TM, Ertelt JM, Rowe JH, Way SS (2010) Regulatory T cell suppressive potency dictates the balance between bacterial proliferation and clearance during persistent Salmonella infection. PLoS Pathog 6(8):e1001043

Jomaa H, Wiesner J, Sanderbrand S, Altincicek B, Weidemeyer C, Hintz M, Türbachova I, Eberl M, Zeidler J, Lichtenthaler HK (1999) Inhibitors of the nonmevalonate pathway of isoprenoid biosynthesis as antimalarial drugs. Science 285(5433):1573–1576

Kaatz GW, McAleese F, Seo SM (2005) Multidrug resistance in *Staphylococcus aureus* due to overexpression of a novel multidrug and toxin extrusion (MATE) transport protein. Antimicrob Agents Chemother 49(5):1857–1864

Kanjilal S, Citorik R, LaRocque RC, Ramoni MF, Calderwood SB (2010) A systems biology approach to modeling Vibrio cholerae gene expression under virulence-inducing conditions. J Bacteriol 192(17):4300–4310

Kaper JB, Morris JG, Jr, Levine MM (1995) Cholera. Clin Microbiol Rev 8(1):48–86

Kirschner DE, Young D, Flynn JAL (2010) Tuberculosis: global approaches to a global disease. Curr Opin Biotechnol 21(4):524–531

Kitano H (2002) Systems biology: a brief overview. Science 295(5560):1662–1664

Konjufca V, Miller MJ (2009) TwoâŁłphoton microscopy of host–pathogen interactions: acquiring a dynamic picture of infection in vivo. Cell Microbiol 11(4):551–559

Kruh NA, Troudt J, Izzo A, Prenni J, Dobos KM (2010) Portrait of a pathogen: the *Mycobacterium tuberculosis* proteome in vivo. PLoS One 5(11):e13938

Kumar D, Nath L, Kamal MA, Varshney A, Jain A, Singh S, Rao KVS (2010) Genome-wide analysis of the host intracellular network that regulates survival of *Mycobacterium tuberculosis*. Cell 140(5):731–743

Kuroda M, Kuroda H, Oshima T, Takeuchi F, Mori H, Hiramatsu K (2003) Two-component system VraSR positively modulates the regulation of cell-wall biosynthesis pathway in *Staphylococcus aureus*. Mol Microbiol 49(3):807–821

Lakshmanan M, Koh G, Chung BK, Lee DY (2014) Software applications for flux balance analysis. Brief Bioinform 15(1):108–122

Latendresse M, Krummenacker M, Trupp M, Karp PD (2012) Construction and completion of flux balance models from pathway databases. Bioinformatics 28(3):388–396

Lew JM, Kapopoulou A, Jones LM, Cole ST (2011) TubercuList–10 years after. Tuberculosis 91(1):1–7

MacPherson DW, Gushulak BD, Baine WB, Bala S, Gubbins PO, Holtom P, Segarra-Newnham M (2009) Population mobility, globalization, and antimicrobial drug resistance. Emerg Infect Dis 15(11):1727–1732

Marcotte EM, Pellegrini M, Ng HL, Rice DW, Yeates TO, Eisenberg D (1999) Detecting protein function and protein–protein interactions from genome sequences. Science 285(5428):751–753

Marino S, El-Kebir M, Kirschner D (2011) A hybrid multi-compartment model of granuloma formation and T cell priming in tuberculosis. J Theor Biol 280(1):50–62

Maus C, Rybacki S, Uhrmacher A (2011) Rule-based multi-level modeling of cell biological systems. BMC Syst Biol 5(1):166

McNicholl JM, Downer MV, Udhayakumar V, Alper CA, Swerdlow DL (2000) Host–pathogen interactions in emerging and re-emerging infectious diseases: a genomic perspective of tuberculosis, malaria, human immunodeficiency virus infection, hepatitis B, and cholera. Annu Rev Public Health 21:15–46

Nelson EJ, Harris JB, Morris JG Jr, Calderwood SB, Camilli A (2009) Cholera transmission: the host, pathogen and bacteriophage dynamic. Nat Rev Microbiol 7(10):693–702

Olszewski KL, Morrisey JM, Wilinski D, Burns JM, Vaidya AB, Rabinowitz JD, Llinás M (2009) Host-parasite interactions revealed by *Plasmodium falciparum* metabolomics. Cell Host Microbe 5(2):191–199

Orth JD, Thiele I, Palsson BØ (2010) What is flux balance analysis? Nat Biotechnol 28(3):245–248

Padiadpu J, Vashisht R, Chandra N (2010) Protein–protein interaction networks suggest different targets have different propensities for triggering drug resistance. Syst Synth Biol 4(4):311–322

Pellegrini M, Marcotte EM, Thompson MJ, Eisenberg D, Yeates TO (1999) Assigning protein functions by comparative genome analysis: protein phylogenetic profiles. Proc Natl Acad Sci U S A 96(8):4285–4288

Pieters J (2008) *Mycobacterium tuberculosis* and the macrophage: maintaining a balance. Cell Host Microbe 3(6):399–407

Pinney JW, Westhead DR, McConkey GA (2003) Petri Net representations in systems biology. Biochem Soc Trans 31(Pt 6):1513–1515

Plata G, Hsiao TL, Olszewski KL, Llinas M, Vitkup D (2010) Reconstruction and flux-balance analysis of the *Plasmodium falciparum* metabolic network. Mol Syst Biol 6:408

Plikat U, Voshol H, Dangendorf Y, Wiedmann B, Devay P, Muller D, Wirth U, Szustakowski J, Chirn GW, Inverardi B et al (2007) From proteomics to systems biology of bacterial pathogens: approaches, tools, and applications. Proteomics 7(6):992–1003

Rachman H, Strong M, Schaible U, Schuchhardt J, Hagens K, Mollenkopf H, Eisenberg D, Kaufmann SHE (2006) *Mycobacterium tuberculosis* gene expression profiling within the context of protein networks. Microbes Infect 8(3):747–757

Raman K, Chandra N (2009) Flux balance analysis of biological systems: applications and challenges. Brief Bioinform 10(4):435–449

Raman K, Bhat AG, Chandra N (2010) A systems perspective of host–pathogen interactions: predicting disease outcome in tuberculosis. Mol Biosyst 6(3):516–530

Raman K, Rajagopalan P, Chandra N (2005) Flux balance analysis of mycolic acid pathway: targets for anti-tubercular drugs. PLoS Comput Biol 1(5):e46

Raman K, Rajagopalan P, Chandra N (2006) Principles and practices of pathway modelling. Curr Bioinform 1(2):147–160

Raman K, Yeturu K, Chandra N (2008) targetTB: a target identification pipeline for *Mycobacterium tuberculosis* through an interactome, reactome and genome-scale structural analysis. BMC Syst Biol 2(1):109

Rappuoli R, Aderem A (2011) A 2020 vision for vaccines against HIV, tuberculosis and malaria. Nature 473(7348):463–469

Ravasz E, Somera AL, Mongru DA, Oltvai ZN, Barabasi AL (2002) Hierarchical organization of modularity in metabolic networks. Science 297(5586):1551–1555

Sassetti CM, Boyd DH, Rubin EJ (2003) Genes required for mycobacterial growth defined by high density mutagenesis. Mol Microbiol 48(1):77–84

Sauro H (2004) An introduction to biochemical modeling using JDesigner. Keck Graduate Institute, Claremont

Schellenberger J, Que R, Fleming RMT, Thiele I, Orth JD, Feist AM, Zielinski DC, Bordbar A, Lewis NE, Rahmanian S (2011) Quantitative prediction of cellular metabolism with constraint-based models: the COBRA Toolbox v2. 0. Nat Protoc 6(9):1290–1307

Schlatter R, Schmich K, Vizcarra IA, Scheurich P, Sauter T, Borner C, Ederer M, Merfort I, Sawodny O (2009) ON/OFF and beyond-a boolean model of apoptosis. PLoS Comput Biol 5(12):e1000595

Schluger NW, Rom WN (1998) The host immune response to tuberculosis. Am J Respir Crit Care Med 157(3):679–691

Schmidt F, Donahoe S, Hagens K, Mattow J, Schaible UE, Kaufmann SHE, Aebersold R, Jungblut PR (2004) Complementary analysis of the *Mycobacterium tuberculosis* proteome by two-dimensional electrophoresis and isotope-coded affinity tag technology. Mol Cell Proteomics 3(1):24–42

Segre D, Vitkup D, Church GM (2002) Analysis of optimality in natural and perturbed metabolic networks. Proc Natl Acad Sci U S A 99(23):15112–15117

Shannon P, Markiel A, Ozier O, Baliga NS, Wang JT, Ramage D, Amin N, Schwikowski B, Ideker T (2003) Cytoscape: a software environment for integrated models of biomolecular interaction networks. Genome Res 13(11):2498–2504

Shin S, Desai SN, Sah BK, Clemens JD (2011) Oral vaccines against cholera. Clin Infect Dis 52(11):1343–1349

Siek JG, Lee LQ, Lumsdaine A (2002) The boost graph library: user guide and reference manual. Addison-Wesley Longman Publishing Co., Inc. Boston

Singh VK, Ghosh I (2006) Kinetic modeling of tricarboxylic acid cycle and glyoxylate bypass in *Mycobacterium tuberculosis*, and its application to assessment of drug targets. Theor Biol Med Modelling 3(1):27

Spagnuolo AM, Dirita V, Kirschner D (2011) A model for *Vibrio cholerae* colonization of the human intestine. J Theor Biol 289:247–258

Spellberg B, Guidos R, Gilbert D, Bradley J, Boucher HW, Scheld WM, Bartlett JG, Edwards J Jr (2008) The epidemic of antibiotic-resistant infections: a call to action for the medical community from the Infectious Diseases Society of America. Clin Infect Dis 46(2):155–164

Stoll G, Viara E, Barillot E, Calzone L (2012) Continuous time Boolean modeling for biological signaling: application of Gillespie algorithm. BMC Syst Biol 6(1):116

Strong M, Graeber TG, Beeby M, Pellegrini M, Thompson MJ, Yeates TO, Eisenberg D (2003) Visualization and interpretation of protein networks in *Mycobacterium tuberculosis* based on hierarchical clustering of genomeâŁłwide functional linkage maps. Nucleic Acids Res 31(24):7099–7109

Szklarczyk D, Franceschini A, Kuhn M, Simonovic M, Roth A, Minguez P, Doerks T, Stark M, Muller J, Bork P (2011) The STRING database in 2011: functional interaction networks of proteins, globally integrated and scored. Nucleic Acids Res 39(Suppl 1):D561–D568

Trautmann L, Sekaly RP (2011) Solving vaccine mysteries: a systems biology perspective. Nat Immunol 12(8):729–731

Tucker PA, Nowak E, Morth JP (2007) TwoâŁłcomponent systems of*Mycobacterium tuberculosis*—structureâŁłbased approaches. Methods Enzymol 423:477–501

Tyagi JS, Sharma D (2004) Signal transduction systems of mycobacteria with special reference to *M. tuberculosis*. Curr Sci 86(1):93–102

Vashisht R, Mondal AK, Jain A, Shah A, Vishnoi P, Priyadarshini P, Bhattacharyya K, Rohira H, Bhat AG, Passi A et al (2012) Crowd sourcing a new paradigm for interactome driven drug target identification in *Mycobacterium tuberculosis*. PLoS One 7(7):e39808

Vodovotz Y, Csete M, Bartels J, Chang S, An G (2008) Translational systems biology of inflammation. PLoS Comput Biol 4(4):e1000014

Wang RS, Albert R (2011) Elementary signaling modes predict the essentiality of signal transduction network components. BMC Syst Biol 5(1):44

Wang JH, Byun J, Pennathur S (2010a) Analytical approaches to metabolomics and applications to systems biology. Seminars Nephrol 30(5):500–511

Wang Y, Cui T, Zhang C, Yang M, Huang Y, Li W, Zhang L, Gao C, He Y, Li Y (2010b) Global protein–protein interaction network in the human pathogen *Mycobacterium tuberculosis* H37Rv. J Proteome Res 9(12):6665–6677

Weckwerth W (2003) Metabolomics in systems biology. Annu Rev Plant Biol 54:669–689

Weston AD, Hood L (2004) Systems biology, proteomics, and the future of health care: toward predictive, preventative, and personalized medicine. J Proteome Res 3(2):179–196

WHO (2012) Fact sheets: infectious diseases. World Health Organganization

Winzeler EA (2005) Applied systems biology and malaria. Nat Rev Microbiol 4(2):145–151

Wittmann DM, Krumsiek J, Saez-Rodriguez J, Lauffenburger DA, Klamt S, Theis FJ (2009) Transforming Boolean models to continuous models: methodology and application to T-cell receptor signaling. BMC Syst Biol 3(1):98

Chapter 9
Systems Pharmacology and Pharmacogenomics for Drug Discovery and Development

Puneet Talwar, Yumnum Silla, Sandeep Grover, Meenal Gupta, Gurpreet Kaur Grewal and Ritushree Kukreti

Abstract Systems pharmacology involves the application of systems biology approaches, integrating high throughput experimental data from different experimental techniques such as genomics and proteomics involving computational analytical approaches, to understand the mechanism of action of drugs, identify potential drug targets, use existing drugs for other disease indications and study adverse drug reactions. The significance of using integrated approach is that it allows drug action and drug response to be studied in the context of whole genome or proteome. Basically, a strong and simplified platform for the development of systems pharmacology is provided by information from genetic studies, disease pathophysiology, pharmacology, protein-protein and protein-drug interactions. Network analyses of interactions involved in disease pathophysiology and drug response will allow the integration of the

"Variability is the law of life, and as no two faces are the same, so no two bodies are alike, and no two individuals react alike and behave alike under abnormal conditions which we know as disease."
Sir William Osler (1849–1919)

R. Kukreti (✉) · S. Grover · M. Gupta · G. K. Grewal · P. Talwar
Lab 403B, Genomics and Molecular Medicine Unit, CSIR-Institute of Genomics and Integrative Biology (IGIB), Mall Road, 110007 Delhi, Delhi, India
Tel.: +91-11-2766 2201; Fax: 91-11-2766471
e-mail: ritus@igib.res.in

Y. Silla
Lab No-201, Genomics and Molecular Medicine Unit, CSIR-Institute of Genomics and Integrative Biology (IGIB), South Campus, Mathura Road, 110025 New Delhi, India
Tel.: +91-9718288936
e-mail: sillayumnam@igib.in; bio.sillayumnam@gmail.com

S. Grover
e-mail: grover.sandeep@gmail.com

M. Gupta
e-mail: meenal002@gmail.com

G. K. Grewal
e-mail: gpkgrewal@gmail.com

P. Talwar
e-mail: talwar.puneet@gmail.com

V. Singh, P. K. Dhar (eds.), *Systems and Synthetic Biology*,
DOI 10.1007/978-94-017-9514-2_9

systems-level understanding of drug action with genetic information enabling personalized medicine. Developments and insights from merging systems pharmacology and pharmacogenomics studies will provide new information on the complexities of disease associated with the identification of multiple targets for drug treatment and understanding adverse events caused by off-targets of drugs. In this chapter, we explored the current and future application of systems biology approaches in integrating large scale data from high-throughput genomic technologies with complex disease phenotypes, drug disposition pathways which might lead to not only newer and more effective therapies, but safer medications with fewer side effects.

Keywords Systems biology · Genome · Network analysis · Adverse drug reactions

9.1 Introduction

Complex disorders such as cancer, type 2 diabetes, depression, stroke, schizophrenia and Alzheimer's disease (AD) are some of the most prevalent, debilitating and yet poorly treated conditions. Although over the past decade several newer drug molecules have been introduced into the market, the drug discovery process has currently slowed down due to several reasons such as the increasing cost and duration of bringing a drug to market, low or variable efficacy and issues of adverse drug reactions (ADR) (Zhao and Iyengar 2012; Hughes et al. 2011; Bhogal and Balls 2008; Kola and Landis 2004). Furthermore, an increase in the failure rate for most new drugs has been reported in Phase II and III clinical trials by the Centre for Medicines Research in the UK (Arrowsmith 2011a, b). This can be attributed to the conventional drug discovery approach which has several drawbacks such as identification of new targets by linking individual cellular components to an tissue/organ-level phenotype which leads to lack of mechanistic understanding of how drug interactions at the molecular and cellular level manifest themselves as alterations in tissue/organ-level function, use of poorly predictable cell-based assay and in vivo animal models which show variable efficacy and may not work for humans and finally an inability to predict adverse events caused by the drug (Zhao and Iyengar 2012). Conventional drug discovery and development approach is depicted in Fig. 9.1a. At present, it is estimated that the average cost and time of bringing a drug from research stage to market via the conventional drug discovery and development (DDD) route is around $ 1 billion and from 12 to 15 years respectively (Hughes et al. 2011). Due to these factors, it is imperative to explore for new approaches which may improve and expedite the process of drug discovery. Since complex disorders are the focus of current drug-discovery efforts, understanding the disease mechanism at systems level using systems biology approach could help in the discovery and development of drug molecules with higher success rate in clinical trials.

An integrated approach to study and understand the function of biological systems and how perturbations such as therapeutic drug administration affect such systems

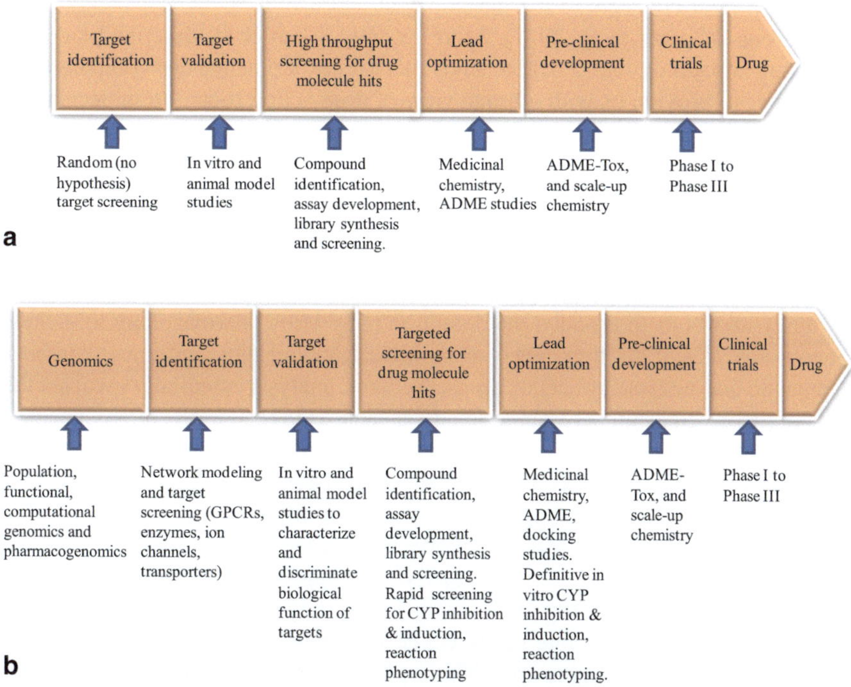

Fig. 9.1 **a** Conventional drug discovery and development approach. **b** Systems approach for drug discovery and development

is provided by systems biology. The biological system can be at the molecular, subcellular organelle, cell, tissue, organ or organism level. Hood et al. originally defined systems biology as "the study of all the elements in a biological system (all genes, mRNAs, proteins, etc) and their relationships one to another in response to perturbations" (Hood 2002). Later, the same group broadened the definition to "systems biology represents an analytical approach to the relationship among elements of a system, with the goal of understanding its emergent properties" (Weston and Hood 2004). More recently, Naylor et al. have combined and modified the definition: "Systems biology is the process of interrogating the genetic, genomic, biochemical, cellular, physiological and clinical properties of a system to define and create a system pathway or network that can be used to predicatively model a biological event(s)" (Naylor and Chen 2010; Naylor and Cavanagh 2004).

In this chapter, systems pharmacology, an emerging systems biology field, is described that may facilitate many of the current attempts to improve the drug discovery and development process. Systems pharmacology refers to the area of systems biology dealing with the representation of disease mechanisms of action (i.e. with the pharmacology of drug targets) (Cucurull-Sanchez et al. 2012). Sorger et al. defined systems pharmacology as 'an approach to translational medicine that combines computational and experimental methods to elucidate, validate and apply new pharmacological concepts to the development and use of small molecule and biologic drugs' (Sorger et al. 2011). More recently, Zhao et al. described the term systems pharmacology as ' a field of study that uses experimental and computational approaches to provide us with a broad view of drug action rooted in molecular interactions between the drug and its targets in the context of such targets interacting with and regulating other cellular components' (Zhao and Iyengar 2012).

In recent years, system biology approaches have been increasingly used in the pharmacology field to understand drug action at the cellular, tissue, organ and organismal levels. The application of computational systems biology approaches along with accumulated experimental genomic and proteomic knowledge to pharmacology allows us to broaden the definition of systems pharmacology to include network analyses at multiple levels of biological organization and to explain both therapeutic and adverse effects of drugs. Network analysis essentially involves the study of the relationship between topology at each scale (i.e., level) of organization (atomic/molecular, cellular/tissue, organ, and organismal) and connections between levels that give rise to organ- and organismal-level functions. This understanding helps us to understand how drugs that interact with different components are able to produce organ- and organismal-level effects, both therapeutic and adverse (Zhao and Iyengar 2012; Csermely et al. 2013). Systems approach for drug discovery and development is depicted in Fig. 9.1b.

A new dimension has come up by connecting genomic status and drug action from the field of pharmacogenomics. This type of integration is important for understanding drug action and effects at different levels of organization. In recent years, genomics has been shown to account for a considerable proportion of inter-individual variability in the drug effect, while showing consistent intra-individual responses (Ma and Lu 2011; Evans and McLeod 2003; Drazen et al. 2000). Pharmacogenomics is the study of the genetic basis of individual variation in response to therapeutic agents (Giacomini et al. 2012). It is an inter-disciplinary field involving molecular biology, human genetics and genomics, bioinformatics, pharmacology, and internal medicine (Yan 2010; Nebert 1999). The investigation of genetic diversity in humans can make it possible to tailor optimal drug prescription and to bring the right drug to the right person. This field may have a deep impact on every step of medical care, from diagnosis to drug prescription and from drug design to clinical trials.

The purpose of this chapter is to provide an overview of the emerging system biology approaches in the field of drug discovery and development with major focus on systems pharmacology and pharmacogenomics.

9.2 Understanding Complex Disorders Using Systems Biology

The advent of high throughput technological platforms (genomics, proteomics, and metabolomics) coupled with rigorous bioinformatic analyses using powerful statistical, computational and network modeling tools have led to the identification and characterization of biomolecules (DNA sequences, transcripts, proteins, lipids, and other metabolites) giving rise to the concept of systems biology (Naidoo et al. 2011; Tyers and Mann 2003). The conventional approaches have failed to completely elucidate the etiopathogenic mechanisms underlying complex disorders such as cancer and neurodegenerative disorders as they have focused on a few selected genes and proteins. Systems biology refers to an integrative analysis approach in which large numbers of biomolecules such as genes, proteins etc are measured simultaneously over time in cells, tissues or whole body (Kitano 2002). It integrates diverse fields encompassing biochemistry and cell biology with genetics, proteomics and bioinformatics to obtain comprehensive understanding of biological systems at various levels (Noorbakhsh et al. 2009). This understanding could lead to improved drug development and targeting, multidrug treatments, adverse drug reaction predictions as well as biomarker discovery.

In recent years, the focus on drug discovery has shifted from a molecular and cellular level to tissue and biological system level i.e., network pharmacology (Berger and Iyengar 2009; Boran and Iyengar 2010) to better understand the complex disorders which are often caused by combined effect of multiple molecular abnormalities rather than being the result of functional defect in a single gene or protein. In contrast to single-target approach, the network biology approach identifies a combination of interacting genes/proteins whose perturbation results in the clinical phenotype observed. In essence, genes/proteins in a biological system form molecular networks which eventually determine physiological or pathological functions within cells and organisms (Oltvai and Barabasi 2002). Methodological workflow describing the steps in network creation and analysis is depicted in Fig. 9.2. Biological systems should therefore be viewed as a web of interacting genes or proteins to permit the understanding of their complexity at 'system level' in terms of quantitative and spatio-temporal changes (Csete and Doyle 2002).

High-throughput studies with focus on the pathogenesis of complex disorders can be categorized into two methodological approaches (Kitano 2002):

1. The first approach involves global analysis of quantitative and/or qualitative changes in biomolecules (association of genetic variants or gene expression changes) followed by establishing pathways linking biomolecules (genes or proteins), pathways (cell signalling or gene regulatory) and disease processes (cancer etc.).
2. The second approach involves analysis of molecular networks or modules with intricate topologies formed at different system levels (e.g. transcription and cell signalling) within a cell.

9.2.1 Elucidating the Role of Systems Biology and Genomics in Complex Disorders

With the completion of the Human Genome Project and advancements in high throughput technologies, the field has entered into the post-genomic era leading to

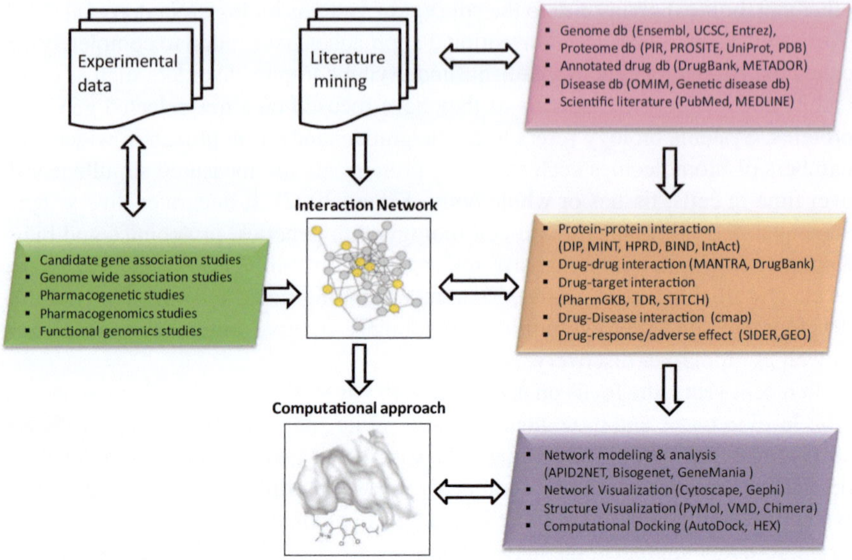

Fig. 9.2 Methodological workflow describing the steps in network creation and analysis: Mining of literature data from databases and integration with experimental data such as genomic and pharmacogenomic can be used to create an interaction network. Network creation and analysis involves the use of various biological tools and further application of computational approaches may help in identification of better drug targets with better efficacy

the advent of whole genome studies. A significant outcome was the systematic identification of single nucleotide polymorphisms (SNPs). SNPs are single nucleotide base changes found commonly in the DNA sequence that can describe variation between individuals (Penrod et al. 2011). A comprehensive catalog of all SNPs gathered from populations with European, Asian, and, African ancestry in the initial phase has been maintained in the HapMap database. The major initial objective of the International HapMap Project was to map genetic variants by comparing the genetic sequences of different individuals and identifying chromosomal regions where genetic variants are shared (International HapMap Consortium 2005). This wealth of information on SNPs has led to a dramatic increase in studies that seek to connect genetic polymorphisms with disease and individual responses to drugs and environmental factors. As the genetic variation among individual's averages to be about 1 in 500–1000 base pairs (Venter et al. 2001), a significant number of genes may contain polymorphisms that contribute to disease and that many may play a role in adverse drug responses (ADR). Although, most of the current research focuses on the association between phenotype/disease and SNPs, SNPs can also serve as biomarkers of ADRs because, unlike other factors such as age, co-morbidity, and environment, an individual's genetic makeup remains static throughout their lifetime. In addition, genetic testing also offers the potential of replacing empirical dose adjustment for many drugs that is based upon therapeutic assessment of pharmacologic or toxic effect after initial

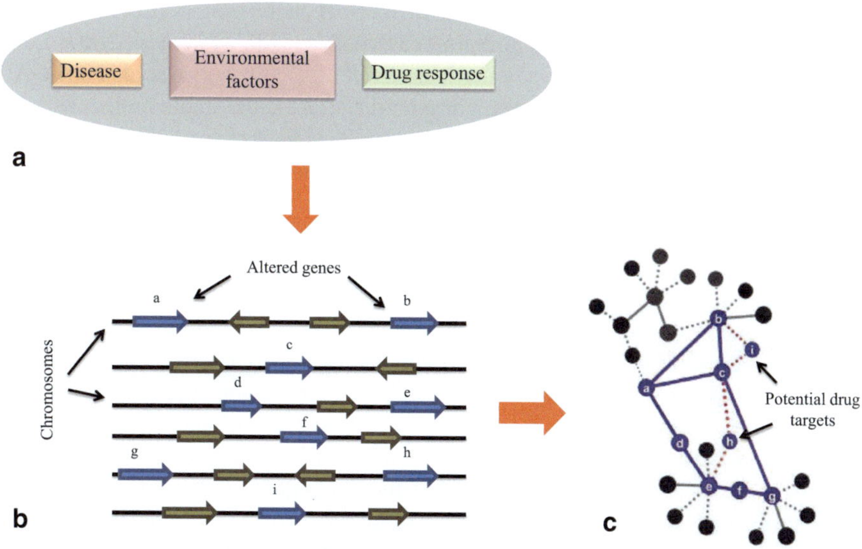

Fig. 9.3 Framework depicting systems approach in drug discovery: **a** Disease condition, environmental factors and drug response phenotypes leads to altered genetic signatures when compared among cases and control individuals. These altered genes can be identified using genotyping SNPs or sequencing methods. **b** These individual genes may not lead to mechanistic understanding of the underlying pathway and potential therapeutic agent. **c** Using network creation and visualization tools, protein-protein interaction or gene interaction network can be built which will provide the information about the altered pathways implicated in the disease conditions. To find the targets in therapies of polygenic, complex diseases network influence strategy is required which targets neighbours of central nodes exerting an indirect influence on the central nodes often representing the 'real targets'

dosing. Furthermore, predictive genetic tests could also be of value in the drug development process by rescuing drugs that failed Phase III clinical trials due to toxicity within a subset of participants (Weiss et al. 2008). Meanwhile, in the last decade, development costs of new drugs has increased tremendously along with high-profile drug withdrawals due to late stage clinical trial failures leading to fewer approvals of new drugs (Caskey 2007). However, with approximately 25,000 genes in the human genome and 20 million SNPs, the analysis and interpretation of huge amount of information has become a compellingly complex problem (Ma'ayan et al. 2005). SNPs are currently used in studies with different study designs including candidate gene association studies, genome wide association studies and pharmacogenomics studies. Framework depicting systems approach in drug discovery is depicted in Fig. 9.3.

Once the genetic variants or SNPs associated with the disease phenotype are identified, the functional effect of predisposing SNP can be identified using bioinformatic approaches which will provide an insight into the mechanisms underlying the disease (Ma'ayan and Iyengar 2006). Several in silico tools have characterized the functional effect of SNPs by assessing their effect on the protein structure or their impact on functional sites at the protein or DNA level (Ma'ayan et al. 2007; Levy et al. 2007; Sayers et al. 2013; Bauer-Mehren et al. 2009; Jegga et al. 2007). All these approaches, although valuable, consider the effect at the single molecule level. In this context, the functional significance of SNPs is better correlated if the evaluation is performed at the system-wide level, for instance by determining their effect on the dynamics of signalling pathways (Cavallo and Martin 2005). Moreover, it is also important to consider the effect of SNPs, in particular, those having an impact at the protein level (non synonymous SNPs, nsSNPs), in the context of biological networks. Although synonymous SNPs and SNPs located in regions that modulate gene expression (e.g. promoters, introns, splice sites, transcription factor binding sites) can also alter gene or protein function and as a consequence lead to disease (Reumers et al. 2006; Kim et al. 2008; Ryan et al. 2009; Klipp et al. 2008), nsSNPs have a more evident effect on the protein function in the biological processes, and are therefore better in explaining disease phenotypes. The study of protein function is usually assessed by experiments aimed at disrupting the activity of the protein, for instance by means of altering the protein sequence at residues suspected to be critical for the function (e.g. *in vitro* mutagenesis experiments). Several databases and web tools gather, manage and provide information about SNPs (Kimchi-Sarfaty et al. 2007; De Gobbi et al. 2006) and their association with diseases (Ma'ayan et al. 2007; Capon et al. 2004) as well as mutations of clinical relevance (Cartegni et al. 2002). Furthermore, several databases also offer information about models of biological networks such as PPI and signalling pathways.

9.2.2 Inferring Pathways from Genetic Association Studies Using Network Biology

Networks have been widely used in many fields of biology to represent the relationships between biological entities. Network is a collection of nodes that are joined together in pairs by edges (Sun 2012). Networks that contain the edges representing relationship with a specified direction are called directed networks whereas network representing relationship between two biological entities (always bidirectional) is called undirected network. In molecular biology and genetics, networks are often used to represent the functional connections among large (e.g., genes, protein) and small molecules (e.g., lipids, drugs) within cells and organisms. Several types of biological networks, such as protein–protein interaction (PPI) networks, metabolic networks and gene interaction networks, have been constructed to illustrate the complex relationship within the biological system. These networks represent the functional or physical connectivity among genes or proteins. Integration of genome wide association studies (GWAS) data with the network-based methods complement the approach of single genetic variant analysis by taking advantage of the available biological knowledge. These approaches play an important role in elucidation of the functional role of the genetic variants, in understanding the molecular mechanism influencing the phenotypic traits and may in turn improve the power to identify phenotype associated genes. The network- and pathway-based methods have been applied successfully in past few years for understanding complex disorders and with increasing availability of GWAS data, these approaches will become more significant to address challenges facing the high throughput studies in the current scenario (Yan 2008).

The availability of pathway databases and curated datasets on the phenotypic effect of genetic variants facilitated the study of genetic factors that contribute to complex disease phenotypes in the context of the structure and dynamics of biological networks. This can have significant consequences for understanding mechanisms of disease and the design of new drug discovery approaches. Several reports have been published detailing the integration of SNP data with protein structural data and pathways (Sherry et al. 2001; Song et al. 2007; Fredman et al. 2002). For instance, DataBins (Song et al. 2007) is a web service for the retrieval and analysis of pathway data from KEGG, and sequence databases such as dbSNP (Kimchi-Sarfaty et al. 2007) with the aim of mapping nsSNPs onto the proteins of a pathway. An interesting strategy for the integration of data retrieved from public resources, such as NCBI dbSNP, UniProt, Reactome and BioModels was put forward by Bauer-Mehren A. et al. The methodology involves generation of attribute files containing phenotypic and genotypic annotations to the nodes of biological networks which are then imported into network visualization tools such as Cytoscape, NAVIGaTOR, Medusa, BioLayout3D, Osprey, ProViz, ONDEX, PIVOT, Pajek (Bauer-Mehren et al. 2009; Agapito et al. 2013). These resources allow the mapping and visualization of interaction among biomolecules and their phenotypic effect on biological networks (e.g. gene interaction networks, protein-protein interaction networks, signalling pathways etc). An example of systems based approach linking disease, genes and drugs interactions through biological pathways is depicted in Fig. 9.4. However, major challenges that exist for the integration of the phenotypic effect of SNPs in the context of biological networks are:

1. Integration of data generated from diverse and heterogeneous experimental technologies.
2. Visualization of information about genetic variations in the context of biological pathways.
3. Incorporation of the effect of the alteration caused by the genetic variation in dynamic models of the pathways.

9.2.3 Integrating Human Diseases, Genetics, Drugs and Drug Targets

Technological advancement led to the rapid identification of disease genes which in turn allows for the construction of a disease-gene network (Goh et al. 2007). This type

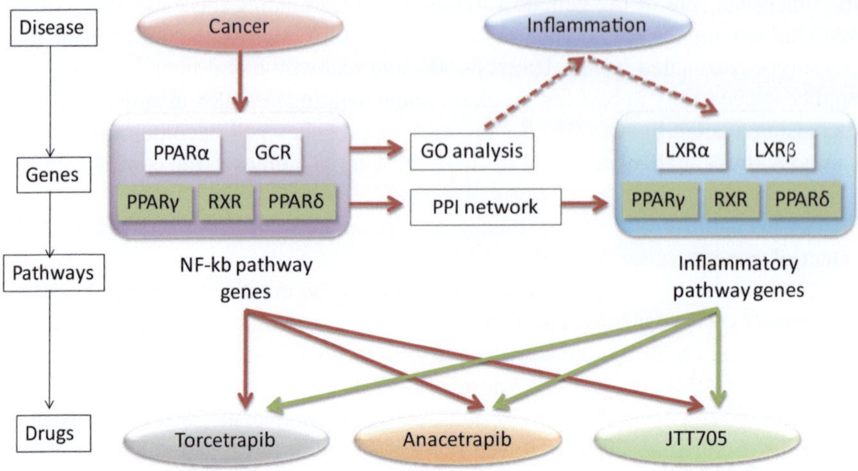

Fig. 9.4 An example of systems based approach linking disease, genes and drugs interactions through biological pathways: Several genes have been implicated in complex disorders such as cancer in genome-wide association and candidate genes studies. When these genes are analyzed for gene ontology terms such as biological process the role of associated significant pathway such as inflammation may become evident. Further integration of protein interaction data from PPI databases can lead to the identification of other significant gene candidates associated with cancer and inflammation. The identification of pathways may help to screen for drug molecules both existing and newer ones. Drug for CETP inhibitors such as Torcetrapib, Anacetrapib, and JTT705 targeting both cancer specific NF-kb pathway and specific genes in inflammatory pathway may be more effective as compared to one targeting a single pathway genes

of network analysis may help identify functional clusters, modules or sub-networks of interacting disease genes and can also be used to predict additional disease gene candidates (Lage et al. 2007; Xu and Li 2006; Franke et al. 2006). Network analysis may play an important role in increasing the power for analyzing the genetic data by combining multiple related genes in a pathway, and to infer the biological function underlying the disease phenotype. Network analysis has been extensively applied to study biological networks including genetic networks (Sun 2012). Thus, new approaches that can integrate multiple biological networks with genetic association study may further bridge the gap between the genetic variants and complex traits. The high-throughput analysis of genomic research has produced a large amount of data to enable network studies. Biological interaction databases such as BioGRID, DIP, HPRD, IntAct, IMID, and MIPS (Aranda et al. 2010; Keshava Prasad et al. 2009; Pagel et al. 2005; Stark et al. 2006; Warde-Farley et al. 2010; Xenarios et al. 2002) provide hundreds of thousands of physical and genetic interactions from a number of organisms including humans. Constructed gene interaction networks from genes within associated loci for complex diseases also showed abundant physical interactions between protein products of associated genes (Rossin et al. 2011). Integrated with the GWAS analysis, the information of PPI can help to interpret the genetic associations with human diseases, and provide hints to plausible functions of the genetic variants (Hannum et al. 2009; Jia et al. 2011).

The biological interaction network based methods essentially represent a framework to incorporate biological knowledge into the genetic studies of complex disorders. Current methods utilize networks to preselect genetic variants for targeted analysis, to enrich the statistical associations and to identify functional modules based on statistical significance, but mostly focus on a single type of integration using one source of biological networks. Furthermore, the analysis of networks of interacting gene products (PPI) and therapeutic agents (drugs) represents a logical and more accurate extension of our understanding of disease, treatments, and their responses (Azuaje et al. 2012). Similarly, networks of drugs and drug targets can also be developed (Ma'ayan et al. 2007; Jia et al. 2011). Combined analysis of such networks can be a valuable initial step towards finding novel approach to identify use of approved drugs for other disease indications and better understand side effects caused by drugs through off target identification in network.

9.3 Understanding Systems Pharmacology, Pharmacogenomics and Drug Development

In the past few years, the field of biomedical science has felt the need for a transformation in approach from reductionism toward a holistic paradigm, from one-drug-fits-all toward personalized medicine. The emerging disciplines, systems biology and pharmacogenomics may help in solving the current problems and guiding the future of drug therapy.

While pharmacogenomics may help achieve personalized medicine, the application of systems biology may help us understand the major issues in pharmacogenomics at different levels. These key issues include the correlations between genotype and phenotype, the associations between structure and function, and the interactions among genes, drugs, and the environment (Yan 2003). Systems biology investigates the roles genes and/or proteins play in the context of complex pathways and interactions and enables the understanding of disease and drug mechanisms at the system level (Kitano 2002). Using computational methods, systems biology may help us simulate large networks of interacting components, organize biological knowledge, and create predictive models. The integration of pharmacogenomics and systems biology may help to understand the disease specific molecular mechanism and mode of drug actions at cellular, molecular and tissue levels and connect information between different levels. For example, variation in genetic structure may cause alterations at the molecular level, which would influence the downstream interactions, pathways, and networks. On the other hand, interactions among genes/proteins, drugs, and the environmental factors at higher levels may also affect the structure and function of genes/proteins at the molecular level, which would in turn change downstream reactions and phenotypes, forming a feedback loop. The analysis of such an interactome may serve as the ultimate key in accurately identifying drug targets, understand drug–response phenotypes and to avoid adverse reactions.

9.3.1 Modelling Drug Action Using Systems Pharmacology Approach

Systems pharmacology is an evolving area that studies drug action across multiple scales of complexity, from molecular and cellular to tissue and organismal levels. The conventional 'one-target one-drug' drug design paradigm initially allowed bringing new drugs to the market. However, a significant decrease in the rate of new drug candidates has been observed due to several reasons:

1. Most drugs interact with multiple targets. For example, anti-diabetic drug rosiglitazone, not only stimulates the peroxisome proliferator activated receptor gamma (PPARγ), but also blocks interferon gamma (INFγ)—induced chemokine expression in Graves's disease or ophthalmopathy (Antonelli et al. 2011).
2. Lack of efficacy and clinical safety or toxicology (Hopkins 2008). For instance, two drugs- cisapride and astemizole have both been withdrawn from markets due to the risk of fatal cardiac arrhythmia associated with their blockade of the hERG potassium ion channel.

There is an urgent requirement for the development of network or pathway based approaches to integrate the accumulating knowledge of chemical biology with systems biology. In silico computational approaches can be used at the various stages of the drug discovery process and, for instance, can involve querying genomic data, running comparative genomics, investigating protein folding, defining protein interaction networks, analyzing the impact of genetic variants, and assisting in clinical trial design, to name only a few (Mah et al. 2011; Pierri et al. 2010; Fernald et al. 2011; Thusberg et al. 2011; Tsai et al. 2009; Tuncbag et al. 2011; Villoutreix 2002; Woollard 2010; Woollard et al. 2011). A practical application of this approach is reported by Yildirim et al. (Yildirim et al. 2007) wherein they combined FDA-approved drugs with a human PPI network (human interactome) in order to analyze the interrelationships between drug targets and disease–gene products i.e. disease–proteins. Identification of therapeutic target and off target using network based approach is depicted in Fig. 9.5.

Key steps in systems pharmacology approach:

1. From a drug or a protein, profiling of multiple annotated (from the literature) or predicted (from the web-tools) targets is performed.
2. Integration of "genomics" data associated with the ensemble of proteins altered by the drug (interactomic, pathway, and genomics) is performed.
3. Finally, analysis of potential clinical effects (therapeutic and adverse effects) associated with the drug is carried out.

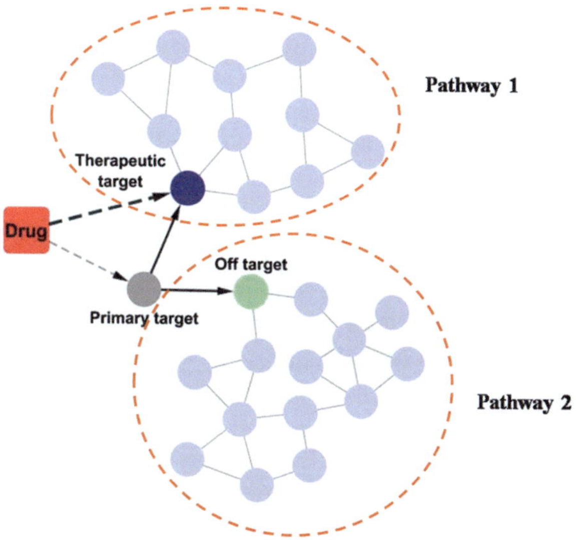

Fig. 9.5 Identification of therapeutic target and off target using network based approach: Pharmacogenomics studies can provide drug response status of individual patients based on their individual genetic architecture leading to identification of poor metabolizer, intermediate metabolizer and ultra metabolizer phenotype. It will also help in elucidation of genes involved in causing adverse drug reactions. Systems pharmacology and network based approach can used to identify off-targets of existing drugs showing variable efficacy and toxicity. For instance, in this figure drug primary target interacts with two proteins—one responsible for therapeutic response and other for adverse reaction. Changing primary target (in *gray*) to existing therapeutic target (*blue* node) may help overcome the side effects of the drug

Systems pharmacology involving network biology approach has emerged as an alternative for accurate predictions of drugs with multiple targets that can cause ADRs (Mendrick 2011). For example, connecting drugs by side effect similarity based on the assumption that drugs with common side-effects are likely to interact with common target proteins can provide insights into the molecular basis of the drug's side effects and allow predicting novel off-targets involved in negative clinical outcomes (Campillos et al. 2008; Yang et al. 2011; Brouwers et al. 2011).

Several databases such as PubChem, CheBI and ChEMBL exist for retrieval of biological information for a large set of chemical compounds. Using PubChem, the ADRs have also been analyzed at the organ level (Pouliot et al. 2011). Several repositories can also be used to extract data related to toxicological effects of small molecules such as Sider (Kuhn et al. 2010), Actor (Judson et al. 2008), or Dailymed (http://dailymed.nih.gov/dailymed). Open Phacts, a public-private partnership will also provide the pharmacological, PK, ADMET (absorption, distribution, metabolism, and excretion—toxicity) and clinical profiles of drugs and small molecules in the near future (http://www.openphacts.org/). In silico approach allows to explore drug repositioning and associations of drugs, targets, and therapeutic responses into an integrated network (Ekins et al. 2011; Kinnings et al. 2009; Oprea et al. 2011; Xie and Bourne 2011). Recent advances include a human disease network (diseasome) linking disorders and disease genes to various known phenotypes (Goh et al. 2007) and a PPI network based on the toxicology of environmental chemicals (Audouze et al. 2010).

9.3.2 Dissecting Variability in Therapeutic Response Through Systems Pharmacology and Pharmacogenomics

Classical pharmacology involves the study of mechanism of drug action at the cell and tissue levels, with major emphasis on identifying targets and understanding drug–target interaction through structure–activity relationships (i.e., the effect of variation in chemical structure on drug activity). However, interindividual variations in drug responses largely form the area of study of pharmacogenomics, a branch of "personalized medicine" that has gained momentum in the post-human project genomic era. It is able to provide explanation for patient-to-patient variation in drug metabolism based on polymorphisms in genes encoding cytochrome P450. In chronic complex diseases such as cancer, variation in drug response at the level of the individual patient is a matter of great concern as large interpatient variability is observed for virtually all targeted and cytotoxic agents, even with drug-naive tumors. This essentially reflects the synergistic effects of the genetic heterogeneity of tumors and common polymorphisms in individual patients. Due to largely unknown origin of genetic variation which appear to be patient specific, their effect on drug response cannot easily be understood and overcome through multi-therapy or differential dosing—at least not without patient-specific rationales, which do not yet exist for most drugs or diseases.

Pharmacogenomics represents a unique opportunity for prediction of drug response by identifying patterns of genetic variation that will guide design of optimal medication regimens in individual patients (Evans and McLeod 2003). In an ideal condition, the application of pharmacogenomics based on the patient's genetic profile would enable the prediction of a patient's response to particular drugs and empower physicians to make right decisions for the treatment. A very useful source of high-quality clinically relevant information about the impact of human genetic variation on drug responses including dosing guidelines, annotated drug labels, and potentially actionable gene-drug associations and genotype-phenotype relationships can be accessed from The Pharmacogenomics Knowledgebase (PharmGKB) (Whirl-Carrillo et al. 2012).

Approaches to understand the variations in drug response can be classified broadly as either correlative or mechanistic. The basic idea of correlative approaches is to match patients with drugs empirically, and to correlate responses with measurable parameters of disease such as histological diagnosis and with clinical factors such as family history, tumor size, and lymph node metastasis. On the other hand, mechanistic studies of drug response seek a detailed understanding of interactions between drugs and their targets, the consequences of binding for downstream proteins, and, ultimately, the impact on cell fate (Yang et al. 2010).

Systems biology can play a major role in understanding precisely how complex cellular networks respond to drugs at a mechanistic level which is extremely challenging. For example, development of MM-121, a therapeutic antibody against the ErbB3 receptor followed careful computational analysis of signaling pathways in tumor cells (Schoeberl et al. 2009). However, physiological drug response is a complex, time-dependent, and probabilistic process at the single-cell level and requires computational tools for elucidating the biochemistry of cell signaling networks, dissecting gene regulatory networks and for advance network-oriented analysis of drug mechanisms. The other challenges include drug response monitoring at multiple time points using quantitative, single-cell assays and integrating this data into pharmacological response genetic signatures (SNPs). In addition, transition from a qualitative level to a quantitative component, including concentration level and kinetic parameters governing the interactions will be another major hurdle.

These challenges can be addressed with a systems approach to pharmacology that is

a. quantitative to predict the behaviours of interacting genes or proteins through knowledge of their individual functional characteristics;
b. mechanistic in explaining disease phenotypes in terms of the comprehensive background of disease genes, genetic variants and drug targets;
c. probabilistic in accounting for the variability between cells and tissues with respect to drug response,
d. postgenomic in analyzing diverse endophenotypes in the light of knowledge of their genetic differences and
e. integrative in assuming that determinants of drug response are multifactorial, that physiological, morphological, and genetic features are important, and that multiple interacting pathways rather than single genes or proteins must be studied.

Overall, systems pharmacology approaches and pharmacogenomics will play a major role in linking drug response phenotype with genetic signatures collected from patients and integrating them with in vitro/in vivo data using system biology approaches, thus contributing to better establish personalized medicine.

Systematic and quantitative studies of adverse side effects have become increasingly important due to rising concerns about the cytotoxicity of drugs in development (Huang et al. 2011). There is an urgent need to design new and accurate models by researchers to assess unwanted side effects and drug actions before initiating costly human clinical trials. Incorporating prior knowledge, including genetic and proteomics can significantly enhance the predictive accuracy of ADR of drugs under development or in clinical trials. Systems approach can play a major role in the analysis of biomolecules and drug entities in a variety of functional network contexts allowing researchers to understand how drugs act in a complex biological system (Li et al. 2009), predict drug safety issues in advance (Butcher et al. 2004; Ekins et al. 2005), identify ADR events early (Mutsumi Fukuzaki et al. 2009; Pouliot et al. 2011), and design personalised diagnostic tests with tailored drug treatments (Barabasi et al.

2011). The use of PPI networks can increase the prediction specificity and the use of GO annotations can increase the prediction sensitivity. The following approach can be followed for the ADR prediction using network pharmacology (Huang et al. 2011):

1. Drug target interactions are expanded in global human PPI networks to build drug target expanding PPI networks.
2. Drug targets are enriched by their gene ontology (GO) annotations to build drug target expanding GO networks.
3. ADR information for each drug is combined with drug target expanding PPI networks and drug target expanding GO networks.
4. Statistics and machine learning are applied to build ADR classification/prediction models.
5. Cross validation and feature selection are used to train prediction models.

9.4 Summary

In the past decade, although a considerable advancement in the high-throughput genome and proteome based experimental technologies has been made, the drug discovery and development slowed down mainly due to high costs, low efficacy and toxicity issues. Systems biology has shown the direction for addressing these problems by showing ability to integrate interdisciplinary research fields such as pharmacology and genomics using network biology. Both genome medicine and systems pharmacology fields are in their infancy but can provide a platform leading to the ultimate goal of personalized medicine i.e. to treat each patient on the basis of the individual's genome. However, success in personalized medicine will require new conceptual and technological developments, integration of different fields using network analysis. This integration will provide a systems-level understanding of drug action and disease complexity ultimately leading to mechanism based understanding of disease and therapy across scales of biological organization. In conclusion, the systems biology applications in future lead to not only newer and more effective therapies, but safer medications with fewer side effects.

Acknowledgements We thank the Director, Institute of Genomics and Integrative Biology (CSIR) for the support. We appreciate Prof. Pawan Dhar for critical evaluation of the manuscript. Financial support from Council of Scientific and Industrial Research (CSIR) (BSC0123) is duly acknowledged. The authors are grateful to Prof. Samir K Brahmachari for his vision and intellectual inputs. PT, YS and SG acknowledge CSIR, Govt. of India and GKG acknowledge DBT, Govt. of India for providing their fellowships.

References

Agapito G, Guzzi PH, Cannataro M (2013) Visualization of protein interaction networks: problems and solutions. BMC Bioinform 14(Suppl 1):S1

Antonelli A, Ferrari SM, Fallahi P, Piaggi S, Paolicchi A et al (2011) Cytokines (interferon-gamma and tumor necrosis factor-alpha)-induced nuclear factor-kappaB activation and chemokine (C-X-C motif) ligand 10 release in Graves disease and ophthalmopathy are modulated by pioglitazone. Metabolism 60:277–283

Aranda B, Achuthan P, Alam-Faruque Y, Armean I, Bridge A et al (2010) The IntAct molecular interaction database in 2010. Nucleic Acids Res 38:D525–531

Arrowsmith J (2011a) Trial watch: phase II failures: 2008–2010. Nat Rev Drug Discov 10:328–329

Arrowsmith J (2011b) Trial watch: phase III and submission failures: 2007–2010. Nat Rev Drug Discov 10:87

Audouze K, Juncker AS, Roque FJ, Krysiak-Baltyn K, Weinhold N et al (2010) Deciphering diseases and biological targets for environmental chemicals using toxicogenomics networks. PLoS Comput Biol 6:e1000788

Azuaje FJ, Dewey FE, Brutsaert DL, Devaux Y, Ashley EA et al (2012) Systems-based approaches to cardiovascular biomarker discovery. Circ Cardiovasc Genet 5:360–367

Barabasi AL, Gulbahce N, Loscalzo J (2011) Network medicine: a network-based approach to human disease. Nat Rev Genet 12:56–68

Bauer-Mehren A, Furlong LI, Rautschka M, Sanz F (2009) From SNPs to pathways: integration of functional effect of sequence variations on models of cell signalling pathways. BMC Bioinform 10(Suppl 8):S6

Berger SI, Iyengar R (2009) Network analyses in systems pharmacology. Bioinformatics 25:2466–2472

Bhogal N, Balls M (2008) Translation of new technologies: from basic research to drug discovery and development. Curr Drug Discov Technol 5:250–262

Boran AD, Iyengar R (2010) Systems pharmacology. Mt Sinai J Med 77:333–344

Brouwers L, Iskar M, Zeller G, van Noort V, Bork P (2011) Network neighbors of drug targets contribute to drug side-effect similarity. PLoS One 6:e22187

Butcher EC, Berg EL, Kunkel EJ (2004) Systems biology in drug discovery. Nat Biotechnol 22:1253–1259

Campillos M, Kuhn M, Gavin AC, Jensen LJ, Bork P (2008) Drug target identification using side-effect similarity. Science 321:263–266

Capon F, Allen MH, Ameen M, Burden AD, Tillman D et al (2004) A synonymous SNP of the corneodesmosin gene leads to increased mRNA stability and demonstrates association with psoriasis across diverse ethnic groups. Hum Mol Genet 13:2361–2368

Cartegni L, Chew SL, Krainer AR (2002) Listening to silence and understanding nonsense: exonic mutations that affect splicing. Nat Rev Genet 3:285–298

Caskey CT (2007) The drug development crisis: efficiency and safety. Annu Rev Med 58:1–16

Cavallo A, Martin AC (2005) Mapping SNPs to protein sequence and structure data. Bioinformatics 21:1443–1450

Csermely P, Korcsmaros T, Kiss HJ, London G, Nussinov R (2013) Structure and dynamics of molecular networks: a novel paradigm of drug discovery: a comprehensive review. Pharmacol Ther 138:333–408

Csete ME, Doyle JC (2002) Reverse engineering of biological complexity. Science 295:1664–1669

Cucurull-Sanchez L, Spink KG, Moschos SA (2012) Relevance of systems pharmacology in drug discovery. Drug Discov Today 17:665–670

De Gobbi M, Viprakasit V, Hughes JR, Fisher C, Buckle VJ et al (2006) A regulatory SNP causes a human genetic disease by creating a new transcriptional promoter. Science 312:1215–1217

Drazen JM, Silverman EK, Lee TH (2000) Heterogeneity of therapeutic responses in asthma. Br Med Bull 56:1054–1070

Ekins S, Nikolsky Y, Nikolskaya T (2005) Techniques: application of systems biology to absorption, distribution, metabolism, excretion and toxicity. Trends Pharmacol Sci 26:202–209

Ekins S, Williams AJ, Krasowski MD, Freundlich JS (2011) In silico repositioning of approved drugs for rare and neglected diseases. Drug Discov Today 16:298–310

Evans WE, McLeod HL (2003) Pharmacogenomics–drug disposition, drug targets, and side effects. N Engl J Med 348:538–549

Fernald GH, Capriotti E, Daneshjou R, Karczewski KJ, Altman RB (2011) Bioinformatics challenges for personalized medicine. Bioinformatics 27:1741–1748

Franke L, van Bakel H, Fokkens L, de Jong ED, Egmont-Petersen M et al (2006) Reconstruction of a functional human gene network, with an application for prioritizing positional candidate genes. Am J Hum Genet 78:1011–1025

Fredman D, Siegfried M, Yuan YP, Bork P, Lehvaslaiho H et al (2002) HGVbase: a human sequence variation database emphasizing data quality and a broad spectrum of data sources. Nucleic Acids Res 30:387–391

Giacomini KM, Yee SW, Ratain MJ, Weinshilboum RM, Kamatani N et al (2012) Pharmacogenomics and patient care: one size does not fit all. Sci Transl Med 4:153ps118

Goh KI, Cusick ME, Valle D, Childs B, Vidal M et al (2007) The human disease network. Proc Natl Acad Sci U S A 104:8685–8690

Hannum G, Srivas R, Guenole A, van Attikum H, Krogan NJ et al (2009) Genome-wide association data reveal a global map of genetic interactions among protein complexes. PLoS Genet 5:e1000782

Hood L (2002) A personal view of molecular technology and how it has changed biology. J Proteome Res 1:399–409

Hopkins AL (2008) Network pharmacology: the next paradigm in drug discovery. Nat Chem Biol 4:682–690

Huang LC, Wu X, Chen JY (2011) Predicting adverse side effects of drugs. BMC Genomics 12(Suppl 5):S11

Hughes JP, Rees S, Kalindjian SB, Philpott KL (2011) Principles of early drug discovery. Br J Pharmacol 162:1239–1249

International HapMap Consortium (2005) A haplotype map of the human genome. Nature 437:1299–1320

Jegga AG, Gowrisankar S, Chen J, Aronow BJ (2007) PolyDoms: a whole genome database for the identification of non-synonymous coding SNPs with the potential to impact disease. Nucleic Acids Res 35:D700–706

Jia P, Zheng S, Long J, Zheng W, Zhao Z (2011) dmGWAS: dense module searching for genome-wide association studies in protein-protein interaction networks. Bioinformatics 27:95–102

Judson R, Richard A, Dix D, Houck K, Elloumi F et al (2008) ACToR–aggregated computational toxicology resource. Toxicol Appl Pharmacol 233:7–13

Keshava Prasad TS, Goel R, Kandasamy K, Keerthikumar S, Kumar S et al (2009) Human protein reference database–2009 update. Nucleic Acids Res 37:D767–772

Kim BC, Kim WY, Park D, Chung WH, Shin KS et al (2008) SNP@Promoter: a database of human SNPs (single nucleotide polymorphisms) within the putative promoter regions. BMC Bioinform 9(Suppl 1):S2

Kimchi-Sarfaty C, Oh JM, Kim IW, Sauna ZE, Calcagno AM et al (2007) A "silent" polymorphism in the MDR1 gene changes substrate specificity. Science 315:525–528

Kinnings SL, Liu N, Buchmeier N, Tonge PJ, Xie L et al (2009) Drug discovery using chemical systems biology: repositioning the safe medicine Comtan to treat multi-drug and extensively drug resistant tuberculosis. PLoS Comput Biol 5:e1000423

Kitano H (2002) Systems biology: a brief overview. Science 295:1662–1664

Klipp E, Herwig R, Kowald A, Wierling C, Lehrach H (2008) Systems biology in practice: concepts, implementation and application. Wiley, Weinheim

Kola I, Landis J (2004) Can the pharmaceutical industry reduce attrition rates? Nat Rev Drug Discov 3:711–715

Kuhn M, Campillos M, Letunic I, Jensen LJ, Bork P (2010) A side effect resource to capture phenotypic effects of drugs. Mol Syst Biol 6:343

Lage K, Karlberg EO, Storling ZM, Olason PI, Pedersen AG et al (2007) A human phenome-interactome network of protein complexes implicated in genetic disorders. Nat Biotechnol 25:309–316

Levy S, Sutton G, Ng PC, Feuk L, Halpern AL et al (2007) The diploid genome sequence of an individual human. PLoS Biol 5:e254

Li J, Zhu X, Chen JY (2009) Building disease-specific drug-protein connectivity maps from molecular interaction networks and PubMed abstracts. PLoS Comput Biol 5:e1000450

Ma Q, Lu AY (2011) Pharmacogenetics, pharmacogenomics, and individualized medicine. Pharmacol Rev 63:437–459

Ma'ayan A, Iyengar R (2006) From components to regulatory motifs in signalling networks. Brief Funct Genomic Proteomic 5:57–61

Ma'ayan A, Blitzer RD, Iyengar R (2005) Toward predictive models of mammalian cells. Annu Rev Biophys Biomol Struct 34:319–349

Ma'ayan A, Jenkins SL, Goldfarb J, Iyengar R (2007) Network analysis of FDA approved drugs and their targets. Mt Sinai J Med 74:27–32

Mah JT, Low ES, Lee E (2011) In silico SNP analysis and bioinformatics tools: a review of the state of the art to aid drug discovery. Drug Discov Today 16:800–809

Mendrick DL (2011) Transcriptional profiling to identify biomarkers of disease and drug response. Pharmacogenomics 12:235–249

Fukuzaki M, Seki M, Kashima H, Sese J (2009) Side Effect Prediction Using Cooperative Pathways. IEEE International Conference on Bioinformatics and Biomedicine, pp 142–147

Naidoo N, Pawitan Y, Soong R, Cooper DN, Ku CS (2011) Human genetics and genomics a decade after the release of the draft sequence of the human genome. Hum Genomics 5:577–622

Naylor S, Cavanagh J (2004) Status of systems biology-does it have a future? Drug Discov Today: BIOSILICO 2:171–174

Naylor S, Chen JY (2010) Unraveling human complexity and disease with systems biology and personalized medicine. Per Med 7:275–289

Nebert DW (1999) Pharmacogenetics and pharmacogenomics: why is this relevant to the clinical geneticist? Clin Genet 56:247–258

Noorbakhsh F, Overall CM, Power C (2009) Deciphering complex mechanisms in neurodegenerative diseases: the advent of systems biology. Trends Neurosci 32:88–100

Oltvai ZN, Barabasi AL (2002) Systems biology. Life's complexity pyramid. Science 298:763–764

Oprea TI, Nielsen SK, Ursu O, Yang JJ, Taboureau O et al (2011) Associating drugs, targets and clinical outcomes into an integrated network affords a new platform for computer-aided drug repurposing. Mol Inform 30:100–111

Pagel P, Kovac S, Oesterheld M, Brauner B, Dunger-Kaltenbach I et al (2005) The MIPS mammalian protein-protein interaction database. Bioinformatics 21:832–834

Penrod NM, Cowper-Sal-lari R, Moore JH (2011) Systems genetics for drug target discovery. Trends Pharmacol Sci 32:623–630

Pierri CL, Parisi G, Porcelli V (2010) Computational approaches for protein function prediction: a combined strategy from multiple sequence alignment to molecular docking-based virtual screening. Biochim Biophys Acta 1804:1695–1712

Pouliot Y, Chiang AP, Butte AJ (2011) Predicting adverse drug reactions using publicly available PubChem BioAssay data. Clin Pharmacol Ther 90:90–99

Reumers J, Maurer-Stroh S, Schymkowitz J, Rousseau F (2006) SNPeffect v2.0: a new step in investigating the molecular phenotypic effects of human non-synonymous SNPs. Bioinformatics 22:2183–2185

Rossin EJ, Lage K, Raychaudhuri S, Xavier RJ, Tatar D et al (2011) Proteins encoded in genomic regions associated with immune-mediated disease physically interact and suggest underlying biology. PLoS Genet 7:e1001273

Ryan M, Diekhans M, Lien S, Liu Y, Karchin R (2009) LS-SNP/PDB: annotated non-synonymous SNPs mapped to Protein Data Bank structures. Bioinformatics 25:1431–1432

Sayers EW, Barrett T, Benson DA (2013) Database resources of the National Center for Biotechnology Information. Nucleic Acids Res 41:D8–D20

Schoeberl B, Pace EA, Fitzgerald JB, Harms BD, Xu L et al (2009) Therapeutically targeting ErbB3: a key node in ligand-induced activation of the ErbB receptor-PI3K axis. Sci Signal 2:ra31

Sherry ST, Ward MH, Kholodov M, Baker J, Phan L et al (2001) dbSNP: the NCBI database of genetic variation. Nucleic Acids Res 29:308–311

Song YC, Kawas E, Good BM, Wilkinson MD, Tebbutt SJ (2007) DataBiNS: a BioMoby-based data-mining workflow for biological pathways and non-synonymous SNPs. Bioinformatics 23:780–782

Sorger PK, Allerheiligen SR, Abernethy DR, Altman RB, Brouwer KL et al (2011) Quantitative and systems pharmacology in the post-genomic era: new approaches to discovering drugs and understanding therapeutic mechanisms. In An NIH white paper by the QSP workshop group (pp 1–48). Bethesda: NIH

Stark C, Breitkreutz BJ, Reguly T, Boucher L, Breitkreutz A et al (2006) BioGRID: a general repository for interaction datasets. Nucleic Acids Res 34:D535–539

Sun YV (2012) Integration of biological networks and pathways with genetic association studies. Hum Genet 131:1677–1686

Thusberg J, Olatubosun A, Vihinen M (2011) Performance of mutation pathogenicity prediction methods on missense variants. Hum Mutat 32:358–368.

Tsai CJ, Ma B, Nussinov R (2009) Protein-protein interaction networks: how can a hub protein bind so many different partners? Trends Biochem Sci 34:594–600

Tuncbag N, Gursoy A, Keskin O (2011) Prediction of protein-protein interactions: unifying evolution and structure at protein interfaces. Phys Biol 8:035006

Tyers M, Mann M (2003) From genomics to proteomics. Nature 422:193–197

Venter JC, Adams MD, Myers EW, Li PW, Mural RJ et al (2001) The sequence of the human genome. Science 291:1304–1351

Villoutreix BO (2002) Structural bioinformatics: methods, concepts and applications to blood coagulation proteins. Curr Protein Pept Sci 3:341–364

Warde-Farley D, Donaldson SL, Comes O, Zuberi K, Badrawi R et al (2010) The GeneMANIA prediction server: biological network integration for gene prioritization and predicting gene function. Nucleic Acids Res 38:W214–220

Weiss ST, McLeod HL, Flockhart DA, Dolan ME, Benowitz NL et al (2008) Creating and evaluating genetic tests predictive of drug response. Nat Rev Drug Discov 7:568–574

Weston AD, Hood L (2004) Systems biology, proteomics, and the future of health care: toward predictive, preventative, and personalized medicine. J Proteome Res 3:179–196

Whirl-Carrillo M, McDonagh EM, Hebert JM, Gong L, Sangkuhl K et al (2012) Pharmacogenomics knowledge for personalized medicine. Clin Pharmacol Ther 92:414–417

Woollard PM (2010) Asking complex questions of the genome without programming. Methods Mol Biol 628:39–52

Woollard PM, Mehta NA, Vamathevan JJ, Van Horn S, Bonde BK et al (2011) The application of next-generation sequencing technologies to drug discovery and development. Drug Discov Today 16:512–519

Xenarios I, Salwinski L, Duan XJ, Higney P, Kim SM et al (2002) DIP, the database of interacting proteins: a research tool for studying cellular networks of protein interactions. Nucleic Acids Res 30:303–305

Xie L, Bourne PE (2011) Structure-based systems biology for analyzing off-target binding. Curr Opin Struct Biol 21:189–199

Xu J, Li Y (2006) Discovering disease-genes by topological features in human protein-protein interaction network. Bioinformatics 22:2800–2805

Yan Q (2003) Pharmacogenomics of membrane transporters: an overview. In: Yan Q (ed) Membrane transporters: methods and protocols, Methods in molecular biology. Humana, Totowa, pp 1–20

Yan Q (2008) The integration of personalized and systems medicine: bioinformatics support for pharmacogenomics and drug discovery. Methods Mol Biol 448:1–19

Yan Q (2010) Bioinformatics for transporter pharmacogenomics and systems biology: data integration and modeling with UML. Methods Mol Biol 637:23–45

Yang R, Niepel M, Mitchison TK, Sorger PK (2010) Dissecting variability in responses to cancer chemotherapy through systems pharmacology. Clin Pharmacol Ther 88:34–38

Yang L, Wang KJ, Wang LS, Jegga AG, Qin SY et al (2011) Chemical-protein interactome and its application in off-target identification. Interdiscip Sci 3:22–30

Yildirim MA, Goh KI, Cusick ME, Barabasi AL, Vidal M (2007) Drug-target network. Nat Biotechnol 25:1119–1126

Zhao S, Iyengar R (2012) Systems pharmacology: network analysis to identify multiscale mechanisms of drug action. Annu Rev Pharmacol Toxicol 52:505–521

Chapter 10
Switching Mechanism in the p53 Regulatory Network

Mohammad Jahoor Alam, Vikram Singh and R. K. Brojen Singh

Abstract p53 is one of the most important signaling molecule which regulates a number of metabolic biochemical pathways. It has a wide role in cellular homeostasis and prevent cellular integrity. p53 prevent the cellular transition from normal to cancer phase. Under the higher cellular stress condition, which is due to the cellular response to different stress inducer viz. Heat shock, DNA damage, rNTP depletion, hypoxia, spindle damage, oncogenic activation, toxic chemicals, the concentration of p53 rises in the cell. In normal state, p53 regulates cell cycle by checking it at G1 phase where cell takes decision either continued the cycle or stop the cycle. Several studies reported that p53 also act as stress suppressor. It helps in transcription of a number of proteins, which has anti-stress effect. It controls the DNA damage by inducing DNA repair proteins. In normal condition, the concentration of p53 is low. The concentration of p53 within the cell fluctuates under the stress condition. p53 has many feedback loops which are responsible for oscillatory behaviour of p53 within the cell. One of major feedback loops which most widely studied is p53-MDM2 feedback loop. P53 positively induced the MDM2 protein but MDM2 act as a negative regulator of p53. In the present study we have shown how the p53 switching behavior at the molecular level is affected by stress induced by Nitric oxide. Nitric oxide is a very important signaling molecule which has a very less half life. We have obtained various transition phases i.e from normal to stress and stress to normal which signify the role of nitric oxide on p53 and cellular dynamics. We have also studied and elaborated the role of Ca^{2+} on the p53-MDM2-NO model. The impact of noise on the system is also studied and well explained.

Keywords p53 · Mdm2 · Ca^{++} oscillations · p53-Mdm2-NO model · p53-Mdm2-NO-Ca model · Chemical Langevin Equation (CLE) · Gillespie's Stochastic Simulation Algorithm (SSA)

R. K. B. Singh (✉)
School of Computational and Integrative Sciences, Jawaharlal Nehru University, New Delhi 110067, India

M. J. Alam
Center for Interdisciplinary Research in Basic Sciences, Jamia Millia Islamia, New Delhi 110025,

V. Singh
Centre for Computational Biology and Bioinformatics, Central University of Himachal Pradesh, Dharamshala 176215, India

V. Singh, P. K. Dhar (eds.), *Systems and Synthetic Biology*,
DOI 10.1007/978-94-017-9514-2_10

10.1 Introduction

*p*53 is an important tumour suppressor protein in the cell. It is a well-conserved phosphoprotein. The human p53 protein consists of 393 amino acids which are structurally and functionally differentiated into four domains: (1) Acidic N-terminal region which contains the 42 amino acids transactivation domain and Hydrophobic proline-rich region (amino acids 64–92) (2) Central sequence-specific DNA-binding domain (amino acids 102–292) (3) Tetramerization domain (amino acids 324–355), and (4)A highly basic C-terminal region regulatory domain (amino acids 363–393). p53 is phosphorylated on different residues which are distributed on different domains of it. The phosphorylation of p53 is generally seen on its serine residues. Many experimental studies suggest that the phosphorylation of p53 leads to the activation and stabilisation of it. It maintains genomic integrity by triggering the production of DNA repair protein. More than 50 % of the human cancers are related with p53. It controlled many key metabolic pathway such as tumor suppression, cell cycle arrest, DNA repair and apoptosis (Lane 1992; Shih 2008). Due to participation in various pathways its concentration level in the cell is frequently varied. Several research work have been done to understand the p53 dynamics and stability. There are several proteins which are directly or indirectly interact with p53. Many experimental studies have proved that a number of protein which interact with p53, either downregulates or upregulates it. The most studied downregulator protein of p53 is MDM2. p53 act as a transcription factor which interact with MDM2 gene due to result of this the sysnthesis of MDM2 increases and this enhances the rate of production of MDM2 protein. MDM2 protein act as a ubiquitin ligase for p53 protein. So, MDM2 leads to the proteosomal degradation of p53 (Lane 1992; Geva-Zatorsky et al. 2006). In an unstressed cell the p53 levels is controlled by $Mdm2$ via a negative feedback loop (Momand et al. 1992). In a stressed cell the activation and stabilisation of p53 is observed. Experimental studies suggest that the p53 is freed from zinc finger domain of MDM2 due to phosphorylation of a serine residue embeded on it.

Nitric oxide (NO) is a short lived ($\sim 1-10$ s) and a bioactive molecule (Schmidt and Walter 1994; Stern 2004). Various experimental studies shown that it can trigger various physiological and pathological processes in mammalian cell types Wang et al. 2002. It is synthesized by various NO synthase enzymes (NOS), namely neuronal (nNOS), inducible (iNOS) (Lowenstein and Padalko 2004) or endothelial (eNOS) (Li et al. 2002; Werner et al. 2003) such that these isoforms convert arginine to NO and citruline (Marletta and Spiering 2003; Dina 2005). Recent experimental studies has reported that NO exhibit two contrast roles in different single cell types, (1)it induces apoptosis (programmed cell death) in some cell types such as macrophages, neurons, pancreatic β-cells, thymocytes, chondrocytes, hepatocytes (Chung et al. 2001; Brune et al. 1999; Kim et al. 2001) etc, whereas (2) it inhibits apoptosis in other cell types such as B-lymphocytes, eosinophils, ovarian follicles, neuronal PC12 cells, embryonic motor neurons (Li and Billiar 1999; Wang et al. 2002; Taylor et al. 2003; Kim et al. 1999) etc. NO has ability to induce cellular stress, activation of $p53$ via DNA damage and disruption of energy metabolism, calcium homeostasis and mitochondrial function which can be taken as toxic action that leads to cell death (Hofseth et al. 2003; Hussain et al. 2003; Chun-Qi and Wogen 2005; Murphy 1999). Various experimental studies reported that NO can upregulats *p*53 (Brune et al. 1999; Messmer et al. 1994) via downregulating *Mdm2* (Wang et al. 2002; Hofseth et al. 2003; Messmer et al. 1994). Extremely excess of *NO* may lead *p*53 to cause cell apoptosis (Brune et al. 1999).

Ca^{2+} is a versatile molecule that plays important role in many biological pathways (Cerella et al. 2003; Samali et al. 2010). Calcium can induces creation of various vasoactive substances in the endothelium which includes nitric oxide, prostacyclin and other prostanoids (Lopez-Jaramillo et al. 1990). Ca^{2+} induces nitric oxide synthase which leads to the production of activated Nitric Oxide Synthase (NOS) (Hansen et al. 2005; Dedkova et al. 2004; Dedon and Tannenbaum 2004; Manser and Houghton 2006). Activated Nitric Oxide Synthase binds with arginine (Jenkins et al. 1995). This interaction leads to the production of nitric oxide and citruline as a by-product (Knowles and Moncada 1994). Recent experimental studies suggest that there are three isoform of nitric oxide synthases, namely, endothelial(eNOS), neuronal(nNOS) and inducible(iNOS) forms (Manser and Houghton 2006; Knowles and Moncada 1994). These nitric oxide synthases are activated through different extra and intra stimuli. Recent studies suggest that NO produced by nNOS and eNOS has a signalling role and are under the strict control of intracellular calcium ions (Silvagno et al. 1996; Wagner et al. 2005).

Recent experimental studies reported that *NO* is an excellent intercellular signaling molecule (Marletta and Spiering 2003; Dina 2005). *NO* is a small and hydrophobic molecule which can pass through cell membrane easily and it is actively and abundantly created inside the cell by the metabolic pathway (Dina 2005; Murphy 1999). Further, it can also diffuse through several cell diameters from its site of synthesis (Murphy 1999; Lancaster 1994, 1997). The diffusion of *NO* leads to various intracellular signal processing and intercellular communication. Moreover, cellular diffusion of nitric oxide and intracellular consumption are supposed to be the two main factors which control *NO* concentration level in cells (Dedkova et al. 2004; Chen and Deen 2001).

Several issues still need to be resolved. For example even if NO induce toxic to cells, how does it activate $p53$ leading which is due to cellular stress induced by nitric oxide and also excess stress cause apoptosis, is still need to be investigated. Further, even if NO is considered as synchronizing molecule, what could be its role in coupling $p53 - Mdm2$ oscillators at different stress conditions, is still need to be investigated and resolved. The roles of Ca^{2+} in providing various state conditions and the role of noise in cellular organization is also important and needs to be investigated systematically. We aim to study an integrated model consisting of two different oscillators, namely calcium and $p53-Mdm2$ oscillators and investigate the influence of ionic calcium to identify the dynamical behaviour of the variables in single cell.

10.2 A p53-MDM2-NO Autoregulatory Network

Nitric oxide (*NO*) is a diffusible molecule (Stern 2004). It is constantly produced in the cell due to cellular metabolism (Stern 2004; Wood and Garthwaite 1994). Experimental studies prove that nitric oxide down regulates the $MDM2$ protein (Wang et al. 2002; Schonhoff et al. 2002). The down regulation of $MDM2$ protein affect the stability of $p53$ protein (Stern 2004) . $MDM2$ protein as well as $p53$ proteins are supposed to be localize in and out of the nucleus (Chen et al. 1995; Liang and Clarke 1999). $p53$ act as a transcription factor. It activates $MDM2$ gene to form *MDM2_mRNA* due to which synthesis of $MDM2$ protein increases in the cells. $MDM2$ binds $p53$ (Proctor and Gray 2008). After complex formation $MDM2$ ubiquitinates $p53$ due to which the $p53$ concentration level decreases in the cell (Haupt et al. 1997; Kubbutat et al. 1997; Momand et al. 2000). *NO* binds with cytosolic $MDM2$ protein and forms *NO_MDM2* complex due to which the concentration level of $MDM2$ protein is decreases (Wang et al. 2002; Schonhoff et al. 2002). The downregulation of $MDM2$ protein, leads to the oscillatory behaviour of $p53$ (Alam et al. 2013). The half life of $p53$ is found to be short (around 30 minutes) (Finlay 1993). Further the half life of $MDM2$ protein, *MDM2_mRNA* and *NO* are very short respectively, 30 min (Finlay 1993; Pan and Haines 1999), $60-120$ min (Hsing and Faller 2000; Mendrysa et al. 2001) and 5–10 s (Wood and Garthwaite 1994; Wang et al. 2002). $p53$ is an integral protein in the cell and constantly synthesize inside the cell (Mcbride et al. 1986). Due its huge network in the biological cell its population inside normal cell stabilized at low level. Moreover, the half life of proteins are varied in biological system which is depends upon the nature of biochemical reaction network. The p53-MDM2-NO model is shown in Fig. 10.1 generated is based upon the above biological interaction. The Table 10.1 shows number of molecule participate in this biochemical network. The molecular species is symbolized in terms of x 's for the shake of simplicity. All reaction channel involved in the model with their respective transition rates are described in Table 10.2.

10.3 A Mathematical Description of the Model

At first we mathematically described the model using deterministic approach. For the deterministic approach, the biochemical reactions shown in Fig. 10.1 can be transformed into a set of coupled ordinary differential equations using simple Mass-action kinetic law. Now, we have following set of coupled equations,

$$\frac{dx_1}{dt} = k_5 - k_7 x_1 x_2 + k_8 x_3 \tag{10.1}$$

$$\frac{dx_2}{dt} = k_1 x_4 - k_4 x_2 + k_6 x_3 - k_7 x_1 x_2 + k_8 x_3 - k_{10} x_5 x_2 \tag{10.2}$$

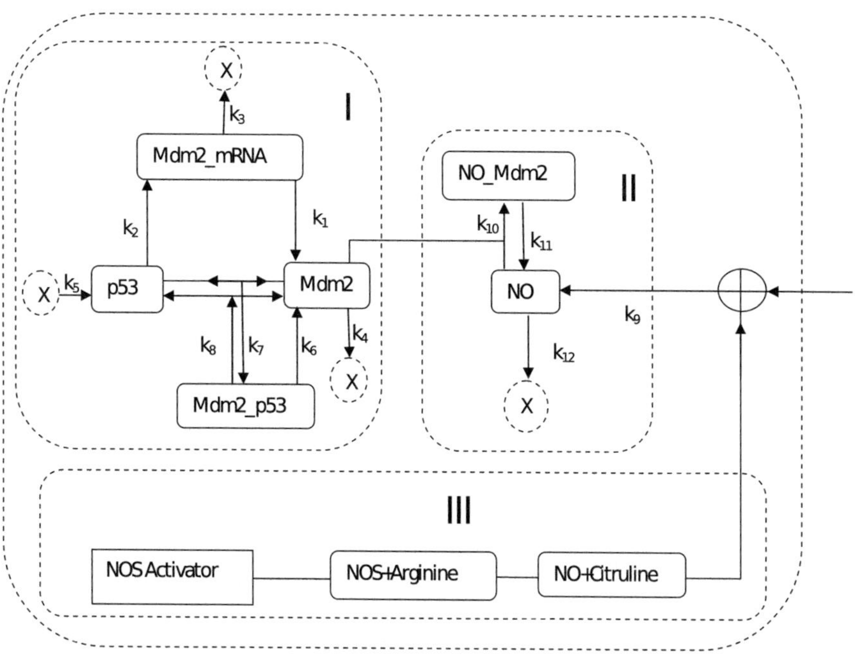

Fig. 10.1 A schematic diagram of p53 network which is induced by nitric oxide

Table 10.1 Molecular species, their description and notation

S. no	Species name	Description	Notation
1.	p53	Unbound p53 protein	x_1
2.	Mdm2	Unbound MDM2 protein	x_2
3.	$MDM2\_p53$	MDM2/p53 complex	x_3
4.	$MDM2\_mRNA$	MDM2 messenger RNA	x_4
5.	NO	Unbound nitric oxide	x_5
6.	NO_MDM2	NO/MDM2 complex	x_6

$$\frac{dx_3}{dt} = -k_6x_3 + k_7x_1x_2 - k_8x_3 \tag{10.3}$$

$$\frac{dx_4}{dt} = k_2x_1 - k_3x_4 \tag{10.4}$$

$$\frac{dx_5}{dt} = k_9 - k_{10}x_5x_2 + k_{11}x_6 - k_{12}x_5 \tag{10.5}$$

$$\frac{dx_6}{dt} = k_{10}x_5x_2 - k_{11}x_6 \tag{10.6}$$

Table 10.2 Chemical reaction, propensity function, rate constant values and references

S.No	Reaction	Propensity function	Values of rate constant	References
1	$x_4 \xrightarrow{k_1} x_4 + x_2$	$k_1 x_4$	$4.95 \times 10^{-4} sec^{-1}$	(Proctor and Gray 2008; Finlay 1993)
2	$x_1 \xrightarrow{k_2} x_1 + x_4$	$k_2 x_1$	$1.0 \times 10^{-4} sec^{-1}$	(Proctor and Gray 2008; Finlay 1993)
3	$x_4 \xrightarrow{k_3} \phi$	$k_3 x_4$	$1.0 \times 10^{-4} sec^{-1}$	(Proctor and Gray 2008; Finlay 1993)
4	$x_2 \xrightarrow{k_4} \phi$	$k_4 x_2$	$4.33 \times 10^{-4} sec-1$	(Proctor and Gray 2008; Finlay 1993)
5	$\phi \xrightarrow{k_5} x_1$	k_5	$0.78 sec^{-1}$	(Proctor and Gray 2008)
6	$x_3 \xrightarrow{k_6} x_2$	$k_6 x_3$	$8.25 \times 10^{-4} sec^{-1}$	(Proctor and Gray 2008)
7	$x_1 + x_2 \xrightarrow{k_7} x_3$	$k_7 x_1 x_2$	$11.55 \times 10^{-4} mol^{-1} sec^{-1}$	(Proctor and Gray 2008)
8	$x_3 \xrightarrow{k_8} x_1 + x_2$	$k_8 x_3$	$11.55 \times 10^{-6} sec^{-1}$	(Proctor and Gray 2008; Finlay 1993)
9	$\phi \xrightarrow{k_9} x_5$	k_9	$1 \times 10^{-2} mol^{-1} sec^{-1}$	(Wang et al. 2002; Alam et al. 2013)
10	$x_5 + x_2 \xrightarrow{k_{10}} x_6$	$k_{10} x_5 x_2$	$1 \times 10^{-3} mol^{-1} sec^{-1}$	(Alam et al. 2013)
11	$x_6 \xrightarrow{k_{11}} x_5$	$k_{11} x_6$	$3.3 \times 10^{-4} sec^{-1}$	(Wang et al. 2002; Alam et al. 2013)
12	$x_5 \xrightarrow{k_{12}} \phi$	$k_{12} x_5$	$1 \times 10^{-3} sec^{-1}$	(Wang et al. 2002; Alam et al. 2013)

Cellular systems are found to be a complex system so that molecular interaction in the system is stochastic or noise induced processes due to random molecular interaction in the system (Rao and Wolf 2002). Moreover, there are several other factors which are responsible to induce noise in the system such as thermodynamics limit etc. (McAdams and Arkin 1997; Blake et al. 2003). The stochastic system is supposed to be real system with qualitative and quantitative prescriptions and it can be well described by taking each and every molecular interaction systematically to find their trajectories in configuration space (Gillespie 1977). One can mathematically described stochastic system by constructing a Master equation of the interaction network, which is mathematically the time evolution of configurational probability $P(\mathbf{x}, t)$ with $\mathbf{x} = \left(x_1, x_2, \ldots, x_6\right)^{-1}$. The Master equation is based on decay and creation of each molecular species at each molecular interaction (Gillespie 1977; McQuarrie 1967). However, Solution of Master equation for a complex system is very difficult to obtained and required huge computational cost. One can compute the trajectory of each and every molecular species in the system using stochastic simulation algorithm (SSA) due to Gillespie (Gillespie 1977) by taking every possible interaction in the complete system. Further, one can simplify this Master equation based on some realistic assumptions which are small time interval of any two consecutive interactions and large molecular population limit (Gillespie 2000). Master equation can be reduce to Chemical Langevin equations (CLE). For our system, we have following CLEs,

$$\frac{dx_1}{dt} = k_5 - k_7 x_1 x_2 + k_8 x_3 + \frac{1}{\sqrt{V}}\left[\sqrt{k_5}\xi_1 - \sqrt{k_7 x_1 x_2}\xi_2 + \sqrt{k_8 x_3}\xi_3\right] \tag{10.7}$$

$$\frac{dx_2}{dt} = k_1 x_4 - k_4 x_2 + k_6 x_3 - k_7 x_1 x_2 + k_8 x_3 - k_{10} x_5 x_2 + \frac{1}{\sqrt{V}}\left[\sqrt{k_1 x_4}\xi_4 - \sqrt{k_4 x_2}\xi_5\right] + \frac{1}{\sqrt{V}}\left[\sqrt{k_6 x_3}\xi_6 - \sqrt{k_7 x_1 x_2}\xi_7 + \sqrt{k_8 x_3}\xi_8 - \sqrt{k_{10} x_5 x_2}\xi_9\right] \tag{10.8}$$

$$\frac{dx_3}{dt} = -k_6 x_3 + k_7 x_1 x_2 - k_8 x_3 - \frac{1}{\sqrt{V}}\left[\sqrt{k_6 x_3}\xi_{10}\right] + \frac{1}{\sqrt{V}}\left[\sqrt{k_7 x_1 x_2}\xi_{11} - \sqrt{k_8 x_3}\xi_{12}\right] \tag{10.9}$$

$$\frac{dx_4}{dt} = k_2 x_1 - k_3 x_4 + \frac{1}{\sqrt{V}}\left[\sqrt{k_2 x_1}\xi_{13} - \sqrt{k_3 x_4}\xi_{14}\right] \tag{10.10}$$

$$\frac{dx_5}{dt} = k_9 - k_{10} x_5 x_2 + k_{11} x_6 - k_{12} x_5 + \frac{1}{\sqrt{V}}\left[\sqrt{k_9}\xi_{15}\right] + \frac{1}{\sqrt{V}}\left[-\sqrt{k_{10} x_5 x_2}\xi_{16} + \sqrt{k_{11} x_6}\xi_{17}\right] - \frac{1}{\sqrt{V}}\left[\sqrt{k_{12} x_5}\xi_{18}\right] \tag{10.11}$$

$$\frac{dx_6}{dt} = k_{10} x_5 x_2 - k_{11} x_6 + \frac{1}{\sqrt{V}}\left[\sqrt{k_{10} x_5 x_2}\xi_{19} - \sqrt{k_{11} x_6}\xi_{20}\right] \tag{10.12}$$

where, V is the system size and ξ_i, $i = 1, 2, \ldots, 20$ are random noise parameters which are given by, $\xi_i(t)\xi_j(t') = \delta_{ij}\delta(t - t')$. The noise term varies with order $O(V^{-1/2})$.

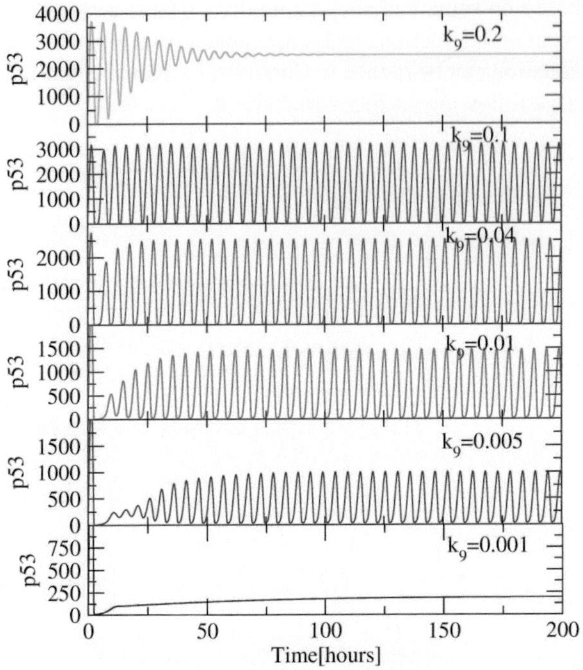

Fig. 10.2 Plots of numerical simulation results showing impact of nitric oxide on p53

10.4 Impact of Nitric Oxide on p53 and Its Network

The deterministic approach is used to show the dynamics of the system when the system size $N = finite$. It assumed the temporal behaviour of the chemically reacting systems is both continuous and deterministic or predictable (Gillespie 1977). In this approach, one can first construct a mathematical model, by translating chemical reaction channels into a set of ordinary differential equations (ODE) using mass action kinetic law. In the present p53-MDM2-NO model, there are six differential equation which is already discuss above Eqs. (1–6). Then solving it by using standard 4th order Runge-Kutta algorithm for numerical integration (Press et al. 1992). The simulation results are plotted in Fig. 10.2. From panels of Fig. 10.2, it is observed that when the concentration of NO is very high $k = 0.2$ level reach to the higher stability. This suggest that at higher level of NO cell moves towards apoptosis due to increase of p53 concentration (Vogelstein et al. 2000). The parameter values taken for this single cell simulation are given in Table 10.2, and the value of k_{NO} $(= k_9)$, creation rate constant, is allowed to vary. Since $NO \propto k_{NO}$, the value of k_{NO} indicates the population of NO in the system. This means that when the value of k_{NO} is small the NO present in the system is low and when the value of k_{NO} increases, NO present in the system is also increased. The results show that at lower value of NO $\left(k_{NO} \leq 0.001\right)$, the two-dimensional plots of pairs of molecular species (proteins and their complexes) show fixed point oscillations indicating stabilization of the dynamics of these molecular species exhibiting normal behaviours of the respective molecular species in the

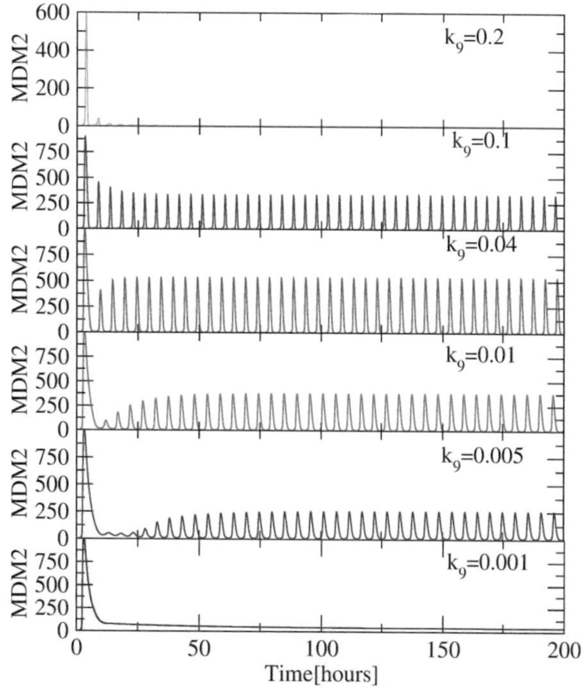

Fig. 10.3 Plots of numerical simulation results showing impact of nitric oxide on p53

system. However, further increase in NO $\left(0.001 \langle k_{NO} \le 0.1\right)$ leads to the transition from fixed point oscillations to nearly limit cycle oscillation (limit cycle oscillation having certain thickness due to fluctuation in the dynamics) takes place. This indicates that $p53$ is activated with the increase in NO showing the enability of NO to cause DNA damage which leads to $p53$ activation (Hofseth et al. 2003). If we further increase NO $\left(k_{NO}\rangle 0.1\right)$, reverse transition i.e transition from the nearly limit cycle oscillations to fixed point oscillations takes place. This could be due to the fact that extremely increase in NO can cause enormous decrease in Mdm2 and increase in p53 correspondingly in the system (i.e. too much toxic to the cell) leading to cell death (Wang et al. 2002; Brune et al. 1999). So we have obtained two stabilization states in $p53$, one for normal like condition and the other for too much toxic leading to killing of cellular functions. In between these two stabilized states we get activated regime of $p53$ which consists of damped and sustained oscillatory behaviours depending on the values of k_{NO}. The term fixed point oscillation means oscillation death dynamics which is different from damped oscillation. Similar behaviour is obtained for dynamics of MDM2 protein as shown in Fig. 10.3.

We next present the stochastic results corresponding to the deterministic results by using the Chemical Langevin Equation (CLE) formalism due to Gillespie (Gillespie 2000) as shown in Fig. 10.4. Here we have fix the stress due to NO (i.e $k_N O$=0.003) and allowed system size to vary. In the upper panel of Fig. 10.4 it is observed that when the system size is very low (i.e., 0.01) we obtained a fix point oscillation. This suggests that when the system size is very low impact of the upon the system is very

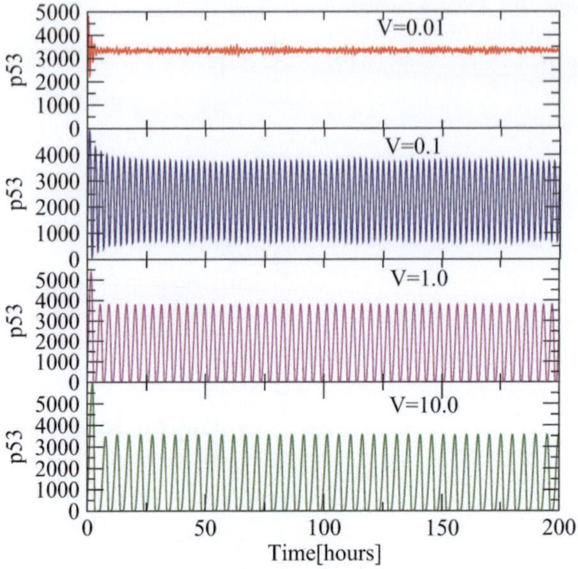

Fig. 10.4 The impact of the noise on the p53-MDM2-NO netrwork. Here, impact of the noise on p53 network is shown

high and noise played a destructive role. Further, when the system size increases, the system moves toward from noisy environment to noise free environment. Hence it is observed from panels that as the system size increases the system attains its normal stress condition.

A similar observation is found when we plot the MDM2 temporal dynamics in Fig. 10.5, by providing similar parameter as in case of p53. This is supposed to be due the counter effect of p53.

10.5 An Integrated p53-MDM2-NO-Ca^{2+} Network Model

Nitric oxide is produced in cell due to enzyme metabolism (Stern 2004; Wood and Garthwaite 1994). Calcium ion acts as a precursor to induce nitric oxide synthetase enzymes (Silvagno et al. 1996; Wagner et al. 2005). The calcium level in an individual cell is considered to be obtained from two sources, one from internal Ca^{2+} pool (from calcium oscillator) (k_7), and the other from extracellular calcium influx (k_6) by the direct diffusion from outside the cell. The overall calcium level binds with nitric oxide synthase (x_{10}) and nitric oxide synthase gets activated (x_{13}). The activated nitric oxide synthase interact with arginine (x_{12}) to produce nitric oxide and citruline as a by-product (Hansen et al. 2005; Dedkova et al. 2004; Manser and Houghton 2006). The level of nitric oxide formed in the cell depends on the level of calcium, and can interact with $p53 - Mdm2$ oscillator via $Mdm2$ protein forming NO_Mdm2 complex (x_5) (Fig. 10.6; Wang et al. 2002; Schonhoff et al. 2002). Even if the half life period of nitric oxide is too short about 5–10 s only (Wood and Garthwaite 1994; Wang et al. 2002), it can move a distance of few hundreds of cells from the site of its

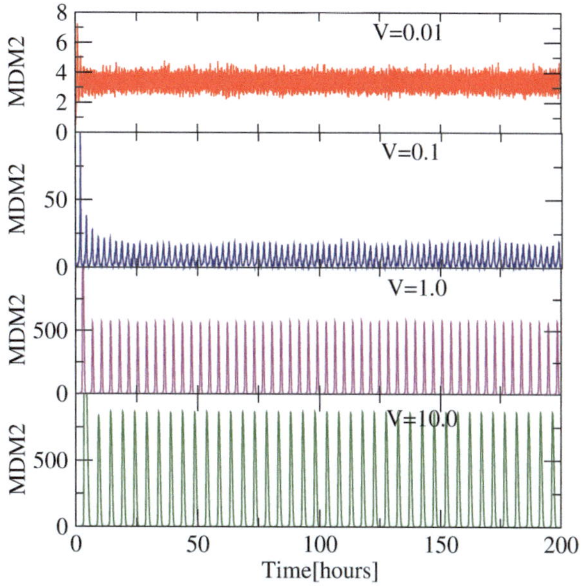

Fig. 10.5 The impact of the noise on MDM2 dynamics is shown

synthesis. Hence, nitric oxide molecule is believed to be one of the most important intracellular and intercellular signaling molecules. In this model, the extracellular influx of nitric oxide molecule is not considered by assuming the amount of nitric oxide created in the cell via calcium is much more as compared to the extracellular influx nitric oxide. Since nitric oxide downregulates $Mdm2$, it eventually affects the dynamics of the $p53$ that leads to the fluctuation of $p53$ level and stabilization (Wang et al. 2002; Schonhoff et al. 2002). Nitric oxide molecule is considered to be unidirectional signaling molecule (from calcium oscillator to $p53 - Mdm2$ oscillator) to study the impact of calcium ion on $p53$ dynamics and regulation. If we consider $\mathbf{S}(t) = \frac{1}{V}\left[X_{10}, \ldots . X_{13}\right]^T = \left[x_{10}, \ldots, x_{13}\right]^T$ as the state of the system that connects the two oscillators (calcium and $p53 - Mdm2$ oscillators) unidirectionally at any instant of time t, the dynamics of the system is given by,

$$\frac{d\mathbf{S}(t)}{dt} = \mathbf{H}\left(x_{10}, \ldots, x_{13}\right) + \frac{1}{\sqrt{V}}\mathbf{H_L}\left(x_{10}, \ldots, x_{13}; \xi_i\right) \tag{10.13}$$

where, the functional vectors $\mathbf{H}$ and $\mathbf{H_L}$ are given by,

$$\mathbf{H}(x_{10}, \ldots, x_{13}) =$$

$$\begin{pmatrix} k_{12} - k14x^* x_{10} \\ -k_{13}x_2x_{11} + k'_{13}x_5 + k_{15}x_{12}x_{13} - k_{16}x_{11} \\ k_{11} - k_{15}x_{12}x_{13} \\ k_{14}x^* x_{10} - k_{15}x_{12}x_{13} \end{pmatrix}$$

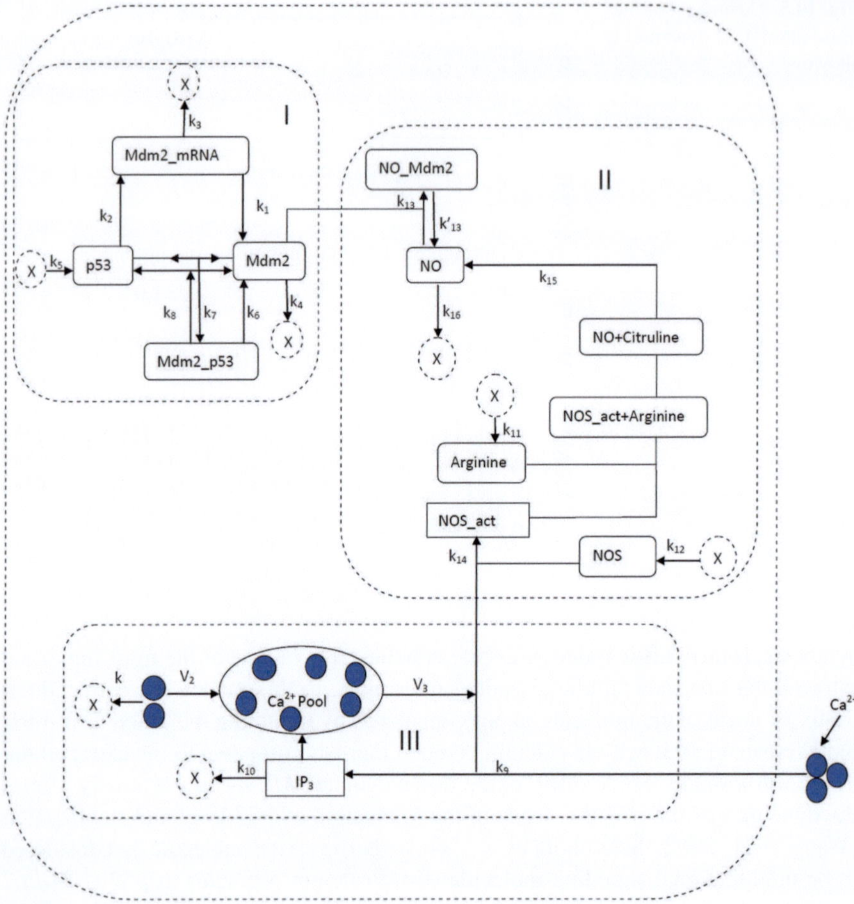

Fig. 10.6 A schematic diagram of the impact of calcium ion upon p53-MDM2 network via nitric oxide

$$\mathbf{H_L}(x_{10}, \ldots, x_{13}; \xi_i) =$$

$$\begin{pmatrix} \sqrt{k_{12}} - \sqrt{k_{14}x^* x_{10}}\xi_{29} \\ [-\sqrt{k_{13}x_2x_{11}}\xi_{30} + \sqrt{k'_{13}x_5}\xi'_{30} + \sqrt{k_{15}x_{12}x_{13}}\xi_{31} \\ -\sqrt{k_{16}x_{11}}\xi_{32}] \\ \sqrt{k_{11}} - \sqrt{k_{15}x_{12}x_{13}}\xi_{33} \\ \sqrt{k_{14}x^*}\xi_{34} - \sqrt{k_{15}x_{12}x_{13}}\xi_{35} \end{pmatrix}$$

Table 10.3 Molecular species, their description and notation

S.No	Molecular species	Description	Notation
1.	p53	Unbound p53 protein	X_1
2.	Mdm2	Unbound Mdm2 protein	X_2
3.	$p53\_Mdm2$	p53/Mdm2 protein	X_3
4.	$Mdm2\_mRNA$	Mdm2 messenger RNA	X_4
5.	NO_Mdm2	Mdm2/NO complex	X_5
6.	Ca_e^{2+}	Extracellular calcium	X_6
7.	Ca_o^{2+}	Released calcium from internal stored calcium	X_7
8.	Calcium-S	Stored calcium in pool	X_8
9.	IP_3	Unbound p53 protein	X_9
10.	NOS	Nitric oxide synthase	X_{10}
11.	NO	Unbound nitric oxide	X_{11}
12.	Ar	Unbound arginine	X_{12}
13.	NOS_act	Activated nitric oxide synthase	X_{13}

The Table 10.3 shows number of molecule participate in this biochemical network. The molecular species is symbolized in terms of x's for the shake of simplicity. All reaction channel involved in the model with their respective transition rates are described in Table 10.4.

10.6 Impact of Calcium Ion on Integrated p53-MDM2-NO Network

The numerical simulation result of the p53 is plotted in Fig. 10.6. In lower panel of Fig. 10.6 we observed that there no activation of p53 molecule. This suggest that when the concentration of calcium ion is very low it has no any effect on system. Further it is noticed as the concentration of calcium ion increases from 0.00001 to 0.05, the behaviour is shifted from normal to stress. Further it is observed that when the concentration of calcium ion is very high i.e. 0.05 the p53 shows steady state. This steady of the p53 signifies that due to higher concentration of calcium ion it trigger synthesis of higher production of NO in the system and due to this cell moved towards the apoptosis. The above observation shows very much agreement with experimental results.

Similarly when we plot the MDM2 temporal dynamics in Fig. 10.7, by providing similar parameter as in case of p53. This is supposed to be due the counter effect of p53.

Next, We have shown the stochastic results corresponding to the deterministic results by using the Chemical Langevin Equation (CLE) formalism due to Gillespie

Table 10.4 Molecular species, their description and notation

S.No	Reaction channel	Description	Kinetic laws	Values of rate constant	References
1	$X_4 \xrightarrow{k_1} X_4 + X_2$	Mdm2 translation	$k_1 X_4$	$4.95 \times 10^{-4} s^{-1}$	(Proctor and Gray 2008; Finlay 1993)
2	$X_1 \xrightarrow{k_2} X_1 + X_4$	Synthesis of Mdm2_mRNA	$k_2 X_1$	$1.0 \times 10^{-4} s^{-1}$	(Proctor and Gray 2008; Finlay 1993)
3	$X_4 \xrightarrow{k_3} \phi$	Degradation of Mdm2_mRNA	$k_3 X_4$	$1.0 \times 10^{-4} s^{-1}$	(Proctor and Gray 2008; Finlay 1993)
4	$X_2 \xrightarrow{k_4} \phi$	Degradation of Mdm2	$k_4 X_2$	$4.33 \times 10^{-4} s-1$	(Proctor and Gray 2008; Finlay 1993)
5	$\phi \xrightarrow{k_5} X_1$	Synthesis of p53	k_5	$0.78\ s^{-1}$	(Proctor and Gray 2008; Finlay 1993)
6	$X_3 \xrightarrow{k_6} X_2$	Decay of p53	$k_6 X_3$	$8.25 \times 10^{-4} s^{-1}$	(Proctor and Gray 2008; Finlay 1993)
7	$X_1 + X_2 \xrightarrow{k_7} X_3$	Synthesis of p53_Mdm2 complex	$k_7 X_1 X_2$	$11.55 \times 10^{-4} mol^{-1} s^{-1}$	(Proctor and Gray 2008; Finlay 1993)
8	$X_3 \xrightarrow{k_8} X_1 + X_2$	Degradation of p53_Mdm2 complex	$k_8 X_3$	$11.55 \times 10^{-6} s^{-1}$	(Proctor and Gray 2008; Finlay 1993)
9	$\phi \xrightarrow{k_9} X_6$	Diffusion of Ca_e^{2+} from extracellular medium to the cell.	k_9	$1 \times 10^{-2} mol^{-1} s^{-1}$	(Adams et al. 1989)
10	$\phi \xrightarrow{V_0} X_7$	Constant input of Ca_o^{2+} inside the cell.	V_0	$2.0\ s^{-1}$	(Houart et al. 1999; Alam et al. 2012)
11	$\phi \xrightarrow{\beta V_1} X_7$	Stimulus-induced influx of calcium from extracellular medium.	βV_1	$\beta = 0.5,\ V_1 = 2.0$	(Houart et al. 1999; Alam et al. 2012)
12	$X_7 \xrightarrow{V_2} X_8$	Pumping of Ca_o^{2+} from cytosol to internal calcium pool.	$V_2 = V_{M2} \frac{X_7^2}{C_2^2 + X_7^2}$	$V_{M2} = 6,\ C_2 = 0.1$	(Houart et al. 1999; Alam et al. 2012)

Table 10.4 (continued)

S.No	Reaction channel	Description	Kinetic laws	Values of rate constant	References
13	$X_8 \xrightarrow{V_3} X_7$	Release of Ca_o^{2+} from calcium pool to cytosol.	$V_3 = V_{M3} \frac{X_7^m}{C_x^m+X_7^m} \frac{X_8^2}{C_y^2+X_8^2} \frac{X_9^4}{C_z^4+X_9^4}$	$V_{M3} = 20, m = 2, C_x = 0.5, C_y = 0.2, C_z = 0.2$	(Houart et al. 1999; Alam et al. 2012)
14	$X_8 \xrightarrow{k_f} X_7$	Release of Ca_o^{2+} from calcium pool to Cytosol due to leakage.	$k_f X_8$	0.01	(Houart et al. 1999; Alam et al. 2012)
15	$X_7 \xrightarrow{k} \phi$	Decay of Ca_o^{2+}.	kX_7	1.0	(Houart et al. 1999, Alam et al. 2012)
16	$\phi \xrightarrow{\beta V_4} X_9$	Stimulus-induced synthesis of IP_3	βV_4	$\beta = 0.5,\ V_4 = 2.0$	(Proctor and Gray 2008; Finlay 1993)
17	$X_9 \xrightarrow{V_5} \phi$	Phosporylation of IP_3 by 3-kinase	$V_5 = V_{M5} \frac{X_9^p}{C_s^p+X_9^p} \frac{X_7^n}{C_d^n+X_7^n}$	$V_{M5} = 5.0,\ p = 2.0,\ C_5 = 1.0,\ n = 4.0,\ C_d = 0.4$	(Houart et al. 1999; Alam et al. 2012)
18	$X_9 \xrightarrow{k_{10}} \phi$	Decay of IP_3	$k_{10}X_9$	$0.01\ s^{-1}$	(Houart et al. 1999; Alam et al. 2012)
19	$\phi \xrightarrow{k_{11}} X_{12}$	Synthesis of Arginine	k_{11}	$0.01\ s^{-1}$	(Wang et al. 2002)
20	$\phi \xrightarrow{k_{12}} X_{10}$	Synthesis of Nitric Oxide Synthase (NOS)	k_{12}	$0.0001\ s^{-1}$	(Wang et al. 2002)
21	$X_{11} + X_2 \xrightarrow{k_{13}} X_5$	Synthesis of Mdm2_NO Complex	$k_{13}X_{11}X_2$	$1.0 \times 10^{-3} s^{-1}$	(Wang et al. 2002)
22	$X_5 \xrightarrow{k'_{13}} X_{11}$	Degradation of Mdm2_NO Complex	$k'_{13}X_5$	$3.3 \times 10^{-4} s^{-1}$	(Wang et al. 2002)
23	$X^* + X_{10} \xrightarrow{k_{14}} X_{13}$	Formation of NOS_act	$k_{14}X^*X_{10}$	$10.0\ s^{-1}$	(Wang et al. 2002)
24	$X_{12}+X_{13} \xrightarrow{k_{15}} X_{11}+ citruline$	Synthesis of nitric oxide and citruline as byproduct	$k_{15}X_{12}X_{13}$	$10.0\ s^{-1}$	(Wang et al. 2002)
25	$X_{11} \xrightarrow{k_{16}} \phi$	Decay of Nitric Oxide	$k_{16}X_{11}$	$0.001\ s^{-1}$	(Proctor and Gray 2008; Finlay 1993)
26	$X_5 \xrightarrow{k_{17}} \phi$	Decay of $Mdm2\_NO$	$k_{17}X_5$	$0.001\ s^{-1}$	(Wang et al. 2002)

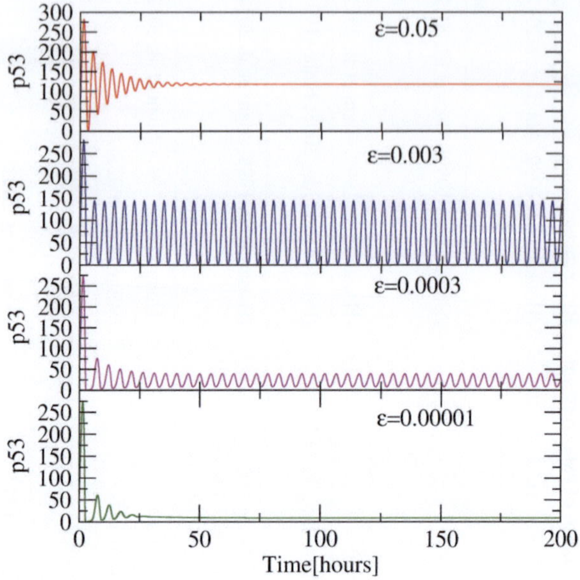

Fig. 10.7 The impact of calcium ion upon p53 in p53-MDM2-NO network is shown at different concentration of calcium ion (i.e., 0.00001, 0.003, 0.003, 0.05)

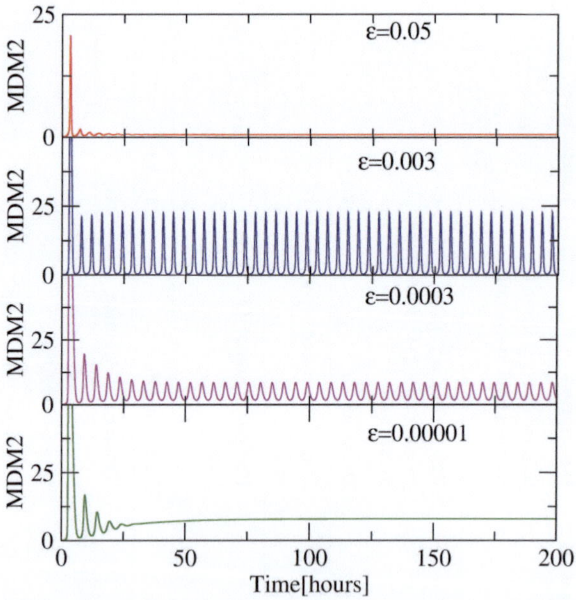

Fig. 10.8 The impact of calcium ion upon MDM2 in p53-MDM2-NO network is shown at different concentration of calcium ion (i.e., 0.00001, 0.003, 0.003, 0.05)

(Gillespie 2000) as shown in Fig. 10.8. Here we have fix the stress due to NO (i.e $k_N O = 0.004$) and allowed system size to vary (100, 300, 500, 1000). In the upper panel of Fig. 10.8 it is observed that when the system size is comparatively very low (i.e., 100) we obtained a fix point oscillation. This suggests that when the system size

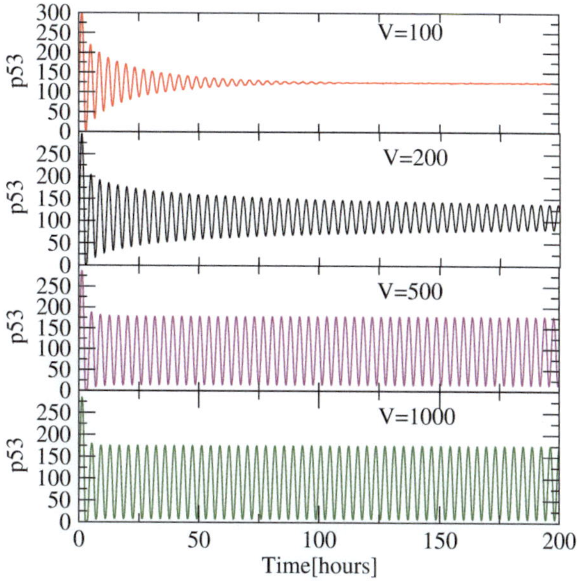

Fig. 10.9 The impact of noise upon p53 in p53-MDM2-NO-Ca network is shown at different system size (i.e., 100, 300, 500, 1000)

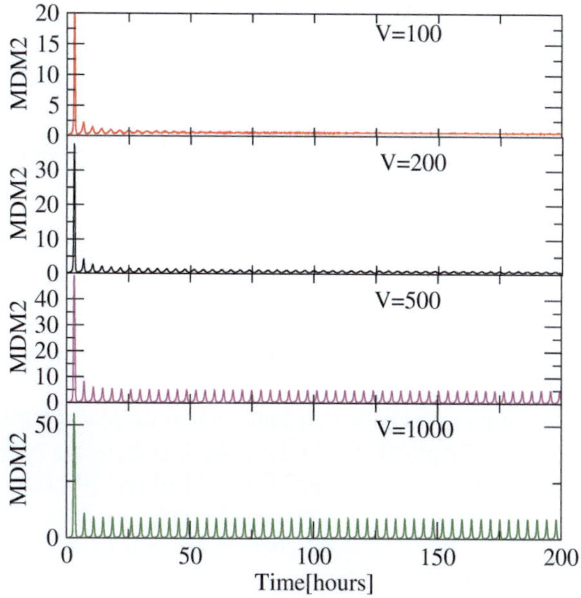

Fig. 10.10 The impact of noise upon p53 in p53-MDM2-NO-Ca network is shown at different system size (i.e., 100, 300, 500, 1000)

is low impact of noise upon the system is very high and noise played a destructive role. Further, when the system size increases, the system moves toward from noisy environment to noise free environment. Hence it is observed from panels in Fig. 10.8 that as the system size increases the system attains its normal stress condition. This

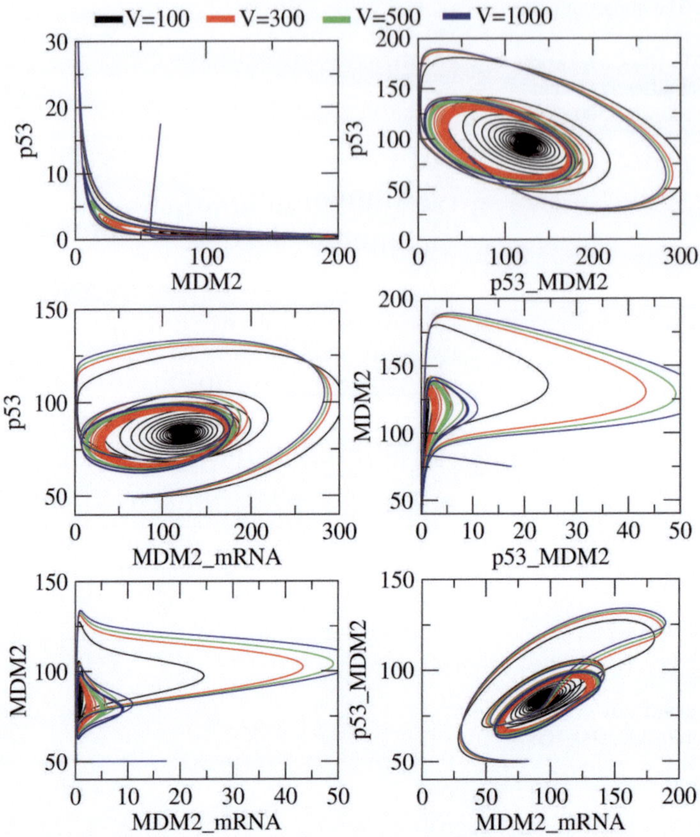

Fig. 10.11 Two dimensional recurrence plot for showing impact of noise upon p53-MDM2-NO-Ca network components are shown at different system size (i.e., 100, 300, 500, 1000)

suggest that when the system size increases system switches from stochastic system to deterministic system.

A similar observation is found when we plot the MDM2 temporal dynamics in Fig. 10.9, by providing similar parameter as in case of p53. This is supposed to be due the counter effect of p53 (Figs. 10.10 and 10.11).

We have also shown a two dimensional recurrence plot for the system for stochastic system. We have varied the system size from low to high as taken above (i.e. 100,300,500,1000). The two dimensional quantitative plot again supported above qualitative plot as shown in Figs. 10.8 and 10.9.

10.7 Conclusion

The impact of nitric oxide on the $p53 - Mdm2$ regulatory network in single cell as well as in coupled cells are studied on a model designed based on various experimental reports. Nitric oxide is being synthesize due to protein-protein interaction inside the cell and is supposed to be toxic in normal cells when its concentration is high. The nitric oxide is maintained at low in normal cell. Nitric oxide directly influence $(p53 - Mdm2)$ network via $Mdm2$. This leads to the oscillatory behaviour of $p53$. In the single cell model, the $p53$ is found to be normal keeping it low and stabilized when NO is low. Again when the NO is increased significantly, $p53$ protein is activated indicated by its oscillatory behaviour. However excess NO leads to the higher stability of p53 and this condition lead to the to cell apoptosis.

The Ca^{2+} ion acts as an activator of $p53$. The simulation results of p53-MDM2-NO-Ca network model suggest that Ca^{2+} activate nitric oxide which in turn affects the $p53 - Mdm2$ network through direct interaction with $Mdm2$. We observed that activation of $p53$ by Ca^{2+} in a cell lifts the cell from normal to stress state. An excess Ca^{2+} level in a cell leads to higher synthesis of nitric oxide switching the cell from normal to apoptotic phase.

The intrinsic noise due to random molecular interaction in the system can be correlated qualitatively with system size such that noise in small system size is large and vice versa (Nandi et al. 2007). The single cell study reveals that the oscillating (damped or sustained) temporal dynamics of $p53$ at negligibly small noise (large V) becomes stabilized (fixed point oscillation) as increase in noise (small V).

There are various issues need to be solved in future for example information transfer among a large number of cells. Since the $p53$ protein is a hub of various biological network so the influences of signaling molecules from various sub-networks need to be considered simultaneously.

Acknowledgement This work is financially supported by University Grant Commission (UGC), India and Ministry of Minority Affair (MOMA), India and carried out in Center for Interdisciplinary Research in Basic Sciences, Jamia Millia Islamia, New Delhi, India.

References

Adams DJ, Barakeh J, Lanskey R, Breemen CV (1989) Ion channels and regulation of intracellular calcium in vascular endothelial cells. FASEB J 3:2389–2400

Alam MJ, Bhayana L, Devi GR, Singh HD, Singh RKB, Sharma BI (2012) Intercellular synchronization of diffusively coupled Ca^{2+} oscillators. J Chem Biol 5:27–34

Alam MJ, Devi GR, Ishrat RR, Agrawal SM, Singh RKB (2013) Switching p53 states by calcium: dynamics and interaction of stress systems. Mol BioSyst 9:508–521

Blake WJ, Kaern M, Cantor CR, Collins JJ (2003) Noise in eukaryotic gene expression. Nature 422:633–637

Brune B, Von Knethen A, Sandau KB (1999) Nitric oxide (NO): an effector of apoptosis. Cell Death Differ 6:969–975

Cerella C, D'alessio M, Nicola MD, Magrini A, Bergamaschi A, Ghibelli L (2003) Cytosolic and endoplasmic reticulum Ca^{2+} concentrations determine the extent and the morphological type of apoptosis, respectively. Ann N Y Acad Sci 1010:74–77

Chen B, Deen WM (2001) Analysis of the effects of cell spacing and liquid depth on nitric oxide and its oxidation products in cell cultures. Chem Res Toxicol 14:135–147

Chen J, Lin J, Levine AJ (1995) Regulation of transcription functions of the p53 tumor suppressor by the mdm-2 oncogene. Mol Med 1:142–152

Chung HT, Pae HO, Choi BM, Billiar TR, Kim YM (2001) Nitric oxide as a bioregulator of apoptosis. Biochem Biophys Res Commun 282:1075J–1079J

Chun-Qi L, Wogen GN (2005) Nitric oxide as a modulator of apoptosis. Cancer Lett 226:1–15

Dedkova EN, Ji X, Lipsius SL, Blatter LA (2004) Mitochondrial calcium uptake stimulates nitric oxide production in mitochondria of bovine vascular endothelial cells. Am J Physiol Cell Physiol 286:C406–C415

Dedon PC, Tannenbaum SR (2004) Reactive nitrogen species in the chemical biology of inflammation. Arch Biochem Biophys 423:12–22

Dina R (2005) Intercellular communication, NO and the biology of Chinese medicine. Cell Comm Sign 3:1–4

Finlay CA (1993) The Mdm2 oncogene can overcome wild-type p53 suppression of transformed cell growth. Mol Cell Biol 13:301–306

Geva-Zatorsky N, Rosenfeld N, Itzkovitz S, Milo R, Sigal A, Dekel E, Yarnitzky T, Liron Y, Polak P, Galit L, Alon U (2006) Oscillations and variability in the p53 system. Mol Syst Biol 2:0033

Gillespie DT (1977) Exact stochastic simulation of coupled chemical reactions. J Phys Chem 31:2340–2361

Gillespie DT (2000) The chemical Langevin equation. J Chem Phys 113:297

Hansen JT, Ferreira A, Yano Y, Kanuparthi D, Romero JR, Brown EM, Chattopadhyay N (2005) Calcium-sensing receptor activation induces nitric oxide production in H-500 Leydig cancer cells. Am J Physiol Endocrinol Metab 288:E1206–E1213

Haupt Y, Maya R, Kazaz A, Oren M (1997) Mdm2 promotes the rapid degradation of p53. Nature 387:296–299

Hofseth LJ, Saito S, Hussain SP, Espey MG, Miranda KM, Araki Y (2003) Nitric oxide induced cellular stress and p53 activation in chronic inflamation. Proc Natl Acad Sci U S A 100:143–148

Houart G, Dupont G, Goldbeter A (1999) Bursting, chaos and biorhythmicity originating from self-modulation of the inositol 1,4,5-trisphosphate signal in a model for intracellular Ca^{2+} oscillations. Bull Math Biol 61:507–530

Hsing A, Faller DV, Vaziri C (2000) DNA-damaging aryl hydrocarbons induce Mdm2 expression via p53-independent post-transcriptional mechanisms. J Biol Chem 275:26024–26031

Hussain SP, Hofseth LJ, Harris CC (2003) Nitric oxide-induced cellular stress and p53 activation in chronic inflammation. Nat Rev Cancer 3:276–285

Jenkins DC, Charles IG, Thomsen LL, Moss DW, Holmes LS, Baylis SA, Rhodes P, Westmore K, Emson PC, Moncada S (1995) Roles of nitric oxide in tumor growth. Proc Natl Acad Sci U S A 92:4392–4396

Kim YM, Chung HT, Kim SS, Han JA, Yoo YM, Kim KM (1999) Nitric oxide protects PC12 cells from serum deprivation-induced apoptosis by cGMP-dependent inhibition of caspase signaling. J Neurosci 19:6740–6747

Kim PK, Zamora R, Petrosko P, Billiar TR (2001) The regulatory role of nitric oxide in apoptosis. Int Immunopharmacol 1:1421–1441

Knowles RG, Moncada S (1994) Nitric oxide synthases in mammals. Biochem J 298:249–258

Kubbutat MHG, Jones SN, Vousden KH (1997) Regulation of p53 stability by Mdm2. Nature 387:299–303

Lancaster JR (1994) Simulation of the diffusion and reaction of endogenously produced nitric oxide. Proc Natl Acad Sci U S A 91:8137–8141

Lancaster JR (1997) A tutorial on the diffusibility and reactivity of free nitric oxide. Nitric Oxide 1:18–30

Lane DP (1992) p53, guardian of the genome. Nature 358:15–16
Li J, Billiar TR (1999) The anti-apoptotic actions of nitric oxide in hepatocytes. Cell Death Differ 6:952–955
Li H, Wallerath T, Munzel T, Forstermann U (2002) Regulation of endothelial-type NO synthase expression in pathophysiology and in response to drugs. Nitric Oxide 7:149–164
Liang SH, Clarke MF (1999) A bipartite nuclear localization signal is required for p53 nuclear import regulated by a carboxyl-terminal domain. J Biol Chem 274:32699–32703
Lopez-Jaramillo P, Gonzalez MC, Palmer RMJ, Moncada S (1990) The crucial role of physiological Ca^{2+} concentrations in the production of endothelial nitric oxide and the control of vascular tone. Br J Pharmacol 101:489–493
Lowenstein CJ, Padalko E (2004) iNOS (NOS2) at a glance. J Cell Sci 117:2865–2867
Manser RC, Houghton FD (2006) Ca^{2+}-linked upregulation and mitochondrial production of nitric oxide in the mouse preimplantation embryo. J Cell Sci 119:2048–2055
Marletta MA, Spiering MM (2003) Trace elements and nitric oxide function. J Nutr 133:1431S–1433S
McAdams HH, Arkin A (1997) Stochastic mechanisms in gene expression. Proc Natl Acad Sci U S A 94:814–819
Mcbride OW, Merry D, Givolt D (1986) The gene for human p53 cellular tumor antigen is located on chromosome 17 short arm (17p13). Proc Natl Acad Sci U S A 83:130–134
McQuarrie DA (1967) Stochastic approach to chemical kinetics. J Appl Probab 4:413–478
Mendrysa SM, McElwee MK, Perry ME (2001) Characterization of the 5' and 3' untranslated regions in murine mdm2 mRNAs. Gene 264:139–146
Messmer UK, Ankarcrona M, Nicotera P, Brune B (1994) p53 expression in nitric oxide-induced apoptosis. FEBS Lett 355:23–26
Momand J, Zambetti GP, Olson DC, George D, Levine A (1992) The mdm-2 oncogene product forms a complex with the p53 protein and inhibits p53-mediated transactivation. Cell 2:1237–1245
Momand J, Wu HH, Dasgupta G (2000) MDM2–master regulator of the p53 tumor suppressor protein. Gene 242:15–29
Murphy MP (1999) Nitric oxide and cell death. Biochim Biophys Acta 1411:401–414
Nandi A, Santhosh G, Singh RKB, Ramaswamy R (2007) Effective mechanisms for the synchronization of stochastic oscillators. Phys Rev E 76:041136
Pan Y, Haines DS (1999) The pathway regulating MDM2 protein degradation can be altered in human leukemic cells. Cancer Res 59:2064–2067
Press WH, Teukolsky SA, Vetterling WT, Flannery BP (1992) Numerical recipe in fortran. Cambridge University, Cambridge
Proctor CJ, Gray DA (2008) Explaining oscillations and variability in p53-Mdm2 system. BMC Sys Biol 2:75
Rao CV, Wolf DM, Arkin AP (2002) Control, exploitation and tolerance of intracellular noise. Nature 420:231–237
Samali A, Fulda S, Gorman AM, Hori O, Srinivasula SM (2010) Cell stress and cell death. Int J Cell Biol 2010:245803
Schmidt HH, Walter U (1994) NO at work. Cell 78:919–925
Schonhoff CM, Daou MC, Jones SN, Schiffer CA, Ross AH (2002) Nitric oxide-mediated inhibition of Hdm2-p53 binding. Biochemistry 41:13570–13574
Shih CT, Roche S, Romer RA (2008) Point mutations effects on charge transport properties of the tumor-suppressor gene p53. Phys Rev Lett 100:018105
Silvagno F, Xia H, Bredt DS (1996) Neuronal nitric-oxide synthase-mu, an alternatively spliced isoform expressed in differentiated skeletal muscle. J Biol Chem 271:11204–11208
Stern JE (2004) Nitric oxide and homeostatic control: an intercellular signalling molecule contributing to autonomic and neuroendocrine integration? Prog Biophys Mol Bio 84:197–215
Taylor EL, Megson IL, Haslett C, Rossi AG (2003) Nitric oxide: a key regulator of myeloid inflammatory cell apoptosis. Cell Death Differ 10:418–430
Vogelstein B, Lane D, Levine AJ (2000) Surfing the p53 network. Nature 408:307–310

Wagner J, Ma L, Rice JJ, Hu W, Levine AJ, Stolovitzky GA (2005) p53-Mdm2 loop controlled by a balance of its feedback strength and effective dampening using ATM and delayed feedback. IEE Proc Syst Biol 152:109–118
Wang X, Michael D, de Murcia G, Oren M (2002a) p53 activation by nitric oxide involves down-regulation of Mdm2. J Biol Chem 277:15697–15702
Wang Y, Vodovotz Y, Kim PK, Zamora R, Billiar TR (2002b) Mechanisms of hepatoprotection by nitric oxide. Ann N Y Acad Sci 962:415–422
Wang X, Michael D, Muricia GD, Oren M (2002c) p53 Activation by nitric oxide involves down-regulation of Mdm2. J Biol Chem 277:15697–15702
Werner ER, Gorren AC, Heller R, Werner-Felmayer G, Mayer B (2003) Tetrahydrobiopterin and nitric oxide: mechanistic and pharmacological aspects. Exp Biol Med (Maywood) 228:1291–1302
Wood J, Garthwaite J (1994) Models of the diffusional spread of nitric oxide: implications for neural nitric oxide signalling and its pharmacological properties. Neuropharmacology 33:1235–1244

Chapter 11
Systems Biology of MicroRNA

Remya Krishnan and Pawan K. Dhar

Abstract MicroRNAs are a class of small non-coding RNAs that has a significant role in regulating almost all life processes. They were first discovered in nematodes and subsequently in flies, fishes, birds, reptiles and mammals. It has been experimentally validated that microRNAs run a parallel control system that determines the genotype and phenotypes of individuals. In humans, microRNAs remotely control physiological process at different levels. Their systematic regulation leads to the up-regulation, down- regulation or gene silencing of specific mRNA sequences that holds the key to several diseases. MicroRNAs have an undisputed role in tumorigenesis and several other diseases in humans; hence an in depth understanding of their regulatory mechanism is the need of the hour. In this review we discuss the biogenesis of microRNAs, what qualifies anmiRNA, biological roles of miRNA, methods to predict a mature miRNA and miRNA targets and a systems biology approach to the microRNA regulatory network in humans.

Keywords MicroRNA · Cancer · Heart diseases · Systems biology

11.1 General

MicroRNAs are single-stranded small regulatory non-coding RNAs of about ~22 nucleotide in length. They belong to the class of small non-coding RNAs that are actively transcribed, having a biologically significant role, but do not code for a protein. In 1993, Victor Ambross, Rosalind Lee and Rhonda Feinbaum discovered in *Caenorhabiditiselegans* that the gene controlling the larval development *lin 4* did not encode for a protein but instead a short nucleotide molecule of about 22 bp that was responsible for regulating the expression of LIN 4 protein. This was the

P. K. Dhar (✉)
Department of Life Sciences, School of Natural Sciences,
Shiv Nadar University, Dadri, UP, India
e-mail: pawan.dhar@snu.edu.in

R. Krishnan
Department of Computational Biology and Bioinformatics,
University of Kerala, Trivandrum, Kerala, India

V. Singh, P. K. Dhar (eds.), *Systems and Synthetic Biology*,
DOI 10.1007/978-94-017-9514-2_11

first discovered micro RNA. However, there was no evidence of small non-coding RNAs in organisms other than nematodes, until another small non-coding RNA was discovered in *C. elegans* (Reinhart et al. 2000). In 2000, Let-7 family of microRNAs was found to be conserved across species for more than 400 million years by the Ruvkun lab. It paved the path to extensive research in the class of small regulatory non-coding, regulatory microRNAs (miRNA), most of which was highly conserved across species. There has been an exponential increase in miRNA research to identify its genomics, biogenesis, mechanism and functions ever since. As of 2012, there are about $\sim$20,000 research articles published on PubMed. It appears that microRNAs serve as regulatory hubs for several physiological functions and disease conditions especially tumorigenesis and feed-forward loops strongly enriched in transcription factor networks.

11.2 Biogenesis

MicroRNAs have been discovered endogenously in animals, plants and viruses (Bartel 2004). Even though they themselves do not encode for a protein, they regulate the translation process by enhancing, degrading or silencing the target expression. MicroRNAs are initially found in their pre-mature state called primary miRNA (**pri-miRNA**) which is a hair-pin like structure. This molecule is later processed by an RNase-III like enzyme **Drosha (Drsh-1)** to form a stem-loop structure called **pre-miRNA**. This molecule is then transported to the cytoplasm by **Exportin-5** and further processed by a **Dicer (Dcr-1)**complex to form the mature single stranded **miRNA**. (Kim 2004). The mature miRNA is incorporated into the RNA induced silencing complex (RISC), which recognizes specific targets in the mRNA sequence and induces post transcriptional gene silencing. The human genome encodes about 1600 microRNA precursor sequences and 2042 mature sequences (Source: mirBasever 19 data release August, 2012; Fig. 11.1).

11.2.1 What Qualifies a MicroRNA?

MicroRNAs are single stranded non-coding RNAs which can be easily mistaken with the other ncRNAs such as rRNA, tRNA and specifically siRNA that are present endogenously in the cells that resemble miRNAs in most functions. Therefore to avoid confusion, a consensus was reached to identify a bona fide microRNA and a uniform system for microRNA annotation was adopted (Ambros et al. 2003).

- RNA hybridization methods like Northern Blotting must detect a 22 nucleotide long RNA transcript
- The transcript thus identified must match with the cDNA Library of the genome of the organism from which they were cloned.

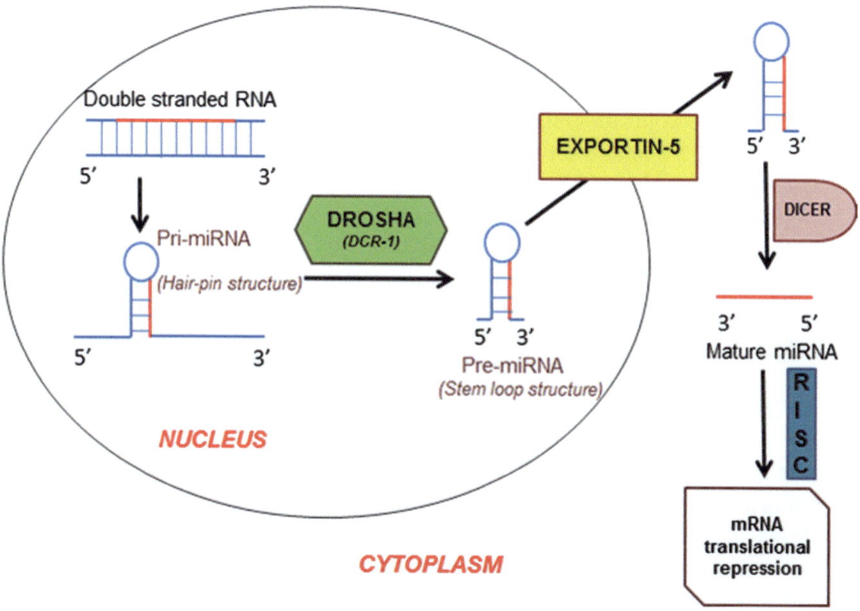

Fig. 11.1 MicroRNA biogenesis

- Prediction of a hair-pin like precursor to the 22 nucleotide long RNA transcript. These precursors can be up to 60–80 nt long in animals and in plants it can go up to several hundreds.
- The 22 nt long microRNA and its precursor must be phylogenetically conserved.
- With the loss or reduction of the Dicer complex function, there must be a detectable accumulation of the miRNA precursor.

11.3 Biological Roles of MicroRNAs in Humans

MicroRNAs act either by exo-nucleolytic mRNA degradation or translational inhibition by imperfect complementarity between micro RNA and 3′ UTR (untranslated region) of mRNA. Due to imperfect complimentary pairing microRNAs are capable of targeting any human mRNA, hence they have pivotal control in almost all biological processes such as cell cycle regulation, cell growth, apoptosis, cell differentiation and stress responses. Deregulationof microRNAs lead to several types of human cancers hence understanding their regulatory mechanism can lead to novel strategies for the prevention and treatment of cancers. MiRNAs can also serve as tumor biomarkers with diagnostic and prognostic implications. Pre-clinical trials in humans for assessing the safety and efficacy of miRNA targeted therapy have already begun.

11.3.1 Role in Thermogenesis

Recently, microRNAs have been discovered to have a regulatory role in thermogenesis in response to cold stimulation. Brown Adipose Tissues (BAT) was considered to be scare/absent in humans until their discovery confirmed its significance in energy dissipation. The regulation of BAT differentiation by miRNAs thus opens up the development of therapeutic strategies in targeting over-weight and obesity.

11.3.2 Role in Circadian Rhythms

In all organisms there is an "internal clock" that adjusts every single physiological activity to a 24 h world. Two microRNAs (miR-213 and miR-132) are found to regulate this "internal time keeping machine" in humans. It has been experimentally determined that miR-213 regulates circadian rhythms of expression while *in vivo* knock-out of miR 132 lengthens the circadian period. Recently miRNA-219 and miRNA-132 show their post-transcriptional roles in the modulation of the circadian rhythm (Liu et al. 2012).

11.3.3 Role in Brain Metastasis

miRNA biogenesis that is highly conserved has been found to be linked to the transport and translatability of mRNAs in neurons. The oncogenic and tumor suppressive potential of microRNAs has been emphasized in several studies especiallybrain tumors. The transcriptional levels of miR10b and miR 21 were found to be elevated in the Cerebro spinal fluid (CSF) of patients with glioblastoma and metastatic brain tumor (Teplyuk et al. 2012). This proves that microRNAs in CSF can serve as biomarkers for brain tumors.

11.3.4 Role in Breast Cancer

Breast cancer is the leading cause of death in women worldwide. Breast cancer subtypes are observed to have deranges microRNA expression signatures. As many as 9 microRNAs namely hsa-miR-21, hsa-miR-365, hsa-miR-181b, hsa-let-7f, hsa-miR-155, hsa-miR-29b, hsa-miR-181d, hsa-miR-98, and hsa-miR-29c were upregulated and 7 microRNAs namely hsa-miR-497, hsa-miR-31, hsa-miR-355, hsa-miR-320, hsa-mir-140, hsa-miR-127 and hsa-miR-30a-3p were down regulated in breast cancer. Specifically miR 21 overexpression could be associated with advanced tumor stage, lymph node metastasis and poor survival of the patients (Yan et al. 2008).

11.3.5 Role in Cardiology

miR208a was discovered to play a pivotal role in cardiac hypertrophy (myocardial thickening) in mice. Recently miR 208a have been found to play a similar role in the human heart namely arrythmiasis, cardiac remodeling, expression of cardiac hypertrophy pathway components and cardiac conduction system. Other micro RNAs like MiR 1 and miR 133 regulate heart development by upregulation in pericardial mesoderm.

11.3.6 miRNAsRegulate Stem Cell Regeneration and Differentiation

microRNAs have been identified to be the key players in stem cell regeneration and differentiation. Stem cells are by nature pluripotent (capable of differentiating to all embryonic lineages), therefore stem cell transplantation is considered as a most promising medical treatment for tissue regeneration. The expression of miR 302–367 clusters which include miR 302 a/b/c/d and miR367 cloned from human somatic cells, increases during stem cell regeneration and decreases during differentiation (Anokye-Danso et al. 2011).

11.3.7 Other Important Functions

miR29 has an important role in regulating innate and adapative immune responses to intracellular bacterial infection. Epidermal Growth Factor (EGFR) over-expression is related to tumorous glioma but miR 146b-5p has been found to suppress the EGFR expression hence their re-constitution is expected to be useful for treatment of invasive tumor (Katakowski et al. 2010). miR-15/107 has been implicated in several functions such as cell division, metabolism, stress response, angiogenesis amd several diseases like human cancers, cardiovascular disease, and neurodegenerative diseases including Alzheimer's disease. Gene regulation networks based in miRNA activities have predicted the role of miRNAs in several brain functions like learning, memory and other neuropsychiatric disorders. Experimental evidence proves the role of miR122 in several hepatic functions and liver including chronic hepatitis and hepatocellular carcinoma. MicroRNAs are so robust and diverse in function that they are predicted to even control the ageing process in humans.

11.4 *In Silico* Approach to MicroRNA Gene Finding and Target Identification

MicroRNAs are the key regulators of gene expression hence *in silico* computational analysis have become indispensable to identify novel miRNAs and their targets. There are several methods and tools currently available to predict novel miRNA precursors, mature miRNA and their targets. A few are briefly described here.

11.4.1 Pre-miRNA Prediction Tools

Novel pre-miRNA sequence prediction methods can be broadly classified as based on comparative genomics, homology and ab-initio methods (Tempel et al. 2012). Data suggests that miRNA is highly conserved across species and that they have an ancient role in gene regulation. Homology based approach uses the existing information regarding sequences and structures available in databases such as miRbase. Some of the tools that use this approach are miRAlign and ERPIN.

Furthermore, mirPara, CID-miRNA, Vmir, miRPred are some of the ab-intio method based tools that predict the precursor miRNA sequences. A more recent tool miRNAFold (http://EvryRNA.ibisc.univ-evry.fr/ .) has been developed to predict the precursor miRNA is found to be more accurate and faster than the above mentioned methods.

11.4.2 Mature miRNA Prediction Tools

To date, predicting miRNA across genomic sequences has been met with limited success due to the vast diversity of premature miRNA sequences and their designated pathways. Even though several filters like phylogenetic conservation, homology searches or machine based methods can be used the question still remains, "*Does nature use such filters in miRNA biogenesis?*" Even though there are no tools that can accurately identify the miRNA sequence and function from a given hair-pin sequence yet there are tools that has about 90 % accuracy in doing so. In humans, a target-centered approach has also been attempted that depends on the highly conserved motifs in 3′-UTRs. Machine learning based approaches that include SVM (Support Vector Machine), HMM (Hidden Markov Model), NBC (Naïve Bayes Classifier) have also been found to be effective. e.g. HMMMir is an algorithm for de novo prediction of miRNA with 88 % accuracy (Kadri et al. 2009). FOMmir is a prediction algorithm with 91 % accuracy on human genome set that uses a Fixed Order Markov model based on secondary structural pattern.

Table 11.1 List of the most commonly used microRNA target prediction tools

S.No.	Tool	URL	Reference
1.	DIANA microT analyzer	http://diana.pcbi.upenn.edu/cgi-bin/micro_t.cgi/	Cai et al. (2004)
2.	MicroInspector	(http://www.imbb.forth.gr/microinspector)	Rusinov et al. (2005)
3.	miRanalyser	http://web.bioinformatics.cicbiogune.es/microRNA/miRanalyser.php	Hackenberg et al. (2009)
4.	miRanda	http://www.microrna.org/microrna/home.do	Enright et al. (2003)
5.	miRTar	http://mirtar.mbc.nctu.edu.tw/human/	Kai-Hsu et al. (2011)
6.	miRTif	http://mirtif.bii.a-star.edu.sg/	Yang et al. (2008)
7.	PicTar	http://pictar.mdc-berlin.de/	Rajewsky et al. (2006)
8.	RegRNA	http://regrna.mbc.nctu.edu.tw/	Huang et al. (2006)
9.	RNA hybrid	http://bibiserv.techfak.uni-bielefeld.de/rnahybrid/	Rehmsmeier et al. (2004)
10.	RNA22	http://cbcsrv.watson.ibm.com/rna22.html	Miranda et al. (2006)
11.	TargetScanS	http://targetscan.org/	Lewis et al. (2003)

11.4.3 miRNA Target Prediction Tools

An overview of the most commonly used tools for identifying microRNA targets are listed in Table 11.1.

11.5 Human MicroRNA Regulatory Network—A Systems Approach

MicroRNAs have been studied extensively ever since it was first discovered in nematodes (Bartel 2004). However a more comprehensive approach is required in order to study the human regulatory network in its fullness. This can be achieved by a systems biology approach which helps in providing the “big picture”. MicroRNAs can act as “regulatory network hubs” underlying several physiological processes and disease conditions. Integrative genomic approaches have proved that miRNAs acts co-operatively or redundantly to regulate a given pathway and also uses feed-back loops to regulate expression of their target gene. However the ENCODE project has proposed a detailed design principle of how the human regulatory information is organized has been put forward which states that microRNAs occupy a bottle neck in the information flow hierarchy by acting in a distal fashion co-regulating with transcription factors. Hence targeting these bottle-necks (with drugs) can affect the regulatory circuit and in turn the flow of information.

Systems Biology takes us from the study of a single miRNA to a more comprehensive understanding of miRNA regulatory network through which hierarchical interactions and functions are integrated along with signaling, metabolic pathway interactions and gene regulatory networks (Iorio et al. 2011). Even though several formalisms are available to date, their application in studying microRNA regulatory networks is limited.

Recent miRNAnetwork studies include microRNA mediated regulation of a human cellular signaling network suggesting that miRNAs frequently target network downstream signaling components than ligands and cell surface receptors. Combinatory regulations between transcription factors (TF) and microRNAs have been studied based on network architecture (Shalgi et al. 2007). The construction and analysis of an integrated regulatory network based on sequencing data has been performed in higher eukaryotes. This integrated regulatory network shows regulation at three levels namely: (a) TF $\rightarrow$ Target Gene (b) TF $\rightarrow$ miRNA (c) miRNA $\rightarrow$ Target gene (Cheng et al. 2011). Functionally distinct classes of microRNAs were identified based on the analysis of network topologies. (Yu et al 2008). According to this study two different classes of microRNAs were identified, the first class was regulated by a large number of TFs while the second class by few TFs. The expression profiles of the two classes suggest that the microRNAs exhibited distinct functionality in embryonic cells and adult tissues and were uniquely wired to TFs in the network topology. A systems biology approach to identify candidate microRNA targets during progression of poly cystic kidney disease (PKD) has been done. Dysregulation of miRNA genes have been identified and experimentally validated by qPCR and a network model has been proposed suggesting the onset of PKD and cyst formation (Pandey et al. 2011).

A study of design principle of a tissue specific regulatory network in eight human tissues was recently reported (Li et al. 2012). In this study, strong hubs, weak hubs and non-hubs that include miRNAs in feed- forward regulation of network motifs were identified individually in all tissues. A comprehensive systems biology approach thus enables the study of microRNA regulatory networks and pathways as a whole and also the distinct possibility of identifying microRNAs and their targets as therapeutic hotspots.

References

Ambros V et al (2003) A uniform system for microRNA annotation. RNA 9(3):277–279

Anokye-Danso F et al (2011) Highly efficient miRNA-mediated reprogramming of mouse and human somatic cells to pluripotency. Cell Stem Cell 8(4):376–388

Bartel D (2004) MicroRNAs: genomics, biogenesis, mechanism and function. Cell 116:281–97

Cai X, Hagedorn CH, Cullen BR (2004) Human microRNAs are processed from capped, polyadenylated transcripts that can also function as mRNAs. RNA 10(12):1957–1966

Cheng C, Yan KK, Hwang W et al (2011) Construction and analysis of an integrated regulatory network derived from high-throughput sequencing data. PLoS Comput Biol 7:e1002190

Enright AJ, John B, Gaul U, Tuschl T, Sander C, Marks DS (2003) miRanda algorithm: microRNA targets in Drosophila. Genome Biol 5:R1

Hackenberg M et al (2009) miRanalyzer: a microRNA detection and analysis tool for next-generation sequencing experiments. Nucleic Acids Res 37(2):W68–W76

Huang HY, Chien CH, Jen KH et al (2006) RegRNA: a regulatory RNA motifs and elements finder. Nucleic Acids Res 34:W429–W434
Iorio MV, Casalini P, Piovan C, Braccioli L, Tagliabue E (2011) Breast cancer and microRNAs: therapeutic impact. Breast 20(Suppl 3):S63–S70
Kadri S, Hinman V, Benos PV (2009) HHMMiR: efficient de novo prediction of microRNAs using hierarchical hidden Markov models. BMC Bioinformatics 10(1):S35
Hsu JB Chiu CM, Hsu SD et al (2011) miRTar: an integrated system for identifying miRNA-target interactions in human. BMC Bioinformatics 12:300
Katakowski M et al (2010) MiR-146b-5p suppresses EGFR expression and reduces in vitro migration and invasion of glioma. Cancer Invest 28(10):1024–1030
Kim VN (2004) MicroRNA precursors in motion: exportin-5 mediates their nuclear export. Trends Cell Biol 14(4):156–159
Lewis BP, Shih IH, Jones-Rhoades MW, Bartel DP, Burge CB (2003) Prediction of mammalian microRNA targets. Cell 115(7):787–798
Li J, Hua X, Haubrock M, Wang J, Wingender E (2012) The architecture of the gene regulatory networks of different tissues. Bioinformatics 28:i509–i514
Liu K, Wang R (2012) MicroRNA-mediated regulation in the mammalian circadian rhythm. J Theor Biol 304:103–110
Miranda KC, Huynh T, Tay T et al (2006) A Pattern-based method for the identification of MicroRNA binding sites and their corresponding hetero-duplexes. Cell 126(6):1203–1217
Pandey P, Quin S, Ho J et al (2011) Systems biology approach to identify transcriptome reprogramming and candidate microRNA targets during the progression of polycystic kidney disease. BMC Syst Biol 5(56):1–23
Rajewsky N, Chen K (2006) Natural selection on human microRNA binding sites inferred from SNP data. Nat Genet 38:1452–1456
Rehmsmeier M, Steffen P, Hochsmann M, Giegerich R (2004) Fast and effective prediction of microRNA/target duplexes. RNA 10(10):1507–1517
Reinhart BJ, Slack FJ, Basson M, Pasquinelli AE, Bettinger JC, Rougvie AE, Horvitz HR, Ruvkun G (2000) The 21-nucleotide let-7 RNA regulates developmental timing in Caenorhabditis elegans. Nature 403(6772):901–906
Rusinov V, Baev V, Minkov IN, Tabler M (2005) MicroInspector: a web tool for detection of miRNA binding sites in an RNA sequence. Nucleic Acids Res 33:W696–W700
Shalgi R, Lieber D, Oren M, Pilpel Y (2007) Global and local architecture of the mammalian microRNA-transcription factor regulatory network. PLoS Comput Biol 3(7):e131
Tempel S, Tahi F (2012) A fast ab-inito method for predicting miRNA precursors in genomes. Nucleic Acids Res 40(11):e80
Teplyuk NM et al (2012). MicroRNAs in cerebrospinal fluid identify glioblastoma and metastatic brain cancers and reflect disease activity. Neuro Oncol 14(6):689–700
Yan LX et al (2008) MicroRNA miR-21 overexpression in human breast cancer is associated with advanced clinical stage, lymph node metastasis and patient poor prognosis. RNA 14(11):2348–2360
Yang Y, Wang PY, Li KB (2008) MiRTif: a support vector machine-based microRNA target interaction filter BMC Bioinformatics 9(12):S4
Yu, X. et al (2008) Analysis of regulatory network topology reveals functionally distinct classes of microRNAs. Nucleic Acids Res 36:6494–6503

Part II
Synthetic Biology

Chapter 12
A Brief Introduction to Synthetic Biology

Mrugainduta Patil and Pawan K. Dhar

Abstract "*What I cannot create, I do not understand.*" -*Richard Feynman*

Organisms are complex systems that run massively parallel and interactive molecular processes. From an H-atom to the whole cell, cells manage information over at least six orders of magnitude in size. To understand a system that integrates contextual, temporal and spatial complexity by default, calls for innovative and massive data gathering efforts. Data gathered over the last two decades has revealed a huge inventory of molecular parts and contextual interactions. Still a good understanding of biology is lacking.

In the early 2000, people asked: can one assemble biological systems from scratch from a standard inventory of parts, instead of relying upon the naturally evolved systems. One of the key foundational papers that hinted towards engineering approach to biology, was a three gene circuit called repressilator that was plugged into *E.coli* as a non-native applet and stably expressed (Elowitz and Leibler, Nature 403:335–338, 2000). This paper accelerated the thought process that we now know as synthetic biology.

Keywords Synthetic biology · Biological engineering · Logic gates · Truth table

12.1 What Is Synthetic Biology?

At a first look the term synthetic biology might give an impression of 'chemical biology'. However, the intended application of the term is engineering inspired biology. Dr. Barbara Hobom was the first to use the phrase 'synthetic biology' in her paper to describe genetically engineered bacteria (Hobom 1980). The bacteria were altered synthetically using recombinant DNA technology. However, the term largely remained synonymous with 'bioengineering'. At the meeting of American Chemical

P. K. Dhar (✉)
Department of Life Sciences, School of Natural Sciences, Shiv Nadar University, Dadri, India
e-mail: pawandhar@gmail.com

M. Patil
Symbiosis School of Biomedical Sciences, Symbiosis International University, Pune, India

V. Singh, P. K. Dhar (eds.), *Systems and Synthetic Biology*,
DOI 10.1007/978-94-017-9514-2_12

Society in San Francisco in 2000, Eric Kool and others used the term "synthetic biology" to describe molecules that are synthesized and can work in living systems (Benner and Sismour 2005).

The term "Synthetic Biology" was formally proposed in the first annual meeting of synthetic biology at MIT (June 2004) to indicate engineering approach in biology as against the classic reductionist approach. The intended meaning of the term "synthetic" was 'non-native' or novel, as against the commonly used connotation of "chemical". The synthetic approach was about making non-natural parts, devices and circuits or making non-natural combinations of existing parts, devices and circuits. For the last several years, the scientific community has extensively debated the wisdom of bringing in a new terminology—given that in practice it looks same old molecular biology that people have been applying for decades. Due to this reason, the MIT people proposed another term: "constructive biology" in place of synthetic biology. However, the term "constructive biology" did not resonate well enough with the scientific community. People wondered if the traditional approach should be called 'destructive biology'! Furthermore, many other terms like: biological technology, biological engineering, biodesign, biomolecular engineering, biosystems engineering were also suggested. Among these terms 'biological engineering' seems to be closest to the intended meaning of "synthetic biology".

The new line of thought is about constructing biology from a set of off-the-shelf parts. A part is defined as a minimal biological unit that interacts. The unit may be a single molecule as in proteins, or DNA sequence as in gene or promoter.

Moving from Reductionism to Integration Over the past couple decades, the focus in molecular biology has been to understand logic operations based on gene, RNA and protein interactions. Most of what we know today has been discovered by studying a particular process in isolation (Philip Ball 2005). Technological advancement has made it easier and less cumbersome to study how genes act in concert to regulate molecular processes that are wired into pathways and networks (Brown and Botstein 1999). This has reintroduced the 'systems' concept from engineering into biology in the form an approach called systems biology (Kitano 2002).

For a biologist to think in 'systems-way' the challenge is to discover and understand molecular networks by reverse engineering, as it would be difficult to understand a network as a function of individual gene expression or protein interaction events (Hartwell et al. 1999).

Moving from Integration to Engineering To understand the design principles of cellular construction, engineers at MIT proposed a 'synthetic' approach that follows a ground up construction of systems.

The de novo designing of complex biological systems is achieved by combining basic input/output units that represent a certain biological behavior. From these sets of standard biological parts, devices are constructed. In the biological setting devices are a set of biochemical reactions like translation, transcription, allosteric regulation, enzymatic reactions and so on. Biological devices produce regulated outputs by interacting with other devices, generating a circuit to perform biosynthetic and metabolic tasks etc. Although given enormous diversity of biological parts, it is very difficult to interface devices; the background setting of design architecture provides options to construct complex systems with rich functionalities (Andrianantoandro et al. 2006).

Though the original MIT definition proposed the construction of novel parts, devices or circuits—in practice a number of variants of synthetic biology have emerged. One of the most widely adopted variant is the installation of an existing pathway in a non-native host. This approach has been popular with the metabolic engineering community. Other activities like making a non-AGTC genome, synthesizing proteins from not-coding DNA, installing circuits as standalone applets, whole genome cloning, engineering inter-cellular signal transmission in microbial consortia, developing technology for long DNA synthesis, evolving standards for data exchange and so on, have also been parked under the term "synthetic biology".

12.2 Synthetic Biology and Engineering

The earliest subtle comparisons between engineering and biology were provided by none other than, Monod and Jacob when they described the operon as a regulatory circuit responding to the **logic** of the cell and signaling by transmitters and receivers (Ball 2005).

The origin of engineering inspired approach stems from the fact that biology and engineering are similar in many aspects. For example, both exhibit similarity in terms of robustness, multi-tasking, fault tolerance, running linear and non-linear processes, analog and digital behavior and run jobs serial and parallel.

However, there are many differences too. -introducing non-native components into the circuit, modification of existing logic gates or overriding them completely to obtain desired results are non-trivial in biology. These differences make it challenging to convert biology into an engineering discipline. Creating a circuit in isolation is one thing but expressing it in a non-native setting is a completely different challenge. Thus, the take home message is: adopt engineering philosophy but have sufficient patience and take sufficient care.

12.3 Key Engineering Concepts Used in Synthetic Biology

12.3.1 Logic Gates

A logic gate is a fundamental building block of an electronic circuit. Graphically logic gates are represented as a combination of logic input and a logic output response. The input and output may be represented as 0 (absence of signal) and 1 (presence of signal). There are several examples of logic gates in biology e.g. lac operon (NOT gate), substrate-enzyme reaction (AND gate), activator-inducer mediated process (AND gate). Even though there are similarities, it is important to recognize that an electronic logic gate does not reverse the information during the passage of information through the gate. However, some biologic gates can lead to partial reversibility of

Fig. 12.1 Bio truth table

Inducer / repressor concentration

Protein concentration		n1	n2	n3	n4	n5
	n1					
	n2					
	n3					
	n4					
	n5					

information during passage e.g. part of the substrate-enzyme complex gets reconstituted back into substrate. Thus, the level of complexity in managing the information in biological circuits is higher than that in the electronic circuitry. To digitize biology and create predictable behaviors in a shorter time span, it is important to develop bio-logic gates (Wang et al. 2011).

12.3.2 Truth Table

A truth table is a logic table used to document input/output response of the system under study. Both the input and output are in the form of logical boolean values 0 and 1. For example, in NOT gate when the input is 0, output is 1 or if the input is 1 and output is zero. This helps design electronic circuits in such a way that the passage of information is efficient and robust.

In comparison a Bio-Truth table is quantitative e.g. given a certain amount of inducer concentration what is the corresponding protein concentration? (Fig. 12.1) Though a bio-truth table may look higher in resolution, it has serious implementation issues. The values in the bio-truth table are not portable, unlike that in engineering systems. This is due to the fact that these values are contextual i.e. dependent upon strain, metabolic state, culture condition and so on.

12.4 Parts Standardization

The kind of redesigning that we see in synthetic biology calls for the need of a bigger and better toolbox, with broader store of fundamental components than just those provided by nature. The current technology of gene manipulation and pathway redesigning is ingenious but also very limiting at the same time. Synthetic biology hence looks at broadening this range of molecular tools (Ball 2005). But along with designing new synthetic materials, comes the need to standardize it. Without exactly knowing the behavior of the particular component under various conditions, or standardizing it to behave similar in various systems, the synthesized component is of less or no use. Same concept applies to the materials provided by nature itself. Unless a particular gene sequence is assured to give the same product in various systems under different conditions, it is of no use to the synthetic biologist. Designing something ambitious is good, but it has to be reproducible and reliable.

The concept of parts standardization also comes from engineering experience where a 'behavioral part-inventory data' is used to assemble devices and complex systems from scratch. In the field of engineering the assembly line construction is so perfect that one goes from computer model to manufacturing in one step e.g. in designing aircrafts. Engineers do not fly thousands of aircrafts to select good designs from the ones that do not crash! But that's exactly what happens in biology. We push a given gene construct in tens of millions of cells with a hope to identify the cells that have correctly received the gene.

Can the situation change if biologists adopt an engineering approach? Is it possible to assemble devices and circuits with reasonable accuracy directly from computational models? To answer this question, pilot experiments are needed where parts-behavior (input/output values) are documented in a range of environmental settings e.g. different culture conditions, various inducer and repressor condition, different types of strains and so on. Data generated from such experiments can help one capture parts-behavior in various contexts. The downside is that it is infeasible to plan every possible experimental variation for every type of part. Thus, given the limited availability of experimental data, we know 'most likely-parts-behavior' instead of 'exact-parts-behavior'.

Once a reasonably comprehensive parts-behavior library of several model organisms is constructed by adopting a uniform protocol across the community, gaps may be filled-in by using (a) training data set from previous experiments and (b) a quantitative modeling formalism. Given budgetary and time constraints, a more realistic version of the parts-behavior inventory will be a database that has 'experimentally determined and computationally predicted' behavioral profile of every functional sequence of DNA part that an organism is made of.

Unlike in engineering, in biological systems one also needs to create a 'device behavior' and a 'circuit behavior' library. Creating parts library does not mean that it can be used to assemble device and circuit library with ease and accuracy. This is due to the fact that biological systems are analog and contextual, in contrast to engineering systems that are digital and deterministic. However, the parts inventory helps reduce the search space of composing stable devices and circuits.

Recently, standard biological parts knowledgebase (SBPkb) was created using synthetic biology open language http://www.sbolstandard.org allowing users to find standard biological parts for compilation (Galdzicki et al. 2011). The issue of standardization is important in this context and has been discussed in detail elsewhere (Müller and Arndt 2012).

12.5 Designing a bioCAD Platform

Composing an organism from scratch comes with a strong need to use computation at various levels. For building a cell from its parts inventory, one needs certain abstraction to reduce systemic complexity to modular units, dynamically linking designs at various levels of abstraction, ability to recruit parts from different sources (databases) on the basis of specifications.

In a typical scenario, a user would want to combine well-annotated parts from various databases and collectively express them as a device or a circuit in a non-native environment. To do so, it would be useful to have a computational capability to navigate different part-inventories, select the right part, develop collective model for expression. Given that high throughput and cost-effective DNA synthesis technologies are rapidly emerging, the output file of BioCAD modeling would be a string of DNA bases, which when chemically synthesized, can generate a design of choice. In some cases, the role of BioCAD platform will be to help find a genome-sequence-equivalent of a metabolic pathway, a regulatory network, a transcriptional cascade, perform sequence edits and assemble the final design for synthesis. Currently, several BioCAD tools are available each offering some unique features (Canton et al. 2008). Several good reviews on the need for standardization in biological engineering are available (Marchisio and Stelling 2009).

12.6 Applications

Synthetic biology has had a positive impact on almost every other biological field. Right from making new synthetic molecules and compounds, to designing of novel new drugs and finding applications to solve many other issues of engineering as well as biology. Synthetic biology is enabling us to gain more understanding of the world around us, at the same time giving us more to make changes and produce new improved products.

12.6.1 Making Genes from Junk DNA

We asked a simple question—Why did nature choose a particular region for making proteins? Did she sample all possible combinations, chose some and rejected others. If she did not sample all possibilities, can we express naturally not-coding DNA into functional peptides and proteins. If yes, would all combinations work? What will be the boundary condition of such an approach and what are the best case scenarios?

To address these questions, we invented a technique to artificially convert not-coding sequences into functional genes (Dhar et al. 2009). Given that genome has

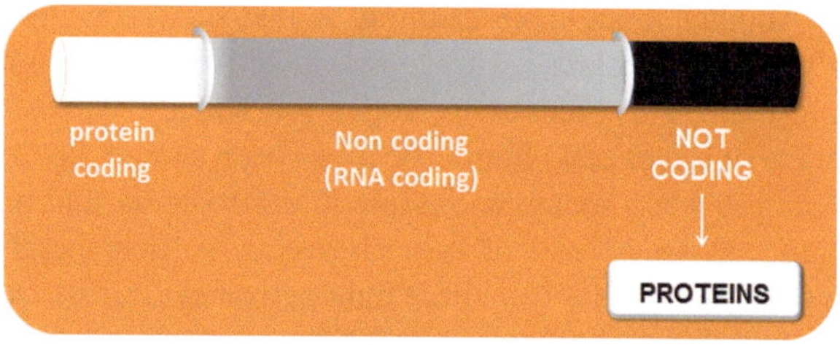

Fig. 12.2 Artificially expressing not coding DNA

three types of sequences: protein coding, RNA coding and not-coding (Fig. 12.2 below), it would be interesting to see what happens when these sequences are artificially expressed. We call these artificially made genes 'eka' (ekam—first, in sanskrit).

Having provided the proof of the concept, the next step was to perform a genome wide scan of intergenic regions and predict what would happen if we expressed a non-coding DNA artificially. To meet this objective, we are currently developing a knowledgebase called EKA. Using EKA Knowledgebase one can find length, sequence composition, structure, function and cellular localization attributes of non-coding DNA, if expressed.

An unexpected offshoot of this work was revisiting the classic question—what makes a gene? Given that (a) naturally non-expressed sequences can be artificially expressed into functional proteins and (b) gene prediction softwares do not identify these sequences as protein coding sequences, the fundamental question remains—what makes a gene? It is clear that community needs to take a fresh look at this question. It is also clear that genome annotation is not over yet! The third offshoot of this work is identification of sequences that are likely to make antimicrobial and anticancer peptides. Our initial experimental findings point towards synthesis of novel peptides towards clinical applications. We would like to express introns, pseudogenes and repetitive sequences and study the outcome.

One of our final goals is to design a novel application-oriented pathway based on proteins made from junk DNA. We are looking at the emergence of a new area that we like to call junkomics i.e., study of the artificially induced expression of junk DNA sequences.

12.6.2 Bionanoscience

The marriage of nanotechnology and synthetic biology is inevitable. The natural nanostructures that exist can be either modified, or mimicked by synthesizing inorganic molecules. Much work has been done in this direction e.g. Audette et al. (2004) found that modified forms pilin proteins that make up the pili of bacterium, can self-assemble into nanotubes about 6 nm in diameter and 100 um long. These proteins were derived by genetically modifying a strain or *P. aeruginosa*. These tubes have been shown to bind to DNA, suggestion nanotubes could be used to construct biocomposite structures (Ball 2005).

Another remarkable example is that of use bacteriorhodopsin immobilized in a polymer to retain its function and actively pump protons against a pH gradient and thereby to "reduce hydrogen ion leakage across the proton exchange membrane of a fuel cell" Similarly, it has been shown that protein-based surfactants can "can provide a membrane-mimetic environment that maintains the integrity of the entire photosystem I of spinach leaves, so that these assemblies of proteins and pigment molecules can effect light-activated electron transport, with applications in photovoltaic technology (Ball 2005).

12.6.3 Regulatory Circuits

All the activities in the cells are controlled by defined circuits made up of genes and proteins. Many of these circuits are either not mapped out or not well understood. These circuits overlap, intersect, or run parallel to bring about the changes required for the cell to survive. To understand how, what happens in the cells, synthetic biology can be used as a tool. Rewiring regulatory circuits and creating new internal circuits to alter the pattern of activity can give a clear idea of what exactly is happening at the molecular level. Artificial gene networks are reality today. Creating regulatory circuits from artificial components whose functions are well understood can be used to control and alter the cells behavior.

One major use of rewiring regulatory circuits is in metabolic engineering. Using these synthetic regulatory circuits, the cells can be modified to have new biosynthetic pathways leading to new products formation, or to increase yield, to correct metabolic dysfunction. A very frequently quoted example is the use of modified bacterium or yeast cells to produce an anti-malarial drug artemisinin. The modified cells produce artemisinic acid, which is a precursor of artemisinin. This method of obtaining artemisinic acid has reduced production costs drastically. The altering of metabolic pathways has not been completely exploited yet. The extreme genetic engineering can allow for natural metabolic and non-natural compounds production.

Another application of rewiring genetic circuits is to elucidate the entire pathways. Synthetic rewiring experiments can be used to test predictions about the role and plasticity of cellular networks. The obvious question, which rises is, that shouldn't we study the real biological system instead of creating amature synthetic networks. The answer to the question is in informal notion of synthetic biology "What I cannot create, I do not understand". Understanding how the connectivity in cellular networks is achieved helps us to create hypothesis driven rewiring experiments. Thus, building biological systems is a useful way to systematically deconstruct underlying biological principles and at the same time "offers a way to understand the underlying designing principle that allow different classes of circuits to be constructed from any type of cellular network." The bottom up approach ensures full control over the circuit design (Sprinzak and Elowitz 2005; Caleb et al. 2010).

12.6.4 *Therapeutics*

Synthetic biology is enabling new therapeutic platform, from the identification of diseases to drug targeting and delivery. Synthetic systems can be designed to study disease mechanism and effect of drugs on that system as well. In fact work in this direction has lead to some amazing discoveries. Researchers developed a synthetic testbed for agammaglobulinaemia by systematically reconstructing the human B-cell receptor signaling pathway in an orthogonal environment. This allowed them to identify various factors that trigger BCR signaling. A rare mutation was detected in a patient and when introduced in the synthetic system showed complete inhibition of assembly of the BCR on the cell surface. Thus this faulty pathway was linked to the disease onset of agammaglobulinaemia.

Once the faulty component has been identified, synthetic biology can be used for drug discovery against it. New drugs can be synthesized against multidrug resistant organisms. Engineered viruses and organisms can be used as drug delivery system for programmed and target specific delivery. Development of the drug against multidrug resistant Mycobacterium tuberculosis by Fussenegger and coworker provides a good example. Ethionamide is currently the best drug against MDR tuberculosis. The mechanism of this drug depends on activation by the enzyme ethA in the organism. However in many cases, the enzyme production is transcriptionally repressed by the protein ethr. The researchers designed a synthetic mammalian gene circuit to address this problem. This circuit was used to identify other inhibitors using an ethr-based transactivator reporter gene. Since the system is cell-based assay, it is easy to enrich the natural inhibitors that are non-toxic and membrane permeable to mammalian cells. (Weber et al. 2008).

12.6.5 *Biosensing*

Cells have a range of sensing and signaling activities running simultaneously. For example, when a cell encounters change in its environment, it activates certain signaling elements to bring about the appropriate stress response. These signals can be detected and synthetically manipulated at transcription, translation and post-translational stages. The basic mechanism by which this is done is by targeting the gene expression controlling elements such as promoter, inducer, transcription factors, natural regulatory RNA molecules such as microRNAs.

The earliest design strategy adopted for biosensing was to engineer the promoter sequence of the gene and place it under synthetic control. This can be achieved by removing, or modifying the activator and repressor sites to fine-tune the promoter's sensitivity to a molecule. Synthetic mammalian biosensors based on this principle have been created for sensing signals such as antibiotics, gases, metabolites, quorum-sensing molecules and even temperature changes. Fussenegger et al, have even created a transgene design, incorporated into mammalian circuits leading to synthetic networks that are responsive to electrical signals. (Weber et al. 2009).

To have true modular genetic parts is inherently difficult due to the interference among natural native and synthetic parts. This calls for careful decoupling of functional modules. Kobayashi et al. (2004) used one such modular design strategy to develop a whole cell *E.coli* biosensor that responded to signals in programmable fashion. In this cell, the sensory module was coupled with a synthetic gene circuit that functions like a central processing unit in the cell. Bayer and Smolke, successfully achieved translational level sensing, when they engineered trans acting ligand responsive riboregulators of gene expression in *S.cerevisiae*. (Bayer and Smolke 2005). The binding of the ligand and aptamer causes a conformational change in the RNA sensor that either signals the continuous translation or inhibition of translation of the output gene reporter. This results in a stable binary like switching which can be detected. The detection threshold can be tuned by altering the RNA sequence. Similarly post translational biosensing has also been made possible (Skerker et al. 2008).

12.7 Critical Appraisal of the Field

Before one accepts the emergence of a new field, it is pertinent to ask: is Synthetic Biology the correct terminology? Are there any alternative terminologies that define the field better? Is designing new organisms faster, easier and cheaper than studying existing organisms? Does the field raise any ethical and legal concerns? What has been the progress of the field so far and what kind of work are we likely to see in future? Looking back, it is clear that synthetic biology approach was somewhat oversold. A number of overhyped promises never saw the light of the day. This is especially true of the BioBrick initiative (Shetty et al. 2008). To my best knowledge there is no "biobrick-based device or circuit" that cannot be made by "non-biobricks". Further, Biobricks do not address the complexity of network dynamics. They are merely abstractions that are "made to feel like modular parts". A large part of synthetic biology remains the same old classic molecular biology and metabolic engineering. Probably the only unique aspect of synthetic biology is the development of long DNA synthesis technologies that may replace routine recombinant DNA methods 1 day. In my opinion, the DNA synthesis technologies would have happened anyway irrespective of emergence of this new approach.

The recent experience of composing and installing user-defined genetic circuits shows that designing new biological systems from scratch and running them as applets is far more challenging than the classic approach. Synthetic biology approach is neither faster, nor easier or cheaper than existing solutions. Maybe once there is sufficient data for every genetic part the assembly process might get slightly less cumbersome. However, given that cells runs massive parallel and massively interactive non-linear and analog molecular processes optimized for a large number of environmental contexts, it is difficult to make an assessment of how much data are enough?

Currently, the community is self-regulated and without any legal framework. However, engineering approach does raise some ethical and safety concerns. Even these have been oversold as the technology to assemble organisms using a conveyer belt like approach, is just not there. The question is: why should anyone adopt a more difficult and less efficient approach to develop a harmful microbe, if cheaper solutions are available. In fact, advances in neurobiology and toxin research are far more worrying than synthetic biology which is still struggling to grapple fundamental issues of biological complexity.

For the last 10 years, the synthetic biology community has designed switches (Gardner 2000), biobricks (Ho-Shing et al. 2012), oscillators (Purcell et al. 2010), quorum sensing (Danino et al. 2010), non-natural DNA (Pinheiro et al. 2012) and minimal synthetic cell (Gibson et al. 2010). In the next 10 years, we hope to see more publications on synthetic chromosomes, alternative genetic codes, several non-bio brick initiatives, genome scale engineering and faster methods to make synthetic cells.

12.8 Summary

Synthetic biology is a bold new approach to construct biological systems with an aim to make organisms from scratch. For installing small networks, the strategy works in some situations. However, making the whole cell from scratch is quite time consuming, effort consuming and exorbitantly expensive. In future, we are likely to see more publications on faster and cheaper methods to make a synthetic chromosome, a synthetic cell or a consortium of user defined cells. The capability to synthesize chromosomes and whole genomes will offer tremendous opportunity for good creative science and also lead to frightening applications.

References

Andrianantoandro E, Basu S, Karig DK, Weiss R (2006) Synthetic biology: new engineering rules for an emerging discipline. Mol Syst Biol 2:2006.0028

Audette GF et al (2004) DNA binding protein nanotubes. Nano Lett 4:1897–1902

Ball P (2005) Synthetic biology for nanotechnology. Nanotech 16:R1–R8

Bayer TS, Smolke CD (2005) Programmable ligand controlled riboregulators of eukaryotic gene expression. Nat Biotech 23:337–343

Benner SA, Sismour AM (2005) Synthetic biology. Nat Rev Genet 6(7):533–543

Brown PO, Botstein D (1999) Exploring the new world of the genome with DNA microarrays. Nat Genet 21:33–37

Caleb JB, Andrew AH, Sergio GP, Wendell AL (2010) Rewiring cells: synthetic biology as a tool to interrogate the organizational principles of living systems. Annu Rev Biophys 39:515–537

Canton B et al (2008) Refinement and standardization of synthetic biological parts and devices. Nat Biotechnol 26:787–793

Danino T et al (2010) A synchronized quorum of genetic clocks. Nature 463:326–330

Dhar PK et al (2009) Synthesizing non-natural parts from natural genomic template. J Biol Eng 3:2

EASAC (2011) European academies science advisory council. Building science into EU policy. Synthetic biology: an introduction 1–11
Elowitz MB, Leibler S (2000) A synthetic oscillatory network of transcriptional regulators. Nature 403:335–338
Galdzicki M et al (2011) Standard biological parts knowledgebase. PLoS ONE 6:e17005
Gardner TS et al (2000) Construction of a genetic toggle switch in *Escherichia coli*. Nature 403:339–342
Gibson DG et al (2010) Creation of a bacterial cell controlled by a chemically synthesized genome. Science 329:52–56
Hartwell LH, Hopfield JJ, Leibler S, Murray AW (1999) From molecular to modular cell biology. Nature 402(6761 Suppl):C47–52
Hobom B (1980) [Gene surgery: on the threshold of synthetic biology]. Med Klin 75(24):834–841
Ho-Shing O et al (2012) Assembly of standardized DNA parts using BioBrick ends in *E. coli*. Methods Mol Biol 852:61–76
Kitano H (2002) Systems biology: a brief overview. Science 295(5560):1662–1664
Kobayashi H et al (2004) Programmable cells: interfacing natural and engineered gene networks. PNAS USA 101:8414–8419
Marchisio MA, Stelling J (2009) Computational design tools for synthetic biology. Curr Opin Biotech 20:479–485
Müller KM, Arndt KM (2012) Standardization in synthetic biology. Methods Mol Biol 813:23–43
Pinheiro VB et al (2012) Synthetic genetic polymers capable of heredity and evolution. Science 336:341–344
Purcell O et al (2010) A comparative analysis of synthetic genetic oscillators. J R Soc Interface 7:1503–1524
Shetty RP et al (2008) Engineering BioBrick vectors from BioBrick parts. J Biol Eng 2:5
Skerker JM et al (2008) Rewiring the specificity of two component signal transduction systems. Cell 133:1043–1054
Sprinzak D, Elowitz MB (2005) Reconstruction of genetic circuits. Nature 438:443–448
Wang B et al (2011) Engineering modular and orthogonal genetic logic gates for robust digital-like synthetic biology. Nat Commun 2:508
Weber W et al (2008) A synthetic mammalian gene circuit reveals antituberculosis compounds. PNAS USA 105:9994–9998
Weber W et al (2009) A synthetic mammalian electrogenetic transcription circuit. Nucleic Acids Res 37:e33

Chapter 13
DNA Structure and Promoter Engineering

Venkata Rajesh Yella, Aditya Kumar and Manju Bansal

Abstract Transcription initiation is the first step in the regulation of gene expression. Promoters are the regions of genomic DNA where transcription initiation machinery assembles and are generally characterized by presence of short nucleotide sequence motifs like TATA-box, Inr element, BRE, etc. However, apart from these motifs, promoter regions have been reported to have structural properties, such as lower stability, lesser bendability and more curvature compared to other genomic regions. Interestingly, these properties are conserved from archaea to mammals, with little differences. Several algorithms have been developed to differentiate promoter regions from non promoters, using DNA structural properties. Here we show that, in *E. coli* and *S. cerevisiae*, genes with different experimentally determined expression levels, differ in their structural features. Promoters of highly expressed or less responsive genes are less stable, less bendable and more curved compared to promoters of lowly expressed or more responsive genes. This suggests that these structural properties can be used to design promoters to modulate gene expression.

Keywords Promoter engineering · DNA structural properties · DNA duplex stability · DNA bendability · Intrinsic curvature · NUCRADGEN · Transcription factor binding sites (TFBSs)

13.1 Introduction

Promoter region can be broadly defined as a small fragment of DNA within a genome that can initiate transcription. Promoter regions are generally composed of small transcription factor binding motifs (of length 4–15 nucleotides), also called

---

M. Bansal (✉) · A. Kumar · V. R. Yella
Molecular Biophysics Unit, Indian Institute of Science,
Bengaluru, Karnataka, India
e-mail: mb@mbu.iisc.ernet.in

A. Kumar
e-mail: aditya@mbu.iisc.ernet.in

V. R. Yella
e-mail: yvrajesh@mbu.iisc.ernet.in

V. Singh, P. K. Dhar (eds.), *Systems and Synthetic Biology*,
DOI 10.1007/978-94-017-9514-2_13

cis-regulatory elements such as TATA box, BRE, Inr element, etc. (Smale and Kadonaga 2003). These sequence motifs are generally identified using Position Weight Matrices (PWMs). These motifs are degenerate, less complex in sequence composition and the probability of finding them in other genomic locations is very high. Early studies on promoter elements were based on small datasets and on model gene systems. Recent high throughput studies on different eukaryotic systems proved that, very few of the promoter regions have the exact consensus motifs (Basehoar et al. 2004; Carninci et al. 2006). The question then arises as to how, in the absence of these base sequences, is the transcription process initiated and regulated. Earlier analyses by Pederson et al. (1998) and Kanhere and Bansal (2005), showed that the promoter regions of both prokaryotic and eukaryotic genomes apparently have different structural features like lower stability, less bendability and higher curvature compared to their flanking regions (Kanhere and Bansal 2005; Pedersen et al. 1998). Stress-induced DNA duplex destabilzation (SIDD) sites were observed to be very closely associated with promoters in the *E. coli* genome (Wang et al. 2004). Recent high-throughput nucleosome position maps for different model systems show that most of the promoter regions are nucleosome free and structurally rigid in nature (Jiang and Pugh 2009). So it is necessary to study promoters not only as simple sequence elements but also as regions with structurally distinct features.

In this chapter, we discuss the structural and compositional properties of DNA in the promoter regions of prokaryotic and eukaryotic genes and the possible role of DNA structural features in promoter engineering. Promoter engineering has many ramifications, here we restrict ourselves only to applications of DNA structural properties to promoter engineering.

13.2 DNA Structural Properties

DNA molecule is highly polymorphic in nature, its structure being dependent on environment, base composition and sequence context (Ghosh and Bansal 2003). B-DNA, A-DNA, Z-DNA and curved or kinked DNA are some of the well characterized double helical polymorphs. Since B-DNA is the most prevalent structure *in vivo*, here we refer to distinct structural property of any particular DNA sequence by its deviation from ideal B-DNA structural parameters or random sequence DNA. DNA structural properties are an outcome of the arrangement of the 4 nucleotide bases, Adenine, Thymine, Guanine and Cytosine which are chemically different and also of the characteristic features of the two grooves (minor and major) arising due to the asymmetric position of glycosidic bonds of base pairs. Hence the DNA structure and properties are expected to vary along its length. The DNA structural properties are broadly divided into two categories, the physico-chemical and conformational properties. Physico-chemical properties such as stability are directly dependent on various inter atomic interactions such as van der Waals, hydrogen bonds and electrostatic. Conformational properties such as wedge angle (for a base paired dinucleotide step), bendability and curvature are dependent on rotational and translational parameters

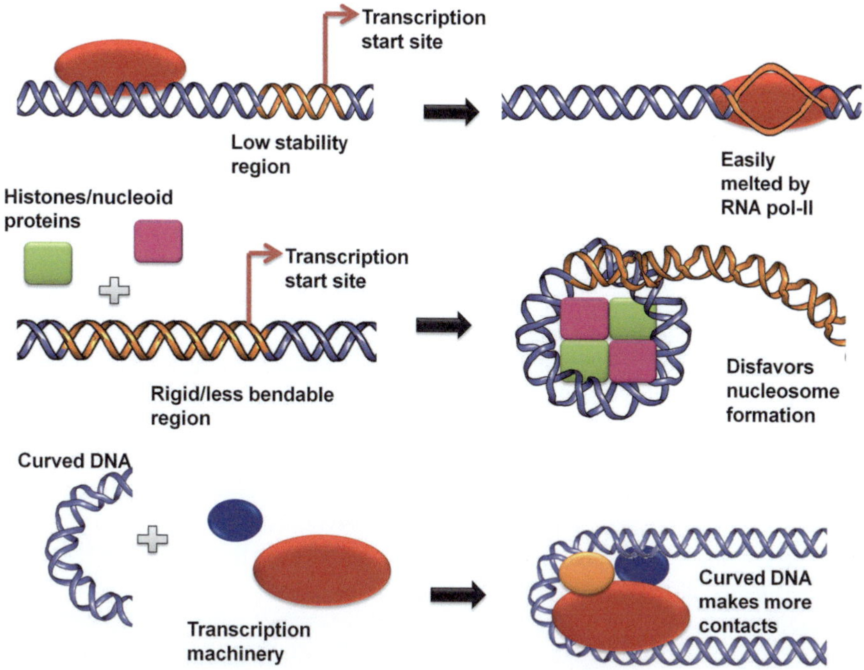

Fig. 13.1 Role of DNA structural properties in transcription initiation

such as roll, tilt, slide and shift (Meysman et al. 2012). DNA stability, bendability and intrinsic curvature are the most well studied and Fig. 13.1 illustrates the biological significance of these structural properties. These three sequence dependent properties differ in their nature and information content and can be studied both at local and global levels, depending on the length of DNA fragment being examined. Intrinsic DNA curvature is however meaningful only if one analyzes relatively long fragments, as a minimum of 30 nucleotides are required to obtain reliable values. Various di, tri and tetra nucleotide models are available to study DNA structural properties at different levels. In addition, supercoil induced DNA destabilization has also been used for identifying the more meltable regions in circular prokaryotic genomes (Bi and Benham 2004). Detailed information about above three structural properties is given in the following box.

DNA Structural Properties

DNA duplex stability: Stability of any chunk of DNA can be expressed in terms of hydrogen bond and stacking interactions, which are short range nearest-neighbor interactions (Allawi and SantaLucia 1997) and depend on identity and orientation of flanking base pairs. Stability primarily depends on GC content with small variations arising due to the actual dinucleotide composition. It is described in terms of standard free energy change, or simply referred to as free energy. Nearest neighbor thermodynamic models can approximately predict the stability of a given fragment of DNA (SantaLucia 1998). The free energy values for a dinucleotide step range from −0.58 kcal/mol (for TA dinucleotide sequence) to −2.24 kcal/mol (for GC dinucleotide sequence), where the higher negative value corresponds to greater stability and is expressed in units of *kcal/mol*.

DNA bendability: The bendability of DNA can be defined as anisotropic bending of DNA in presence of DNA binding factors. DNA bendability can be calculated using di, tri and tetranucleotide models. Here we have used trinucleotide models based on DNase1 sensitivity (Brukner et al. 1995) and Nucleosome Positioning Preference (Satchwell et al. 1986) to estimate bendability, since they are more reliable as compared to dinucleotide models and are derived from experimental data. DNase1 sensitivity model is derived from cutting studies of oligonucleotides by DNase1 enzyme. The bending propensity values in this model range from −0.281(= AAT/ATT) to 0.194(= TCA/TGA). This model differentiates bendability of trinucleotides in terms of ease of bending towards the major groove with higher negative value corresponding to lower bendability. Satchwell's Nucleosome Positioning Preference (NPP) model gives preferences of all possible trinucleotides in the DNA duplex, for their minor or major groove face to be towards the histone core. This model provides relative values for major groove face preferring or minor groove preferring as well as trinucleotides with no rotational position preference, on an absolute scale. The values range from 45(for GCC/GGC) to 2(for CAG/CTG). According to this model, trinucleotides with strong preference for their major groove or minor groove to face the histone core are rigid, whereas trinucleotides without any rotational preference are flexible. The trinucleotide models calculate bendability in arbitrary units and hence give an indication of relative bendability of various DNA sequences.

Intrinsic curvature: Curvature is defined as the anisotropic bending of DNA in solution in the absence of any external forces. Curvature is depends on the order of dinucleotide sequence and the common helical and inter-base pair parameters for each base paired dinucleotide step. Two sets of dinucleotide parameters are available to predict curvature values for a given sequence, which are based on crystal structure data (Bhattacharya and Bansal 1988) and gel mobility data (Bolshoy et al. 1991). Curvature can be represented in different units such as radius of curvature or d/lmax (the ratio of minimum end to end distance 'd' to the contour length of a DNA fragment, 'lmax'). Here we use d/lmax, which ranges from 0 to 1, where 0 corresponds to a completely closed circle (highly curved) and 1 to a perfectly linear DNA fragment (Bansal et al. 1995).

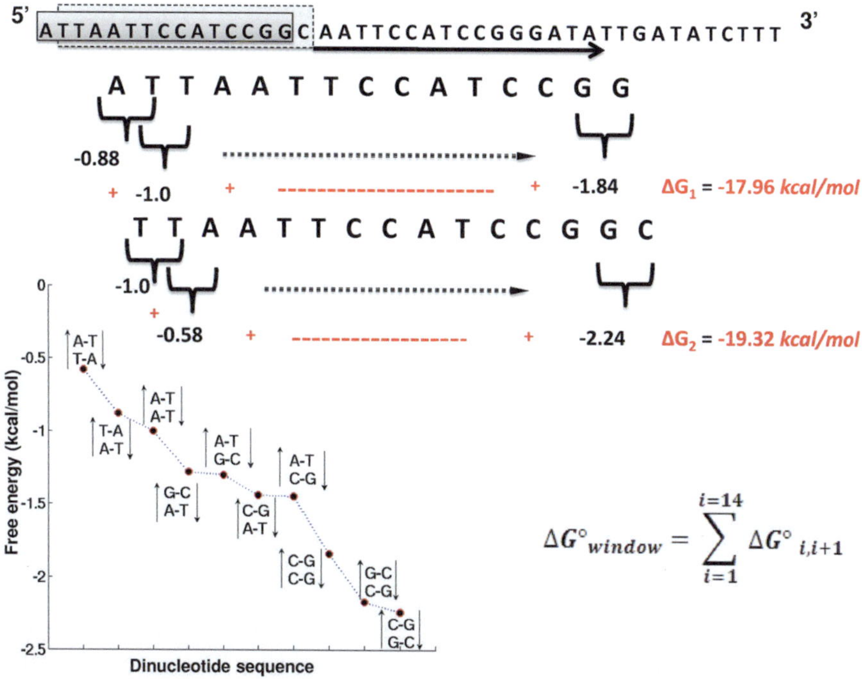

$$\Delta G^{\circ}{}_{window} = \sum_{i=1}^{i=14} \Delta G^{\circ}{}_{i,i+1}$$

Fig. 13.2 DNA stability of a given region, represented by the total free energy of the fragment, which is calculated by summing the free energy of the constituent dinucleotides (values shown in inset at *bottom left*)

Structural properties can be calculated using the various di, tri, tetranucleotide models reported in the literature. These models provide lookup tables, using which one can easily obtain the numerical profiles, representing the various properties, of a given fragment of DNA. To calculate the structural property, first each dinucleotide or trinucleotide in a sequence is replaced with corresponding structural feature value and then summed or averaged over a predefined window. Structural property calculations are generally carried out using a single nucleotide sliding window, as these depend on nearest-neighbor effects, context features and cooperative effects[1]. However, different window sizes can be chosen based on the property/feature being examined and whether one is interested in its manifestation at the local level or global level. For local level structure analysis, calculations are generally carried out for window size of less than 10 nucleotides (one helical turn of DNA). Window size between 10 and 30 nucleotides can be assumed to be a good representative of both local and global level features, while greater than 30 nucleotides is generally considered as global level. Figure 13.2 shows an example of protocol that can be followed for calculation of average free energy using a 15 nucleotide (or 14 dinucleotide steps) sliding window.

[1] Bendability and curvature are cooperative in nature, whereas stability is only restricted to nearest neighbor effects.

13.3 Promoter Regions have Unique Structural Properties

Figure 13.3 shows average structural properties of promoter regions of three different model systems, *E. coli*, *S. cerevisiae* and mouse along with human. These model systems differ in their genomic GC content and nucleotide composition, are well studied and their experimentally validated Transcription Start Site (TSS) data has been published[2]. To obtain the structural profiles, all promoter sequences are aligned, relative to their TSSs and then the structural property profiles are calculated, as described in the previous section. The numerical values obtained for all sequences are averaged at each nucleotide position, to get the mean structural property for each system. Optimal window sizes for each property are different, 15 nucleotide window (14 dinucleotide steps) for stability, 30 nucleotide window (28 trinucleotide steps) for the two bendability models and 75 nucleotide window (74 dinucleotide steps) for curvature calculation (Kanhere and Bansal 2005). The length of promoter sequences are 1001 nucleotides in case of *E. coli* and *S. cerevisiae* and 2001 nucleotides in case of mouse and human. The structural features in mouse and human are extending beyond 500 nucleotides upstream and downstream relative to TSS position.

The average stability profiles in Fig. 13.3, of all four systems, show less stable regions, but the span of the region varies. In *E. coli*, the low stability region extends from −150 to +50 nucleotide region, with a sharp peak close to the TSS. In *S. cerevisiae* the low stability region is observed from around −200 to +50 region with respect to TSS, with three split peaks. The gray lines in Fig. 13.3 show the numerical profiles of shuffled sequences corresponding to the upstream and downstream regions with respect to the TSS. Although shuffling the sequences leads to the context effects vanishing, the overall upstream shuffled sequences are less stable compared to downstream shuffled regions (the effect is very marginal in mouse and human). The lower stability of *E. coli* and *S. cerevisiae* promoters is due to higher AT content in these regions as compared to neighboring regions. In mammalian systems, the promoter regions are GC rich, but two small low stability peaks are observed around −30 and −1 region, which are flanked by more stable GC rich regions. The high stability of core promoter regions (except −30 and −1 region) in mammals is attributed to the presence of CpG islands. The structural property, stability is however a common feature of all classes of bacteria and lower eukaryotes. An algorithm has been developed that uses the relative free energy values to predict promoter regions in whole genome sequences (Rangannan and Bansal 2010). Presence of this low stability feature in promoter regions is very important for genome transcription, as it is then easy for transcription machinery to transiently melt the core promoter DNA and initiate transcription, as shown schematically in Fig. 13.1.

The bendability profiles of the promoter regions of the four systems analyzed differ for two models as seen in Fig. 13.3. Satchwell's nucleosome positioning model shows

---

[2] Human and mouse TSS data were downloaded from DBTSS database (Wakaguri et al. 2008). *E. coli* data downloaded from RegulonDB version 7.0 (Gama-Castro et al. 2011). *S. cerevisiae* data downloaded from Xu et al. transcriptome study (Xu et al. 2009).

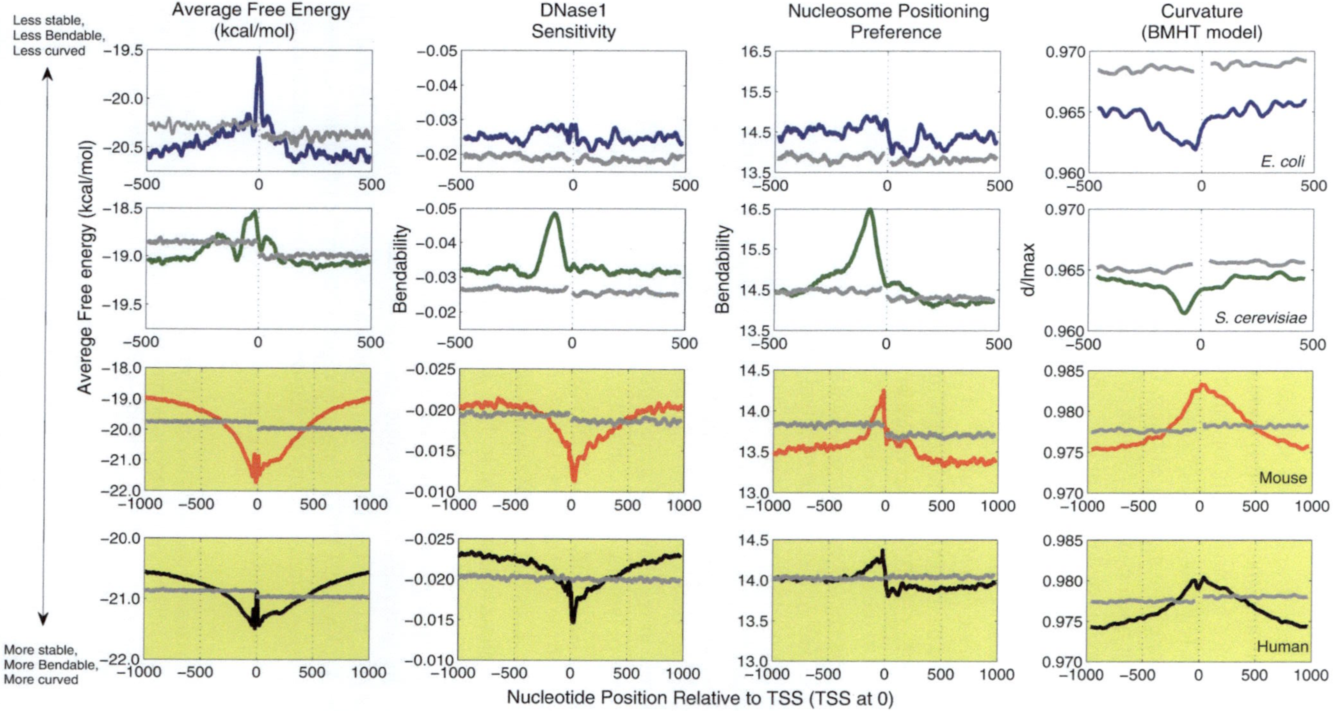

Fig. 13.3 Sequence dependent structural properties in the promoter regions of different model systems. The figure shows distribution of different structural properties, Average Free Energy, DNase1 sensitivity, Nucleosome Positioning Preference and Curvature (from *left* to *right* respectively) in promoter regions of four different systems. The colors *blue*, *green*, *red* and *black* represent structural features of *E. coli* (1597), *S. cerevisiae* (4911), mouse (17451) and human (29456) promoters respectively. *Gray lines* indicate the average values for corresponding shuffled sequences

that promoter regions are less bendable[3] in all systems, whereas DNase1 sensitivity model indicates that while promoter regions of *E. coli* and *S. cerevisiae* are rigid, the core promoter regions of mouse and human are flexible. In *E. coli*, the bendability does not seem to be a distinguishing feature, but it is a distinctive feature of promoter regions in *S. cerevisiae*. The DNase1 sensitivity model indicates that, mammalian promoters are more flexible, since this model calibrates rigidity on the basis of bendability preference towards major or minor groove. There are suggestions that the rigidity of DNA in promoter regions provides greater scope for sliding of DNA binding proteins along its length. In addition, the higher cost of DNA bending may play a major role in open complex formation during transcription, by making the DNA resistant to bending and facilitating escape of the transcription machinery from promoter region. It is also now well accepted that promoter regions should be devoid of nucleosomes for easy access by basal transcription machinery. Rigid DNA can hence act as an antinucleosomal barrier for transcription initiation.

The average intrinsic curvature profiles[4] of promoter regions of *E. coli* and *S. cerevisiae*, show that the near upstream region is more curved compared to downstream region, as shown in Fig. 13.3. The curvature of genomic promoter sequences is also higher as compared to the shuffled sequences. Conversely, the core promoter regions of mammals are less curved compared to flanking regions. The curvature values of shuffled sequences are almost similar to actual genomic sequences. The importance of curvature was recognized very early for promoter regions of pathogenic bacteria and thermo-sensing bacteria (Falconi et al. 1998; Prosseda et al. 2004), but it is less apparent in eukaryotic promoters. It has been proposed that the intrinsically curved DNA segments in the promoter region of prokaryotes act as transcriptional signals and thus play a role in regulation of gene expression. Curved DNA can make more contacts with RNA polymerase and can cause juxtaposition of transcription factors binding to far apart regions of DNA, as shown in Fig. 13.1.

The differences in the structural properties of core promoter regions and the neighboring regions can arise due to differential base composition or due to presence of some select oligonucleotides. Table 13.1 lists the over represented and under represented hexanucleotides in the −150 to −50 regions, relative to TSS, of the four systems. The core promoter regions have characteristic nucleotide composition and they are generally rich in some specific AT rich oligonucleotides. The trinucleotides AAA, TTT, ATA, the tetranucleotides AAAA, TTTT, TATA and the hexanucleotides AAAAAA, TTTTTT are over represented in the promoter regions of *E. coli* and *S. cerevisiae*, whereas promoter regions of human and mouse are enriched

[3] The terms bendability, rigidity and flexibility have been used interchangeably. DNA flexibility is of two types, torsional flexibility (due to variations in twist about the axis) and bending flexibility (or bendability, due to variations in roll, tilt, slide and shift). In present context rigidity or flexibility refers to only bending flexibility.

[4] Curvature values shown here are calculated using BMHT dinucleotide step parameters (Bolshoy et al. 1991) and in-house software NUCRADGEN (http://nucleix.mbu.iisc.ernet.in/nucradgen/index.htm).

Table 13.1 Over represented and under represented hexameric sequences in promoter regions (−150 to −50 with respect to TSS)

Organism	Over represented hexamers	Under represented hexamers
E. coli	AAAAAA, TTTTTT, TAAAAA, CTGGCG, AAAAAC	ACCTAG, CCTAGA, CCTAGG, CTAGGA, CTAGGG
S. cerevisiae	TTTTTT, AAAAAA, TTTTTC, GAAAAA, ATTTTT	GGGTCG, CGACCC, CGGGAC, CGGGGG, GCCCCG
Mouse	GGGCGG, GGCGGG, CCGCCC, CCCGCC, GGGGCG	TCGTAT, TATACG, ATATCG, CGTATA, ATACGA
Human	GGGCGG, GGCGGG, TTTTTT, CCGCCC, CCCGCC	TACGAT, ATACGA, ATATCG, TATCGA, TATACG

with GC rich nucleotide sequences, such as GGG, CCC, GGGG, CCCC, GGCG, GGGC, GGGCGG, GCGGGG along with oligoA-tracts[5], though CG repeats are surprisingly not found in our dataset. The high occurrence of AT rich sequences, particularly A-tracts, in promoter regions of *E. coli* and *S. cerevisiae* is reflected in their lower stability, lesser bendability and higher curvature. Higher eukaryotes, like mammals seem to favor other structural elements (such as G4-motifs (Huppert and Balasubramanian 2007) and oligo G-tracts) along with oligo A-tracts.

13.4 Applications of DNA Structural Properties in Promoter Engineering

Gene expression is the most fundamental biological process, in which the genetic information is used to create a phenotype. Gene expression can be regulated at various levels, such as during transcription, RNA processing, translation and post translational events. The initiation of transcription is the first and probably the most important step in regulation of gene expression. Promoter regions are the elements where the transcription machinery assembles. So the level of gene expression depends on promoter architecture, along with other external factors. It is known that promoters with optimal *cis*-regulatory elements (such as a TATA-box) are sufficient for normal gene expression to occur. The promoter regions can be designed and engineered. To design a synthetic promoter, a general approach is to construct a promoter sequence with naturally occurring regulatory elements such as transcription factor binding sites (TFBSs) and enhancers. Another approach is to design promoters with special DNA structural features such as curved DNA, which have been found to enhance gene expression. Sometimes promoters are designed by incorporating both TFBSs and DNA with special structural features.

Several studies have addressed the importance of DNA structural properties (mainly curvature and bendability) in specific DNA-protein interactions. Some attempts were also made to engineer better promoters. Classical work done by (Bracco et al. 1989), showed that the hybrid promoters (with an upstream curved region

[5] A-tracts consist of stretches of minimum four consecutive A:T base pairs without a flexible TA step.

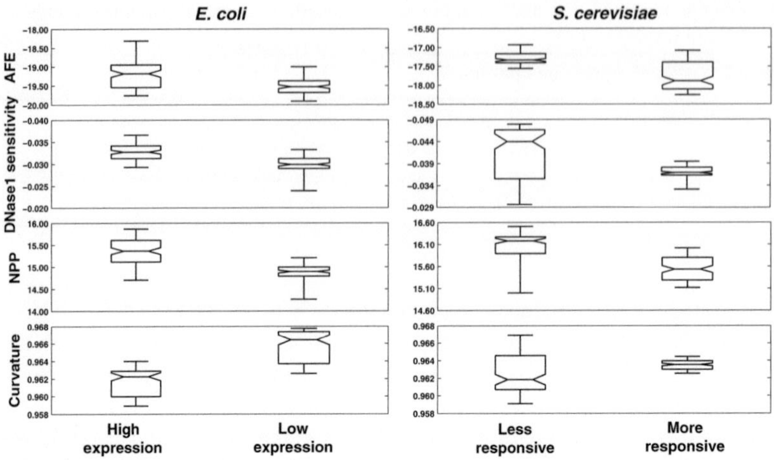

Fig. 13.4 Box plot showing structural properties of promoter regions of highly expressed and lowly expressed genes in *E. coli* and less responsive and more responsive genes in *S. cerevisiae*

arising due to phased A-tracts) are transcriptionally more efficient than wild-type *gal* promoter, with a CAP binding site. Similarly (Gartenberg and Crothers 1991) showed that curved sequences increased transcription initiation tenfold as compared to wild-type *lac* promoter. Very recently Segal and his colleagues (Raveh-Sadka et al. 2012), showed that the rigid poly (dA:dT) tracts can act as tunable components in regulation of gene expression. To understand the role of nucleosome-disfavoring sequences, they systematically studied 70 different variants of wild-type yeast his3 promoter, with poly (dA:dT) tracts of varying length, composition and distance from the transcription factor binding sites (TFBSs). These rigid sequences increase the gene expression with the increase of purity of A-tracts, length, proximity to TFBSs and the effects are predictable at single-cell level. In another study, by the same group, on 777 designed promoters, they found that, the effects of the nucleosome-disfavoring, rigid sequences on gene expression regulation are consistent with their earlier results (Sharon et al. 2012). A study in plants showed that, a few of the engineered promoters with enhanced activity had lower stability as compared to other sequences (Ranjan et al. 2012). Many studies use A-tracts to design promoters, as they have very distinct local structure (Haran and Mohanty 2009), which is also reflected in their special properties like low stability, less bendability and higher curvature (when occurring as phased elements). So, these and other sequences with distinctly different structural features as compared to random sequence DNA (presumed to have ideal B-form properties) can act as modulators of gene expression.

An example of how structural properties play a role in differential gene expression is discussed here. Figure 13.4 shows the structural properties of promoter regions of

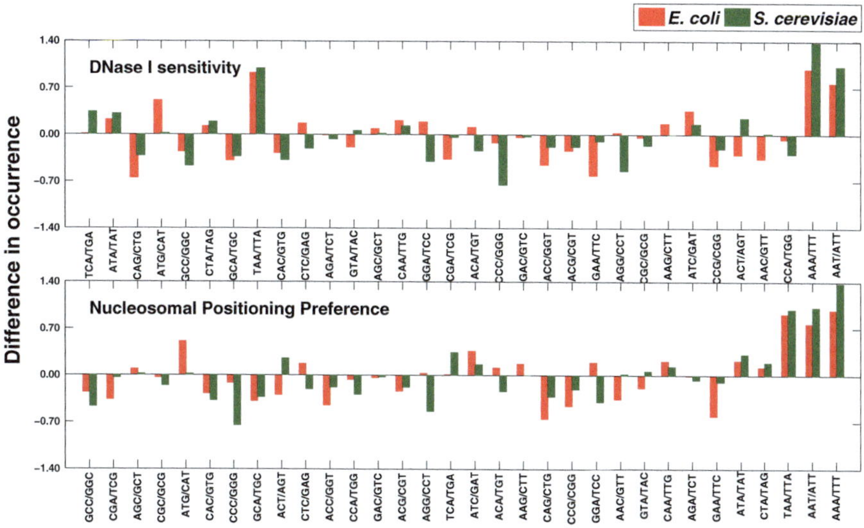

Fig. 13.5 *Bar diagrams* showing the differences in frequency between the highly expressed and lowly expressed gene promoters, for the 32 unique nucleotide trimers. The *top* figure corresponds to difference in trimer frequencies in promoter regions of *E. coli* and *S. cerevisiae*, sorted according to DNase1 sensitivity values. The *bottom* figure corresponds to difference in trimer frequencies sorted according to Nucleosome Positioning Preference values i.e., major groove preferring trimers are to the *left* and minor groove preferring ones to the *right*. In both cases the *first bar* corresponds to *E. coli* and *second bar* to *S. cerevisiae*. Positive values correspond to higher occurrence in highly expressed or less responsive gene promoters

highly and lowly expressed[6] genes in *E. coli* and less responsive[7] and more responsive genes in *S. cerevisiae*. The boxes in this figure represent the distribution of averaged property value of the promoter regions of two differently expressed classes of genes[8]. The core promoter regions, in this case correspond to the -150 to -50 region, where the structural features are observed to be more distinctly different. The promoter regions of highly and lowly expressed genes in *E. coli* or less responsive and more responsive genes in *S. cerevisiae* show significant differences in their properties. The characteristic properties of the promoter regions are even more pronounced for highly expressed or less responsive genes, being less stable, more rigid and more curved compared to lowly expressed genes or more responsive genes (Fig. 13.4).

[6] RNA-seq data was downloaded from (Gama-Castro et al. 2011) and estimation of gene expression was done by analyzing RPKM (Mortazavi et al. 2008) for *E. coli*.

[7] Responsiveness is the gene expression variability measured from curated datasets representing various conditions. Responsiveness data for *S. cerevisiae* was downloaded from (Choi and Kim 2009).

[8] The number of promoter sequences considered for this analysis are 100 in the datasets corresponding to highest and lowest gene expression in case of *E. coli*. The two datasets corresponding to low and high responsiveness contaians 200 sequences for *S. cerevisiae*.

In order to understand the underlying differences in oligonucleotide composition that could play a role in these differential structural features, we analyzed the composition of promoter regions in the two datasets. Figure 13.5 shows the difference in trinucleotide composition of core promoter regions (−150 to −50 regions with respect to TSS) of highly expressed or less responsive and lowly expressed or more responsive genes. The rigid and minor groove face towards histones preferring trinucleotides[9], such as AAA/TTT, AAT/ATT, are over represented in highly expressed gene promoters of *E. coli* and in less responsive gene promoters *S. cerevisiae*, whereas flexible or major groove bend preferring trinucleotides like GCC/GGC, CGA/TCG are under represented. Hence an optimal choice of these trimer sequences in promoter regions of a gene could help engineer its expression level by modulating its structural properties.

13.5 Conclusions

Promoter regions in both prokaryotes and eukaryotes differ in their structural properties when compared to other genomic regions. They are less stable, less bendable and more curved, compared to flanking regions. The special structural features of promoter regions are due to differential base composition, in particular the occurrence of AT rich sequences (at least in prokaryotes and yeast). These structural features are utilized by transcriptional machinery to identify promoter regions from other regions of the genome. Many experimental studies have used the special properties of A-tracts to modulate gene expression. We find that all the promoter specific structural properties are more pronounced in highly expressed or less responsive genes as compared to lowly expressed or more responsive genes in both *E. coli* and *S. cerevisiae*. The highly expressed or less responsive gene promoters are less stable, less bendable and more curved compared to promoters of lowly expressed or more responsive genes. These differences in structural properties can therefore be used to design promoters and modulate gene expression. Although our qualitative study clearly shows that structural properties modulate gene expression, it is not clear if there is a linear relationship between the property value and gene expression level. The role of these structural properties in mammals is also not clear and they may use different structural features like G4-motifs or GC-rich rigid sequences, in their structure mediated regulation of gene expression.

Acknowledgements AK acknowledges CSIR, INDIA for scholarship. MB is a recipient of J. C. Bose National Fellowship of DST, India. We thank Asmita Gupta for assistance in the preparation of Fig. 13.1.

[9] The trinucleotides sorted using NPP model are not according to absolute flexibility values, but on the basis of rotational preference for minor or major groove.

References

Allawi HT, SantaLucia J (1997) Thermodynamics and NMR of internal G.T mismatches in DNA. Biochemistry 36(34):10581–10594

Bansal M, Bhattacharyya D, Ravi B (1995) NUPARM and NUCGEN: software for analysis and generation of sequence dependent nucleic acid structures. Comput Appl Biosci 11(3):281–287

Basehoar AD, Zanton SJ, Pugh BF (2004) Identification and distinct regulation of yeast TATA box-containing genes. Cell 116(5):699–709

Bhattacharya D, Bansal M (1988) A general procedure for generation of curved DNA molecules. J Biomol Struct Dyn 6(1):93–104

Bi C, Benham CJ (2004) WebSIDD: server for predicting stress-induced duplex destabilized (SIDD) sites in superhelical DNA. Bioinformatics 20(9):1477–1479

Bolshoy A, McNamara P, Harrington RE, Trifonov EN (1991) Curved DNA without A-A: experimental estimation of all 16 DNA wedge angles. Proc Natl Acad Sci U S A 88(6):2312–2316

Bracco L, Kotlarz D, Kolb A, Diekmann S, Buc H (1989) Synthetic curved DNA sequences can act as transcriptional activators in *Escherichia coli*. EMBO J 8(13):4289–4296

Brukner I, Sanchez R, Suck D, Pongor S (1995) Sequence-dependent bending propensity of DNA as revealed by DNase I: parameters for trinucleotides. EMBO J 14(8):1812–1818

Carninci P, Sandelin A, Lenhard B, Katayama S, Shimokawa K, Ponjavic J, Semple CA, Taylor MS, Engstrom PG, Frith MC, Forrest AR, Alkema WB, Tan SL, Plessy C, Kodzius R, Ravasi T, Kasukawa T, Fukuda S, Kanamori-Katayama M, Kitazume Y, Kawaji H, Kai C, Nakamura M, Konno H, Nakano K, Mottagui-Tabar S, Arner P, Chesi A, Gustincich S, Persichetti F, Suzuki H, Grimmond SM, Wells CA, Orlando V, Wahlestedt C, Liu ET, Harbers M, Kawai J, Bajic VB, Hume DA, Hayashizaki Y (2006) Genome-wide analysis of mammalian promoter architecture and evolution. Nat Genet 38(6):626–635

Choi JK, Kim YJ (2009) Intrinsic variability of gene expression encoded in nucleosome positioning sequences. Nat Genet 41(4):498–503

Falconi M, Colonna B, Prosseda G, Micheli G, Gualerzi CO (1998) Thermoregulation of Shigella and *Escherichia coli* EIEC pathogenicity. A temperature-dependent structural transition of DNA modulates accessibility of virF promoter to transcriptional repressor H-NS. EMBO J 17(23):7033–7043

Gama-Castro S, Salgado H, Peralta-Gil M, Santos-Zavaleta A, Muniz-Rascado L, Solano-Lira H, Jimenez-Jacinto V, Weiss V, Garcia-Sotelo JS, Lopez-Fuentes A, Porron-Sotelo L, Alquicira-Hernandez S, Medina-Rivera A, Martinez-Flores I, Alquicira-Hernandez K, Martinez-Adame R, Bonavides-Martinez C, Miranda-Rios J, Huerta AM, Mendoza-Vargas A, Collado-Torres L, Taboada B, Vega-Alvarado L, Olvera M, Olvera L, Grande R, Morett E, Collado-Vides J (2011) RegulonDB version 7.0: transcriptional regulation of *Escherichia coli* K-12 integrated within genetic sensory response units (Gensor Units). Nucleic Acids Res 39(Database issue):98–105

Gartenberg MR, Crothers DM (1991) Synthetic DNA bending sequences increase the rate of in vitro transcription initiation at the *Escherichia coli* lac promoter. J Mol Biol 219(2):217–230

Ghosh A, Bansal M (2003) A glossary of DNA structures from A to Z. Acta Crystallogr D Biol Crystallogr 59(4):620–626

Haran TE, Mohanty U (2009) The unique structure of A-tracts and intrinsic DNA bending. Q Rev Biophys 42(1):41–81

Huppert JL, Balasubramanian S (2007) G-quadruplexes in promoters throughout the human genome. Nucleic Acids Res 35(2):406–413

Jiang C, Pugh BF (2009) Nucleosome positioning and gene regulation: advances through genomics. Nat Rev Genet 10(3):161–172

Kanhere A, Bansal M (2005) Structural properties of promoters: similarities and differences between prokaryotes and eukaryotes. Nucleic Acids Res 33(10):3165–3175

Meysman P, Marchal K, Engelen K (2012) DNA structural properties in the classification of genomic transcription regulation elements. Bioinform Biol Insights 6:155–168

Mortazavi A, Williams BA, McCue K, Schaeffer L, Wold B (2008) Mapping and quantifying mammalian transcriptomes by RNA-Seq. Nat Methods 5(7):621–628

Pedersen AG, Baldi P, Chauvin Y, Brunak S (1998) DNA structure in human RNA polymerase II promoters. J Mol Biol 281(4):663–673

Prosseda G, Falconi M, Giangrossi M, Gualerzi CO, Micheli G, Colonna B (2004) The virF promoter in Shigella: more than just a curved DNA stretch. Mol Microbiol 51(2):523–537

Rangannan V, Bansal M (2010) High-quality annotation of promoter regions for 913 bacterial genomes. Bioinformatics 26(24):3043–3050

Ranjan R, Patro S, Pradhan B, Kumar A, Maiti IB, Dey N (2012) Development and functional analysis of novel genetic promoters using DNA shuffling, hybridization and a combination thereof. PLoS ONE 7(3):e31931

Raveh-Sadka T, Levo M, Shabi U, Shany B, Keren L, Lotan-Pompan M, Zeevi D, Sharon E, Weinberger A, Segal E (2012) Manipulating nucleosome disfavoring sequences allows fine-tune regulation of gene expression in yeast. Nat Genet 44(7):743–750

SantaLucia J (1998) A unified view of polymer, dumbbell, and oligonucleotide DNA nearest-neighbor thermodynamics. Proc Natl Acad Sci U S A 95(4):1460–1465

Satchwell SC, Drew HR, Travers AA (1986) Sequence periodicities in chicken nucleosome core DNA. J Mol Biol 191(4):659–675

Sharon E, Kalma Y, Sharp A, Raveh-Sadka T, Levo M, Zeevi D, Keren L, Yakhini Z, Weinberger A, Segal E (2012) Inferring gene regulatory logic from high-throughput measurements of thousands of systematically designed promoters. Nat Biotechnol 30(6):521–530

Smale ST, Kadonaga JT (2003) The RNA polymerase II core promoter. Annu Rev Biochem 72: 449–479

Wakaguri H, Yamashita R, Suzuki Y, Sugano S, Nakai K (2008) DBTSS: database of transcription start sites, progress report 2008. Nucleic Acids Res 36(Database issue):97–101

Wang H, Noordewier M, Benham CJ (2004) Stress-induced DNA duplex destabilization (SIDD) in the *E. coli* genome: SIDD sites are closely associated with promoters. Genome Res 14(8): 1575–1584

Xu Z, Wei W, Gagneur J, Perocchi F, Clauder-Munster S, Camblong J, Guffanti E, Stutz F, Huber W, Steinmetz LM (2009) Bidirectional promoters generate pervasive transcription in yeast. Nature 457(7232):1033–1037

Chapter 14
Synchronous Sequential Computations with Biomolecular Reactions

Vishwesh V. Kulkarni, Hua Jiang, Evgeny Kharisov, Naira Hovakimyan, Mark Riedel and Keshab Parhi

Abstract We present a methodology for implementing synchronous sequential computation using molecular reactions. Such systems perform computations in terms of molecular concentrations, i.e., *molecules per unit volume*, whereas the traditional electronic systems perform computations in terms of voltages, i.e., *energy per unit charge*. Thus far, several researchers have already proposed molecular reactions to implement static logical and arithmetic functions such as addition, multiplication, exponentiation, square root, and logarithms. In this paper, we propose two mechanisms to implement a multi-phase clock using molecular reactions. In addition, we synthesize memory by transferring concentrations between molecular types in the alternating phases of the clock. We illustrate how our methodology can be used to construct *finite impulse response* (FIR) filter, an *infinite impulse response* (IIR) filter and a four-point, two-parallel *fast Fourier transform* (FFT). We also show how these molecular reactions can be translated into DNA strand displacement reactions and validate our designs through chemical kinetics simulations at the DNA reactions level. Our proposed methodology is conceptual but has potential in developing synthetic biological constructs for biochemical sensing and drug delivery.

Keywords Infinite impulse response (IIR) filter · Finite impulse response (FIR) filter · Molecular computation · Synchronous sequential computation · Mass-action

---

V. V. Kulkarni (✉) · H. Jiang · M. Riedel · K. Parhi
University of Minnesota, Minneapolis, USA
e-mail: vkulkarn@umn.edu

H. Jiang
e-mail: hua@umn.edu

K. Parhi
e-mail: parhi@umn.edu

E. Kharisov · N. Hovakimyan
University of Illinois at Urbana-Champaign, Champaign, IL, USA
e-mail: evgeny@illinois.edu

N. Hovakimyan
e-mail: nhovakim@illinois.edu

M. Riedel
e-mail: mriedel@umn.edu

V. Singh, P. K. Dhar (eds.), *Systems and Synthetic Biology*,
DOI 10.1007/978-94-017-9514-2_14

kinetics · Molecular clock · Memory · Fast Fourier Transform(FFT) · Counter · Binary counter

14.1 Introduction

The field of molecular computation has attracted a lot of research activity in recent years (Qian et al. 2010; Seelig et al. 2006; Soloveichik et al. 2010; Yurke et al. 2000). Broadly, the field strives to implement computations using biomolecular processes. Some of the early work in the field discussed molecular solutions to challenging combinatorial problems such as the *Hamiltonian Path Problem* and *Boolean Satisfiability* (Adleman 1994). In spite of the claims of "massive parallelism"—100 Teraflop performance in a test tube—molecular computations are, today, a significantly inferior alternative to conventional silicon computers for number crunching tasks since the chemical reaction networks are inherently slow and complex. As a result, the broad impetus of the field is not on massive number crunching tasks but, rather, on the design of "embedded controllers", i.e., chemical reactions that are engineered into biological systems, such as viruses and bacteria, to perform useful molecular computation *in situ* where it is needed. For example, consider a system for chemotherapy drug delivery with an engineered bacteria. Here, the objective is to ensure that the bacteria reaches the tumors and selectively produce a drug to kill the cancerous cells. In this scenario, an embedded control of the bacteria is needed to decide the amount of the drug and the locations where it needs to be delivered. The computation could be as simple as: "If the chemical type X is present then produce the chemical type Y", where X is a protein marker of cancer and Y is the chemo drug. It could also be more complicated: for example, "Produce Z if X is present and Y is not present or vice-versa", i.e., an *exclusive-or* (XOR) function. It could be even more complicated: for example, "Produce Z if the rate of change of X is within certain bounds", i.e., a band-pass filter function. Interesting results on along these lines are given in (Anderson et al. 2006) and (Venkataramana et al. 2010).

Concepts from digital circuit design have been applied in molecular computations (see (Anderson et al. 2007; Arkin and Ross 1994; Benenson et al. 2004; Weiss et al. 1999; Weiss 2003; Win; Smolke 2007; Win et al. 2009)) to implement specific computational constructs such as logical operations such as copying, comparing and incrementing/decrementing (Senum; Riedel 2011); programming constructs such as "for"and "while" loops (Shea et al. 2010); arithmetic operations such as multiplication, exponentiation and logarithms (Senum and Riedel 2011; Shea et al. 2010); and signal processing operations such as filtering (Jiang et al. 2010, 2011; Samoilov et al. 2002). Building on this work, we now present a general methodology for implementing *synchronous sequential* computation. In our method, global synchrony is achieved by a molecular oscillator providing synchronous clock signals. All computations are carried out under the control of this global clock.

14.1.1 Computational Model

A molecular system consists of a set of chemical reactions, each specifying a rule for how types of molecules combine. For instance,

$$X_1 + X_2 \xrightarrow{k} X_3, \tag{14.1}$$

specifies that one molecule of X_1 combines with one molecule of X_2 to produce one molecule of X_3. Here, the *rate constant* k governs the rate at which this reaction proceeds. We model the molecular dynamics in terms of *mass-action kinetics* (Érdi and Tóth 1989; Horn and Jackson 1972): reaction rates are proportional to (1) the concentrations of the participating molecular types; and (2) the rate constant. Accordingly, for the reaction above, the rate of change in the concentrations of X_1, X_2 and X_3 is

$$-\frac{\widetilde{X_1}}{dt} = -\frac{\widetilde{X_2}}{dt} = \frac{d\widetilde{X_3}}{dt} = k\widetilde{X_1}\widetilde{X_2},$$

where $\widetilde{x}$ denotes the concentration of a given chemical type x. Most prior molecular computation schemes assume that the rate constants are constant valued. As a result, these have a rather limited applicability since, in practice, the rate constants are not time-invariant: these depend on several factors, including cell volume and temperature. Therefore, the molecular computation results implemented by these schemes are not robust.

We aim for robust constructs. Our methodology requires only two coarse values for the rate constants: k_{fast} for the *fast* reactions and k_{slow} for the *slow* reactions. Given such coarse values for these constants, the computation is exact and, furthermore, the exact rates the "fast"reactions or the "slow"reactions do not matter so long as all fast reactions fire sufficiently faster than slow reactions. As the experimental platform, we implement these DNA-based computations using DNA strand displacement. Our contribution can be viewed as the *front-end* of a design flow with the output of our methodology being a set of abstract molecular reactions. Soloveichik et al. have recently developed a "DNA assembler" (Soloveichik et al. 2010); this constitutes the *back-end*. They have shown that the kinetics of molecular reactions can be *emulated* with DNA strand displacements. Reaction rates are controlled by designing sequences with different binding strengths. The binding strengths are controlled by the length and sequence composition of "toehold"sequences of DNA. Different reaction rates can be easily realized by designing DNA strands with different toehold lengths (Soloveichik et al. 2010). They have shown that that *any* system consisting of unimolecular reactions (i.e., those with a single reactant) and bimolecular reactions (i.e., those with two reactants) can be emulated by such DNA strand displacement reactions.

In DNA strand displacement systems, the reaction rates for unimolecular reactions and bimolecular reactions are different. The rates for unimolecular reactions depend on the initial concentration of auxiliary complexes. For design simplicity, all of our designs consist of bimolecular reactions. We map these to DNA strand displacement reactions, using similar experimental parameters as (Soloveichik et al. 2010).

We generate differential equations corresponding to the DNA reactions and obtain transient solutions. Such simulations of the chemical kinetics provide a reasonably accurate prediction of the actual *in vitro* behavior.

14.1.2 Synchronous Sequential Computation

The general structure of our design is illustrated in Fig. 14.1a. As in an electronic system, our molecular system has separate constructs to implement *computation* and *memory*. A clock signal synchronizes transfers between computation and memory. Prior results on the computational reactions are given in (Jiang et al. 2010; Senum and Riedel 2011; Shea et al. 2010). Operations such as addition and scalar multiplication are straightforward. Operations such as multiplication, exponentiation, and logarithms are trickier. These can be implemented with reactions that implement iterative constructs analogous to "for"and "while"loops. Our main contributions include a new method to generate the clock signals and a new method to implement memory.

14.2 Results

14.2.1 Clock

In electronic circuits, a clock signal is generated by an oscillatory circuit that produces periodic voltage pulses. For a molecular clock, we choose reactions that produce sustained oscillations in the chemical concentrations. With such oscillations, a low concentration corresponds to logical value of zero; a high concentration corresponds to a logical value of one. Techniques for generating chemical oscillations are well established in the literature. Classic examples include the Lotka-Volterra, the "Brusselator"and the Arsenite-Iodate-Chlorite systems (Epstein and Pojman 1998; Kepper et al. 2008). Unfortunately, none of these schemes are quite suitable for synchronous sequential computation: we require that the clock signal be perfectly symmetrical, with abrupt transitions between the phases.

We now present two new designs for generating multi-phase clocks. Our first design is a stand alone design in which an external input is not required to build the oscillations. Our second design is a modification of the coupled oscillators recently derived by Kim and Winfree (see Kim and Winfree 2011) and requires an external input. A complete discussion of the second design is beyond the scope of this manuscript and will shortly be presented separately.

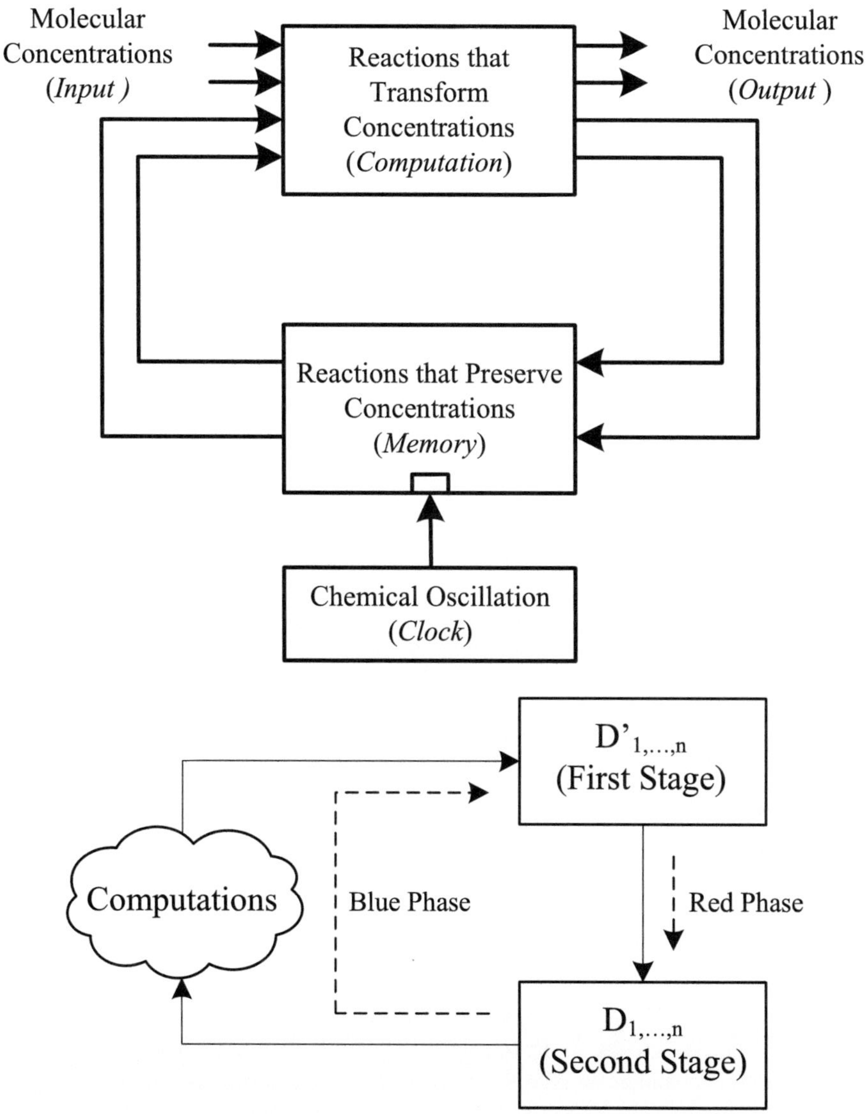

Fig. 14.1 **a** A high level block diagram of a synchronous sequential computational system implemented using molecular reactions. **b** A 2-phase memory transfer scheme used to help implement this architecture

A Stand-Alone Multi-Phase Clock

Let us consider a 4-phase clock. Let these phases be represented by the molecular types labeled R, G, B, V: these labels indicate the color codes red, green, blue, and violet, respectively. Each of these molecular types has an *absence* indicator: r is the absence indicator for R, g is the absence indicator for G, and so on. First consider the reactions:

$$\varnothing \xrightarrow{\text{slow}} r,\ \varnothing \xrightarrow{\text{slow}} g,\ \varnothing \xrightarrow{\text{slow}} b,\ \varnothing \xrightarrow{\text{slow}} v \tag{14.2}$$

$$R + r \xrightarrow{\text{fast}} R,\ G + g \xrightarrow{\text{fast}} G,\ B + b \xrightarrow{\text{fast}} B,\ V + v \xrightarrow{\text{fast}} V. \tag{14.3}$$

As (14.2) shows, the absence indicators r, g, b, and v are generated at a slow rate at all time instants. Here, $\varnothing$ indicates "no reactants"—in other words, such products are generated from a large or replenishable source. As (14.3) shows, the types R, G, B, and V quickly consume the types r, g, b, and v, respectively. The reactions

$$R + v \xrightarrow{\text{slow}} G,\ G + r \xrightarrow{\text{slow}} B,\ B + g \xrightarrow{\text{slow}} V,\ V + b \xrightarrow{\text{slow}} R, \tag{14.4}$$

transfer one phase signal to another, in the absence of the previous one. The essential aspect is that, within the $R - G - B - V$ sequence, the full quantity of the preceding type is transferred to the current type before the transfer to the succeeding type begins. To achieve sustained oscillation, we introduce a positive feedback as follows:

$$2G \underset{\text{fast}}{\overset{\text{slow}}{\rightleftharpoons}} I_G,\ R + I_G \xrightarrow{\text{fast}} 3G,\ 2B \underset{\text{fast}}{\overset{\text{slow}}{\rightleftharpoons}} I_B,\ G + I_B \xrightarrow{\text{fast}} 3B,$$

$$2V \underset{\text{fast}}{\overset{\text{slow}}{\rightleftharpoons}} I_V,\ B + I_V \xrightarrow{\text{fast}} 3V,\ 2R \underset{\text{fast}}{\overset{\text{slow}}{\rightleftharpoons}} I_R,\ V + I_R \xrightarrow{\text{fast}} 3R. \tag{14.5}$$

Consider the first two reactions. Two molecules of G combine with one molecule of R to produce three molecules of G. The first step in this process is reversible: two molecules of G can combine, but in the absence of any molecules of R, the combined form will dissociate back into G. So, in the absence of R, the quantity of G will not change much. In the presence of R, the sequence of reactions will proceed, producing one molecule of G for each molecule of R that is consumed. Due to the first reaction $2G \xrightarrow{\text{slow}} I_G$, the transfer will occur at a rate that is super-linear in the quantity of G; this speeds up the transfer and thereby provides a positive feedback. We choose two nonadjacent phases, R and B, as the clock phases. This approach is easy to implement and, as Fig. 14.2 shows, it works reasonably well *in silico*.

14.2.1.1 An Externally Forced Multi-Phase Clock

Even though the clock described in Sect. 14.2.1 is quite simple, its robustness to uncertainties and disturbances, which it is bound to encounter in wet-lab implementations, is difficult to quantify. As a result, a more sophisticated approach to synthesize feedback is required. For simplicity, let us assume that a possibly vector-valued periodic signal u can be injected in the system externally as an excitation

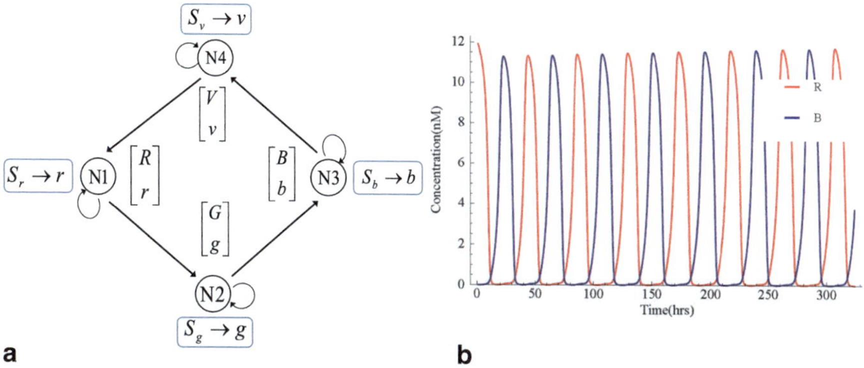

Fig. 14.2 **a** An interconnection network to implement the 4-phase clock. Each node N_i represents a chemical and its absence indicator. By choosing the strength of the interconnections appropriately, oscillations are triggered in each of the 4 nodes in response to the initial concentrations of these chemicals. The absence indicators are generated continually at a slow rate using the sources $S_{(\cdot)}$. **b** ODE simulation of the chemical kinetics of the proposed clock network shows that oscillations are triggered in R and B phase; for this simulation the other two phases were turned off

input. Now, our objective is to ensure that the system output y, comprising concentrations of possibly more than one chemical entities, tracks this signal as closely as possible, i.e., our objective is to synthesize a feedback controller which ensures that a suitable norm $\|y - u\|$ is minimized. We perform the controller synthesis in two stages: in the first stage, we develop a feedback controller and in the second stage, we develop a method of generating the exciting input. The feedback controller is to ensure that (1) the system tracks the reference command satisfactorily, and (2) rejects the modeling uncertainties and exogenous disturbances satisfactorily. For this purpose, we choose $\mathcal{L}_1$ adaptive controller, which enables fast adaptation and provides guaranteed transient performance while preserving robustness of the control system. The $\mathcal{L}_1$ adaptive control theory was originally developed for the systems with fast computing capability (Hovakimyan and Cao 2010), which allow complicated mathematical calculations at relatively large speeds; however some of the $\mathcal{L}_1$ adaptive architectures can be suitable for implementation in chemical reactions. Namely, for the problem in this paper, we choose $\mathcal{L}_1$ adaptive controller with switching adaptation laws (Kharisov and Hovakimyan 2012). This architecture has adaptation laws with simple structure and does not require large values of any of the parameters or signals.

Let us consider a 2-phase clock. This can be realized as an interconnection network of two oscillators. We now describe how such a network can be synthesized as a modification of the Design 1 of the oscillators recently proposed by Kim and Winfree in (Kim and Winfree 2011). In (Kim and Winfree 2011), the wet-lab implemetation aspects of this design is discussed in detail. Hence, in this paper, we focus on the mathematical modifications only; a detailed description of the wet-lab implementation of our proposed clock network will shortly be presented separately.

Let x_1 and x_2 denote the outputs of the two oscillators. Then, the objective of the adaptive controller is to ensure that x_1 and x_2 track the given reference signals r_1 and r_2 with the performance specifications given by the *desired system*

$$\dot{x}_{\text{des}_1}(t) = \frac{1}{\tau^*}\big(r_1(t) - x_{\text{des}_1}(t)\big), \quad \dot{x}_{\text{des}_2}(t) = \frac{1}{\tau^*}\big(r_2(t) - x_{\text{des}_2}(t)\big), \tag{14.6}$$

where $x_{\text{des}_1}(t)$ and $x_{\text{des}_2}(t)$ are the desired values of the states x_1 and x_2 at time instant t, and τ^* is a nominal value of the uncertain system parameter τ. Since $\mathcal{L}_1$ adaptive controller ensures closeness of the control system to the desired system (14.6), we design the oscillation excitation scheme for the system (14.6). In this paper, to demonstrate feasibility of such approach we use simple nonlinear oscillation excitation of the following form:

$$r_1(t) = \mu_1(t) - k_\mu \mu_2(t) \quad r_2(t) = \mu_2(t) + k_\mu \mu_1(t),$$

$$\mu_1(t) = \int_0^t k_r \nu_1(\tau) d\tau \quad \mu_2(t) = \int_0^t k_r \nu_2(\tau) d\tau,$$

$$\dot{\nu}_1(t) = \begin{cases} 0, & \text{if } |\nu_1(t)| > \nu_{\max} \text{ and} \\ & \text{sign}(\nu_1(t)(x_1(t) - b)) > 0, \\ x_{\text{des}_1}(t) - b, & \text{otherwise}, \end{cases}$$

$$\dot{\nu}_2(t) = \begin{cases} 0, & \text{if } |\nu_2(t)| > \nu_{\max} \text{ and} \\ & \text{sign}(\nu_2(t)(x_2(t) - b)) > 0, \\ x_{\text{des}_2}(t) - b, & \text{otherwise}. \end{cases}$$

The block diagram of this scheme is shown in Fig. 14.3a. In this structure, the gain k_μ defines the relative phase of the oscillations. The magnitude and frequency depend on the choice of saturation level $\nu_{\max}$ and the gain k_r. The bias b defines the value around which the oscillations occur. The $\mathcal{L}_1$ adaptive controller is comprised of the state predictor, switching adaptation laws, and the control law. The *state predictor* is given by

$$\dot{\hat{x}}_1(t) = \frac{1}{\tau^*}\Big(\nu_1^* \Omega\big(r I_2, K_I, n\big) + \hat{\sigma}_1(t) - \hat{x}_1(t)\Big),$$

$$\dot{\hat{x}}_2(t) = \frac{1}{\tau^*}\Big(\nu_2^*(1 - \Omega\big(r A_1, K_A, m\big)\Big) + \hat{\sigma}_2(t) - \hat{x}_2(t)\Big),$$

where $\hat{x}_1(t)$ and $\hat{x}_2(t)$ are the predictions for $x_1(t)$ and $x_2(t)$ respectively, ν_1^* and ν_2^* are constants, and Ω is a Hill-type nonlinearity (see Design 1 of (Kim and Winfree 2011) for a complete description); here, $\hat{\sigma}_1(t) \in \mathbb{R}, \hat{\sigma}_2(t) \in \mathbb{R}$ are the adaptive estimates governed by the following *adaptation laws*:

$$\dot{\hat{\sigma}}_1(t) = -\Delta_\sigma \text{sgn}\left[\text{dz}_{\epsilon_\sigma}(\tilde{x}_1(t))\right] \quad \dot{\hat{\sigma}}_2(t) = -\Delta_\sigma \text{sgn}\left[\text{dz}_{\epsilon_\sigma}(\tilde{x}_2(t))\right],$$

where $\tilde{x}_1(t) \triangleq \hat{x}_1(t) - x_1(t)$, $\tilde{x}_2(t) \triangleq \hat{x}_2(t) - x_2(t)$; $\mathrm{sgn}(\cdot)$ and $\mathrm{dz}(\cdot)$ stand for sign and dead–zone functions; $\epsilon_\sigma \in \mathbb{R}^+$ is the dead–zone interval; and $\Delta_\sigma \in \mathbb{R}^+$ is a design constant. In $\mathcal{L}_1$ adaptive control theory the control signal performs compensation for the system uncertainty within the bandwidth of a lowpass filter. Notice that in our case the plant contains input nonlinearity. This nonlinearity is invertible within admissible control input $\left(K_I > 0 \text{ and } K_A > 0\right)$. Therefore to allow compensation for the system uncertainty, we use a virtual control signals $v_1(t)$ and $v_2(t)$ and define the systems control signals using the *nonlinear inversion compensation*:

$$K_I(t) = \frac{[rI_2](t)}{\left(\frac{1}{v_1(t)} - 1\right)^{\frac{1}{n}}}, \quad K_A(t) = \frac{[rA_1](t)}{\left(\frac{1}{1-v_2(t)} - 1\right)^{\frac{1}{m}}}. \tag{14.7}$$

The system uncertainty is compensated with the help of the following *control law*:

$$v_1(s) = k_{g_1} r_1(s) - C(s)\hat{\sigma}_1(s), \quad v_2(s) = k_{g_2} r_2(s) - C(s)\hat{\sigma}_2(s), \tag{14.8}$$

where $k_{g_1} = 1/v_1^*$, $k_{g_2} = 1/v_2^*$, and $C(s)$ is a stable strictly proper transfer function with unit dc gain $C(0) = 1$. As Fig. 14.3b shows, the system performs well in the absence of modeling uncertainties and exogenous disturbances. In the face of such uncertainties and disturbances, the system continues to function well, as will be seen in the Simulations section. One of the drawbacks of this approach to build a multi-phase clock is the necessity of an exciting periodic signal. In itself, that is not a major limitation since the consumption of entities in chemical reactions implies that there has to be a mechanism to inject chemicals externally in a synthetic biological circuit. We are implementing such a system as a modification of the oscillators derived by Kim and Winfree in (Kim and Winfree 2011) and those results will be presented separately.

14.2.2 Memory

To implement sequential computation, it is necessary to store and transfer signals across the clock cycles. In electronic systems, storage is typically implemented with flip-flips. In our molecular system, we implement the storage and transfer using a protocol that is synchronized on the two phases of the clock described earlier. Every memory unit S_i is assigned two molecular types D'_i and D_i. Here, D'_i is the first stage and D_i the second. The blue phase reactions are:

$$\begin{aligned} B + D_i &\xrightarrow{\text{fast}} \mathit{Computations} + B \\ \mathit{Computations} &\xrightarrow{\text{fast}} D'_j. \end{aligned} \tag{14.9}$$

Every unit S_i releases the signal it stores in its second stage D_i. The released signal is operated on by reactions in *computational modules*. These generate results and push them into the first stages of succeeding memory units. Note that D'_j molecules

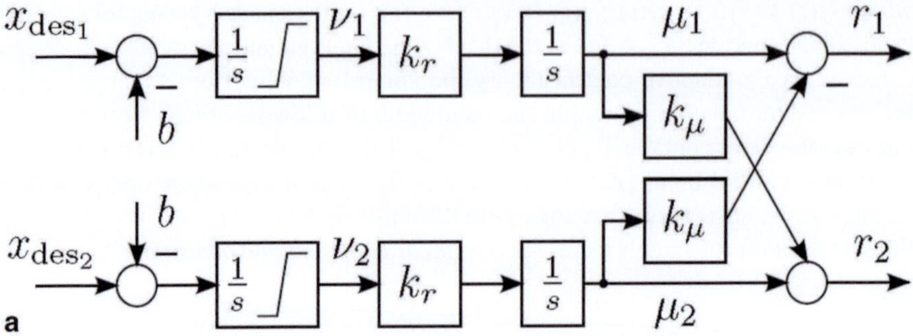

a

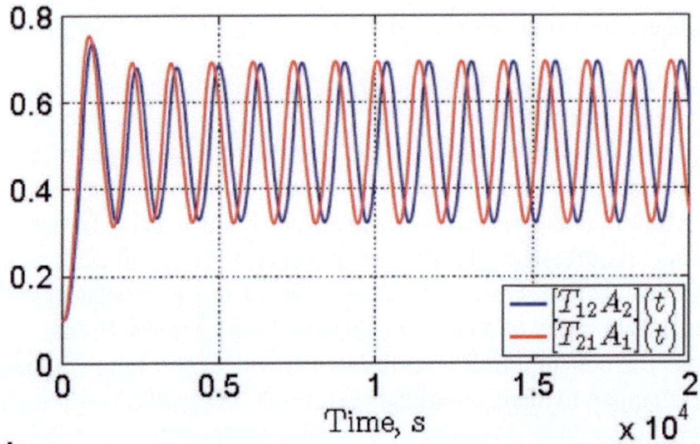

b

Fig. 14.3 **a** A block diagram of the reference signal generator for our adaptive controller to generate the mutli-phase clock. The gain k_μ defines the relative phase of the oscillations. The magnitude and frequency depend on the choice of saturation level $\nu_{\max}$ and the gain k_r. The bias b defines the value at which the oscillations occur. **b** Simulation results for the nominal system show that the 2-phase clock performs well in the absence of modeling uncertainties and exogenous disturbances. The clock performance in the face of modeling uncertainties and disturbances is discussed in the simulations section

will be the first stage of any succeeding memory unit S_j along the signal path from S_i. The red phase reactions are:

$$R + D'_j \xrightarrow{\text{fast}} D_j + R. \tag{14.10}$$

Every unit S_j transfers the signal it stores in D'_j to D_k, preparing for the next cycle. For the equivalent of delay (D) flip-flops in digital logic, $j = k$. For other types of memory units, j and k can be different. For example, for a toggle (T) flip-flop, S_k is the complementary bit of S_j: $D'_j \longrightarrow D_k$ and $D'_k \longrightarrow D_j$ toggle the pair of bits in each clock cycle. The transfer diagram for our memory design is shown in Fig. 14.1b.

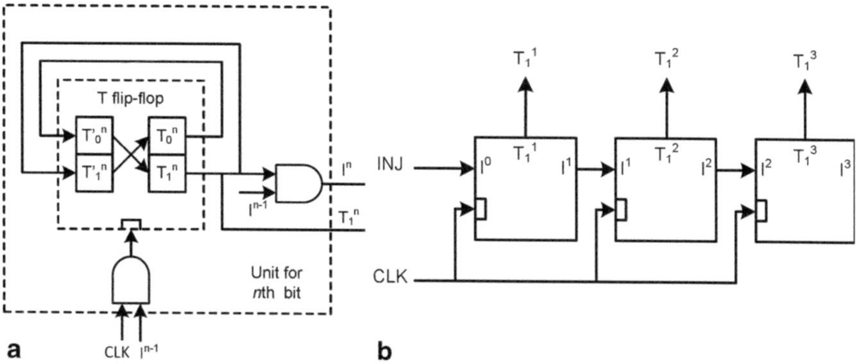

Fig. 14.4 The binary counter. **a** Block diagram of one stage of the counter. **b** Connections between the stages

14.2.3 A Binary Counter

We next present a three-bit binary counter; this design can readily be generalized to an arbitrary number of bits. The counter consists of three identical stages, each of which processes a single bit. The block diagram of a stage is shown in Fig. 14.2a. In the n-th stage, there are two memory units T_1^n and T_0^n that form a T flip-flop. The presence of molecules of T_1^n indicates that this bit is logical one; the presence of molecules of T_0^n indicates that this bit is logical zero. If we provide a non-zero initial concentration to one of the two types, then either T_0^n or T_1^n will always be present. Using the memory implementation discussed earlier, we have types T'^n_1 and T'^n_0 as the first stages of the memory units. The red phase reactions are:

$$R + T'^n_1 \xrightarrow{\text{fast}} T_0^n + R, \quad R + T'^n_0 \xrightarrow{\text{fast}} T_1^n + R. \tag{14.11}$$

These toggle each bit. The blue phase reactions are:

$$I^{n-1} + T_0^n \xrightarrow{\text{fast}} T'^n_0, \quad I^{n-1} + T_1^n \xrightarrow{\text{fast}} T'^n_1 + I^n. \tag{14.12}$$

These feed the output of each T flip-flop back to its input. Note that the T flip-flops transfer molecules only when there are molecules of I^{n-1} injected from the previous stage. If the bit is logical one, i.e., T_1^n is present, then molecules of I^n are injected into the next stage. Figure 14.2a illustrate three connected stages. The reaction

$$B + Inj \xrightarrow{\text{fast}} I^0 + B \tag{14.13}$$

transfers the external injection Inj to I^0 in the blue phase. Since this reaction is the very first of all computational reactions, all injection signals I^n are generated in the blue phase. Accordingly, B is not required in the reaction (14.12), so long as I^n is present among the reactants.

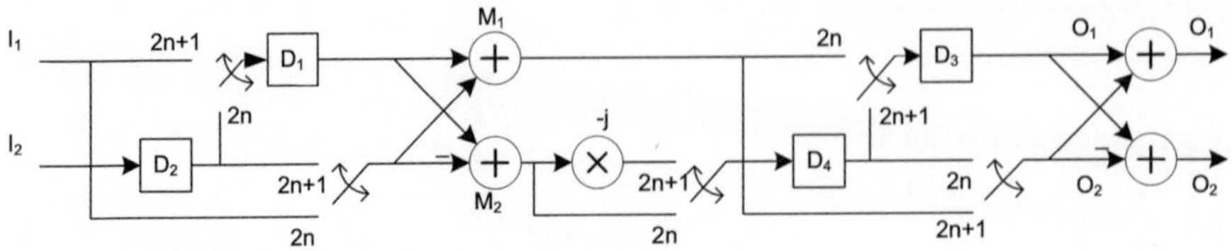

Fig. 14.5 Block diagram of a 4-point pipelined FFT design

14.2.4 A Two-Parallel FFT Design

We next present a molecular implementation of a four-point two-parallel FFT. The FFT operation is commonplace in signal processing. It can have a parallel pipelined architectures for high throughput (Parhi 1999). A block diagram is shown in Fig. 14.5. Assume that the system starts at clock cycle 1. The first two inputs are sampled in cycle 1; the last two inputs are sampled in cycle 2. The system generates the first and third outputs in cycle 3; it generates the other two outputs in cycle 4. There are four switches in this design. Each selects one of the two incoming signals alternatively in different cycles. To achieve this switching functionality in our molecular design, we use two alternating selection signals. We generate these with a pair of D-flip-flops, as shown in Fig. 14.6a. If there is a non-zero initial concentration of S'_1, then S'_0 and S'_1 will be "turned on"once every two cycles, in alternating fashion, starting with S'_1. Note that it is S'_1 and S'_0, not S_1 and S_0, that enable the switches, because they are generated in the blue phase. We could implement this computation with the following reactions, based on the signal transfer principles discussed above:

$$R + S'_0 \xrightarrow{k_{slow}} S_0 + R, R + S'_1 \xrightarrow{k_{slow}} S_1 + R,$$
$$B + S_1 \xrightarrow{k_{slow}} S'_0 + B, B + S_0 \xrightarrow{k_{slow}} S'_1 + B. \tag{14.14}$$

However, there is a problem with this implementation. Enabling signals S'_0 and S'_1 are transferred to S_0 and S_1 by R. It takes time to finish these transfers. Therefore, there will always be overlapped R and S'_0/S'_1. Since S'_0 and S'_1 should only be present during B phase, this overlapping causes computation errors. To cope with this problem, we change the set of reactions given by (14.14) to:

$$V + S'_0 \xrightarrow{k_{fast}} S_0 + V, V + S'_1 \xrightarrow{k_{fast}} S_1 + V,$$
$$B + S_1 \xrightarrow{k_{slow}} S'_0 + B, B + S_0 \xrightarrow{k_{slow}} S'_1 + B. \tag{14.15}$$

Here, V is the clock phase signal following B. S'_0 and S'_1 are quickly transferred to S_0 and S_1 right after B phase ends. With this modification, overlapping between S'_0/S'_1 and R is minimized. Clock phases and enabling switch signals are illustrated in Fig. 14.6b.

The transfer reactions enabled by S'_1 or S'_0 implement the switches. In this system, signals are complex numbers. Both the real and the imaginary parts can be negative

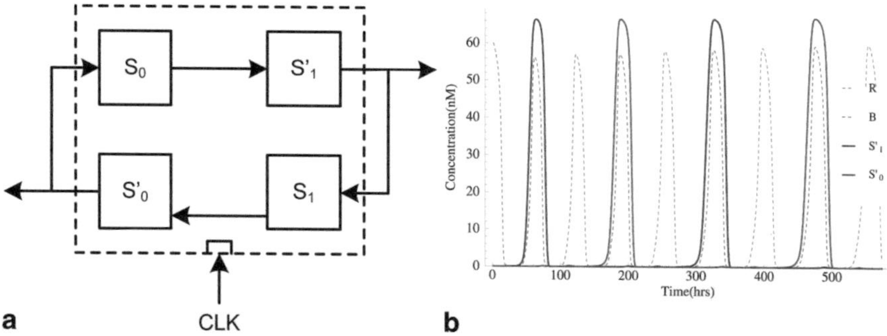

Fig. 14.6 Implementation of FFT using molecular reaction. **a** Generating the selection signals. **b** Clock phases and selection signals

numbers. To represent the signals, each number X is assigned four molecular types X_p, X_n, X_p^*, and X_n^*. The first two are assigned to the real parts: X_p represents the positive component and X_n the negative component. The last two are assigned to the imaginary parts: X_p^* represents the positive component and X_n^* the negative component. Therefore, $X = [X_p] - [X_n] + j([X_p^*] - [X_n^*])$. Adders are implemented by assigning input edges and output edges to the same molecular type (Jiang et al. 2010). Note that there are two negative input edges in the lower two adders. Signals from the negative input edge will be transferred to the opposite component. For example, at the $n+1$st clock cycle, M_1 is transferred to O_1 and O_2 as:

$$S'_1 + M_{1,p} \xrightarrow{\mathrm{k_{slow}}} O_{1,p} + O_{2,n} + S'_1, \quad S'_1 + M_{1,n} \xrightarrow{\mathrm{k_{slow}}} O_{1,n} + O_{2,p} + S'_1,$$

$$S'_1 + M^*_{1,p} \xrightarrow{\mathrm{k_{slow}}} O_{1,p} + O^*_{2,n} + S'_1, \quad S'_1 + M^*_{1,n} \xrightarrow{\mathrm{k_{slow}}} O_{1,n} + O^*_{2,p} + S'_1.$$

or, alternatively, as:

$$S'_1 + M_1 \xrightarrow{\mathrm{k_{slow}}} O_1 + O_2^- + S'_1.$$

Also, for each number X, the reactions

$$X_p + X_n \xrightarrow{\mathrm{k_{fast}}} \varnothing, \quad X_p^* + X_n^* \xrightarrow{\mathrm{k_{fast}}} \varnothing$$

are required. These cancel out equal concentrations of positive and negative components by transferring them to an external sink. There is a $-j$ multiplication in the system. It is implemented by

$$S'_1 + M_{2,p} \xrightarrow{\mathrm{k_{fast}}} D'^*_{4,n} + S'_1, \quad S'_1 + M_{2,n} \xrightarrow{\mathrm{k_{fast}}} D'^*_{4,p} + S'_1,$$
$$S'_1 + M^*_{2,p} \xrightarrow{\mathrm{k_{fast}}} D'_{4,n} + S'_1, \quad S'_1 + M^*_{2,n} \xrightarrow{\mathrm{k_{fast}}} D'_{4,p} + S'_1,$$

or, alternatively, as

$$S'_1 + M_2 \xrightarrow{\mathrm{k_{fast}}} D'^{-*}_4 + S'_1$$

which transfers real/imaginary parts to imaginary/real parts with opposite polarity. Based on the computational operations discussed above, we have the blue phase reactions as:

$$I_1 + S'_1 \xrightarrow{\mathrm{k_{fast}}} D'_1 + S'_1, \quad I_1 + S'_0 \xrightarrow{\mathrm{k_{fast}}} M_1 + M_2^- + S'_0, \quad I_2 + B \xrightarrow{\mathrm{k_{fast}}} D'_2 + B,$$

$$D_2 + S'_0 \xrightarrow{\mathrm{k_{fast}}} D'_1 + S'_0, \quad D_2 + S'_1 \xrightarrow{\mathrm{k_{fast}}} M_1 + M_2^- + S'_1,$$

$$D_1 + B \xrightarrow{\mathrm{k_{fast}}} M_1 + M_2 + B,$$

$$M_2 + S'_1 \xrightarrow{\mathrm{k_{fast}}} D'^{-*}_4 + S'_1, \quad M_2 + S'_0 \xrightarrow{\mathrm{k_{fast}}} D'_4 + S'_0, \quad M_1 + S'_0 \xrightarrow{\mathrm{k_{fast}}} D'_3 + S'_0$$

$$M_1 + S'_1 \xrightarrow{\mathrm{k_{fast}}} O_1 + O_2^- + S'_1, \quad D_3 + B \xrightarrow{\mathrm{k_{fast}}} O_1 + O_2 + B,$$

$$D_4 + S'_0 \xrightarrow{\mathrm{k_{fast}}} O_1 + O_2^- + S'_0, \quad D_4 + S'_1 \xrightarrow{\mathrm{k_{fast}}} D'_3 + S'_1.$$

Note that S'_0 and S'_1 are generated in the blue phase. It is not necessary to list B if a reaction is enabled by S'_0 or S'_1. The red phase reactions are

$$R + D'_1 \xrightarrow{\mathrm{k_{fast}}} D_1 + R, \quad R + D'_2 \xrightarrow{\mathrm{k_{fast}}} D_2 + R$$

$$R + D'_3 \xrightarrow{\mathrm{k_{fast}}} D_3 + R, \quad R + D'_4 \xrightarrow{\mathrm{k_{fast}}} D_4 + R.$$

So the full design of the four-point, two-parallel FFT consists of the above reactions together with the positive/negative canceling reactions as well as the clock generation reactions.

14.2.5 Simulation Case Studies

We present simulation results for the clock, binary counter, and the FFT. For each design, wherever necessary, we list our choice of kinetic constants corresponding to “slow”and “fast”as well as the initial concentrations of the molecular types. We assume that an external source sets the concentrations of the input types to new values at specific intervals. We setup ordinary differential equations corresponding to the mass-action kinetics of the reactions and solve these numerically with MATLAB.

For the simulations we consider the system dynamics with the following parameters $k_p = 0.04$, $k_d = 0.002$, $\tau^* = 500$, $n = m = 5$, $\left[T_{12}^{tot}\right]^* = \left[T_{21}^{tot}\right]^* = 100$; this choice is consistent with Design 1 of (Kim and Winfree 2011). We choose the following $\mathcal{L}_1$ adaptive controller parameters: $\Delta_\sigma = 0.3$, $dz_{\epsilon_\sigma} = 10^{-5}$ along with the following lowpass filter:

$$C(s) = \frac{1}{2s+1}\frac{1}{1s+1}\frac{1}{0.1s+1}.$$

We also set the following parameters of the oscillation excitation scheme: $k_\mu = 0.1$, $k_r = -10^{-5}$, $\nu_{\max} = 2$, $b = 0.5$. For the simulations, we choose the following initial conditions of the plant: [0.1 0.1 0.1 0.1] and of the state predictor $\hat{x}(0) = [0.1\ 0.1]$. Consider the following three scenarios of parametric uncertainties:

- Scenario 1: Nominal system, i.e., no uncertainty. Here, $\tau = \tau^*$, $\nu_1 = \nu_2 = \nu_1^* = \nu_2^*$.
- Scenario 2: Uncertainty is quatified as $\tau = 600$, $\nu_1 = 130$, $\nu_2 = 70$.
- Scenario 3: Uncertainty is quantified as $\tau = 450$, $\nu_1 = 70$, $\nu_2 = 120$.

First, we demonstrate performance of the oscillation excitation scheme. For this, we consider Scenario 1 (nominal case) and turn off the $\mathcal{L}_1$ adaptive controller, i.e., in this case, $\hat{\sigma}(t) \equiv 0$. The simulation results are shown in Fig. 14.7. We can see that the closed loop system achieves sustainable oscillations with desired magnitude, frequency and phase. The influence of the uncertainty to the system performance in the absence of $\mathcal{L}_1$ adaptive controller is illustrated in Figs. 14.8a, and 14.9b. In both cases, the system is not able to maintain synchronized oscillations and its performance is significantly affected by the plant uncertainty. The reason for that is in poor robustness of the oscillation excitation scheme in the absence of $\mathcal{L}_1$ adaptive controller.

Figures 14.8b, and 14.9b show the simulation results for Scenarios 2 and 3 with $\mathcal{L}_1$ adaptive controller turned on. It can be observed that when the $\mathcal{L}_1$ adaptive controller is turned on, the clock performance is very close to the performance of the ideal system, i.e., the nominal system in Scenario 1. This demonstrates that the $\mathcal{L}_1$ adaptive controller well compensated for the existing plant uncertainty and recovered the robustness and performance of the clock network.

14.2.5.1 Counter

For the three-bit counter, we set the initial concentrations of T_0^0, T_0^1, and T_0^2 to 10 (corresponding to bits "000") and R to 100. We set the initial concentrations of all the other molecular types to 0. We set the concentration of type Inj to 10 at time points 50, 500, 1000, 1500, 2000, 2500, and 3000. We set k_{fast} to 100; and k_{slow} to 1. The results of a MATLAB ODE simulation are shown in Fig. 14.10.

We see that the bit signals T_1^0, T_1^1, and T_1^2 toggle at the correct time points. The system counts the number of injection events from "000" to "111" correctly.

One observation from Fig. 14.10 is that the concentrations of T_1^0, T_1^1, and T_1^2 when the corresponding bit is "1"degrade slowly over time. This is because of slightly overlapped clock phases.

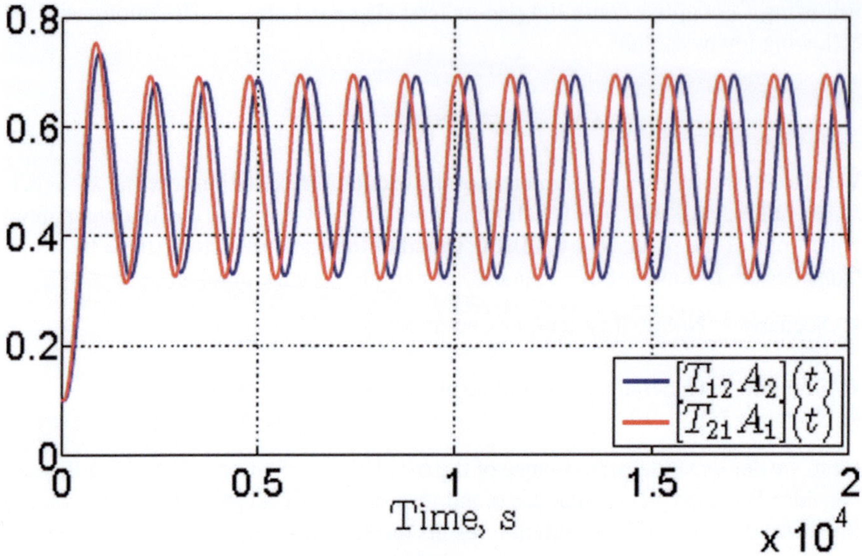

a **System output**

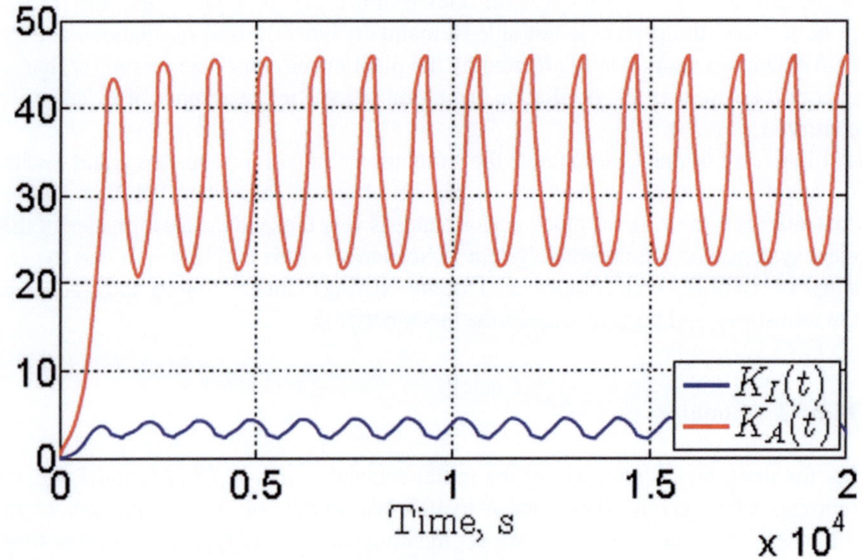

b **Control history**

Fig. 14.7 Simulation results for Scenario 1 with $\mathcal{L}_1$ adaptive controller turned off

There is always a slight leakage in the amount of R in the blue phase and a slight leakage in the amount of B in the red phase. This error accumulates over time, due

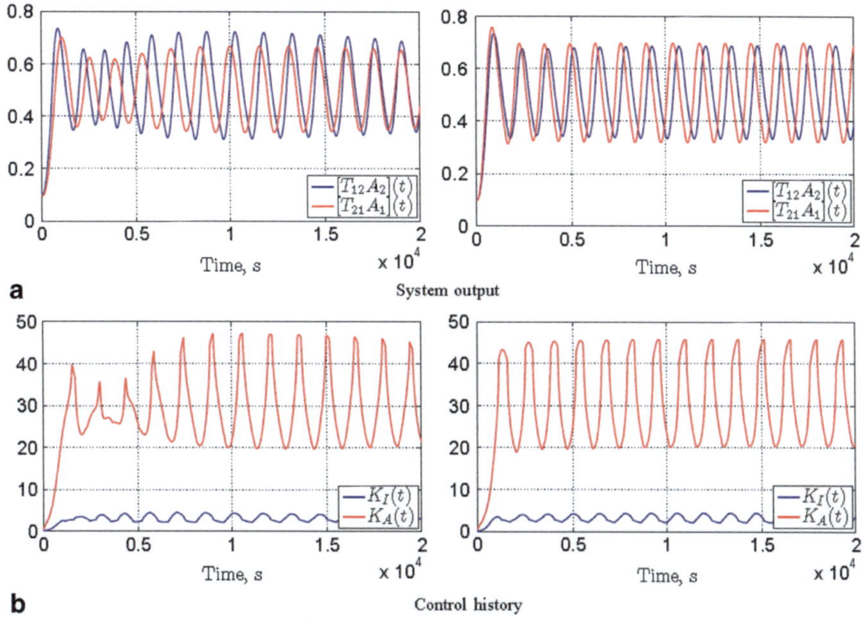

Fig. 14.8 **a** Simulation results for Scenario 2 with $\mathcal{L}_1$ adaptive controller turned off. **b** Simulation results for Scenario 2 with $\mathcal{L}_1$ adaptive controller turned on

to the feedback loop in each stage of the counter. To mitigate against this, we could select a higher ratio of $\lambda = \frac{k_{\text{fast}}}{k_{\text{slow}}}$.

14.2.5.2 FFT

For our FFT design, we the set initial concentration of S'_0 to 50 and that of R to 100. Recall that S'_0 is transferred to S_0 in the first red phase and S_0 is transferred to S'_1 in first blue phase. So the computation begins with S'_1. We set the initial concentrations of all the other types to 0. We inject I_1 and I_2 in the red phase. The output types O_1 and O_2 are produced in clock cycle 3, in a blue phase. We clear them out in following red phase. We set k_{fast} to 100; and k_{slow} to 1. The results of a MATLAB ODE simulation are shown in Fig. 14.11. The inputs are a sequence of real numbers {10, 15, 10,0}. The outputs are {35, −15j, 5, 15j}, as shown in the figure.

Since there is no feedback in the FFT architecture, no error due to the leakage of R and B accumulates. We analyzed the computational errors of the counter and the FFT design in terms of the fast-to-slow ratio λ. For the counter, the error is defined as the differences of concentrations from a perfect "0"or a perfect "1". We consider the average error accumulated in one clock cycle for one bit. For the FFT design, we consider the relative error of the simulated outputs compared to theoretical outputs. These errors are listed in Table 14.1. As expected, we see that the error decreases

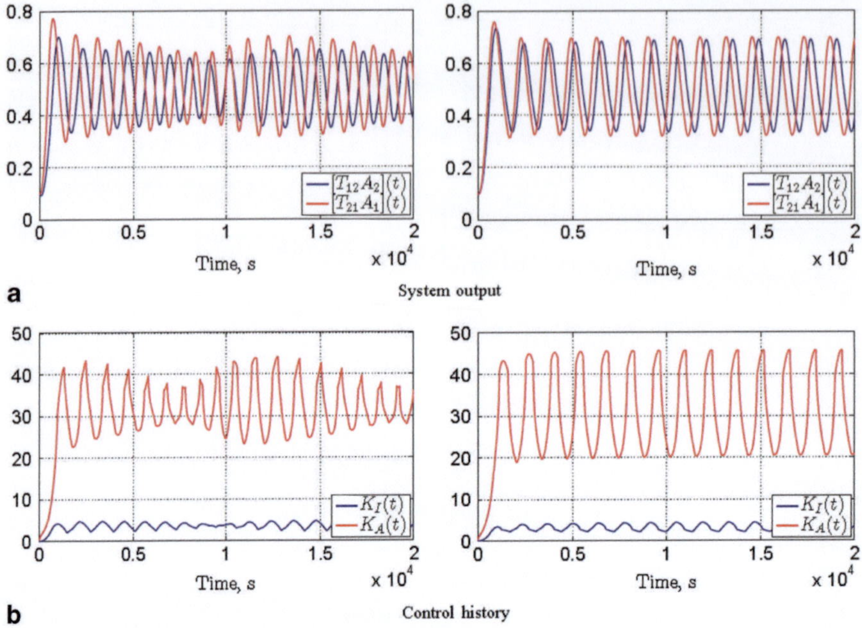

Fig. 14.9 **a** Simulation results for Scenario 3 with $\mathcal{L}_1$ adaptive controller turned off. **b** Simulation results for Scenario 3 with $\mathcal{L}_1$ adaptive controller turned on

Table 14.1 Relative error in simulations

λ	Counter (Average error per cycle per bit) (%)	FFT (%)
10	0.9871	28.107
100	0.2078	3.5428
1000	0.0169	0.2691

as λ increases since a higher fast-to-slow ratio causes fewer reactions to fire in the incorrect clock phase.

14.3 Conclusion

We have presented a robust, rate-independent methodology for implementing synchronous sequential computations using molecular reactions. Our methodology is rate independent in the sense that the computations performed by our molecular system are exact over a broad range of values for the kinetic constants and, furthermore, are independent of the specific values for the reaction rates. The results in this paper are complementary to the results on *self-timed* asynchronous methodologies for molecular computation presented in (Jiang et al. 2010). As in electronic circuit

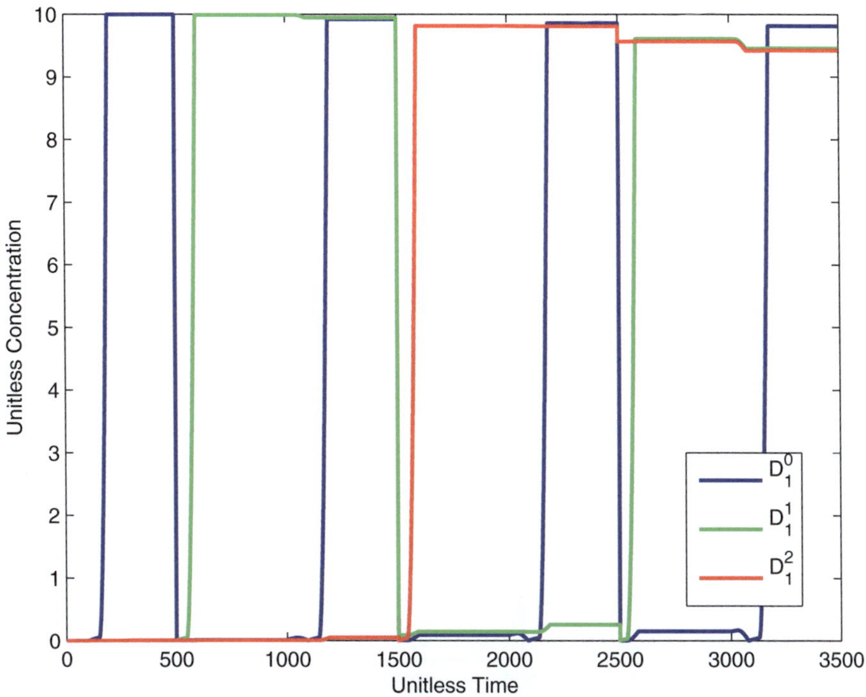

Fig. 14.10 Transient simulation result of the counter

design, there are advantages and disadvantages to asynchronous and synchronous design styles for molecular computing. A synchronous implementation leads to simpler designs with fewer reactions but suffers from the drawback of the error accumulations across the clock cycles. We have presented two different mechanisms for synthesizing multi-phase clocks. Our first mechanism is a stand-alone mechanism in the sense that the clock signal can be generated in response to the initial conditions of the chemicals in the system. This mechanism is very simple to implement but suffers from a lack of robustness and tunability. In addition, since the chemicals in a closed system degrade with time, this approach cannot be relied upon to generate a clock signal that can last for arbitrarily long periods. Our second mechanism overcomes these limitations by using an $\mathcal{L}_1$-adaptive controller. We have demonstrated that a well-known oscillator network recently synthesized by Kim and Winfree (see Design 1 of Kim and Winfree 2011)) can be rendered tunable and robust using our $\mathcal{L}_1$ adaptive controller. However, this approach requires an exogenous periodic signal as an input to the clock network. We are separately presenting details on the required wet-lab implementation set-up to realize such a network of clocks. In addition, we have proposed molecular reactions to implement memory and binary counters, and have validated our designs *in silico*. We have illustrated the utility of these building blocks by synthesizing a four-point two-parallel FFT using molecular reactions.

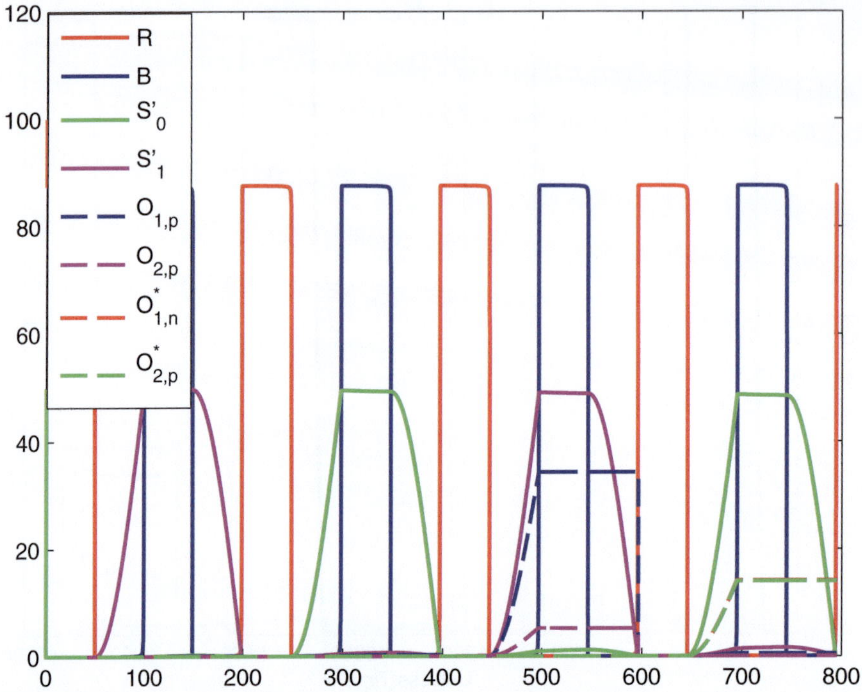

Fig. 14.11 Transient simulation of the FFT design

We are exploring the mechanism of DNA strand-displacement as an experimental chassis (Soloveichik et al. 2010). DNA strand-displacement reactions can emulate chemical reactions with nearly any rate structure. Reaction rates are controlled by designing sequences with different binding strengths. The binding strengths are controlled by the length and sequence composition of "toehold"sequences. With the right choice of toehold sequences, reaction rates differing by as much as 10^6 can be achieved. Our contribution can be positioned as the "front-end"of the design flow – analogous to technology-independent design. DNA assembly can be considered the "back-end"– analogous to technology mapping to a specific library.

Acknowledgments This research is supported, in parts, by the NSF CAREER Award 0845650, NSF CCF 0946601, NSF CCF 1117168, and AFOSR.

Appendix: $\mathcal{L}_1$ Controller to Synthesize a Multi-Phase Clock

Let the functions sgn(ξ) and dz(ξ) denote the "sign"and the "dead–zone"functions respectively. An instance of the dead–zone function is shown in Figure 14.12a; the theory used by us is valid for some other definitions of the dead–zone function as well.

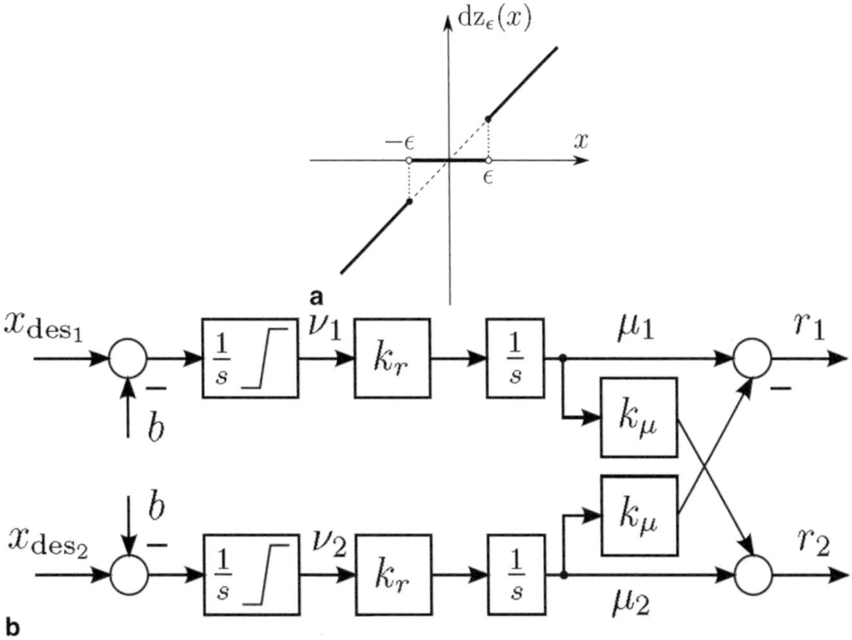

Fig. 14.12 **a** A deadzone nonlinearity. **b** A block diagram of the reference signal generator for our adaptive controller to generate the mutli-phase clock. The gain k_μ defines the relative phase of the oscillations. The magnitude and frequency depend on the choice of saturation level $\nu_{\max}$ and the gain k_r. The bias b defines the value at which the oscillations occur

If ξ is a vector, then we assume that these functions are applied to each component of the vector independently. For example

$$\text{sgn}\Big(\big[\xi_1\ \xi_2\ \xi_3\big]^\top\Big) = \Big[\text{sgn}(\xi_1)\ \text{sgn}(\xi_2)\ \text{sgn}(\xi_3)\Big]^\top .$$

With a slight abuse of notation, we shall interchangeably uses time–domain and frequency–domain notation to refer to signals. For example, $\xi(t)$ denotea ξ as a function of time and $\Xi(s)$ denotes its Laplace transform. We approach controller synthesis via a 2-stage process. In the first stage, we develop a feedback controller. In the second stage, we develop a method of exciting sustainable oscillations. The feedback controller in our design plays two roles: (1) it is used to ensure that the system tracks the reference commands generated by oscillation excitation block, and (2) it is used to compensate for the environmental uncertainties and disturbances to reduce their effect on the closed-loop system behavior. For this purposes we choose $\mathcal{L}_1$ adaptive controller, which enables fast adaptation and provides guaranteed transient performance while preserving robustness of the control system.

The $\mathcal{L}_1$ adaptive control theory was originally developed for the systems with fast computing capability (Hovakimyan and Cao 2010), which allow complicated mathematical calculations at relatively large speeds; however some of the $\mathcal{L}_1$ adaptive architectures can be suitable for implementation in chemical reactions. Namely, for the problem in this paper, we choose $\mathcal{L}_1$ adaptive controller with switching adaptation laws (Kharisov and Hovakimyan 2012). This architecture has adaptation laws with simple structure and does not require large values of any of the parameters or signals. Thus, the adaptive controller is designed to ensure that the system outputs x_1, x_2 track the given reference signals $r_1(t)$ and $r_2(t)$ with the performance specifications given by the *desired system*

$$\dot{x}_{\text{des}_1}(t) = \frac{1}{\tau^*}\big(r_1(t) - x_{\text{des}_1}(t)\big), \;\; \dot{x}_{\text{des}_2}(t) = \frac{1}{\tau^*}\big(r_2(t) - x_{\text{des}_2}(t)\big), \tag{14.16}$$

where $x_{\text{des}_1}(t)$ and $x_{\text{des}_2}(t)$ are the desired values of the states x_1 and x_2, respectively, and τ^* is a nominal value of the uncertain system parameter τ.

$\mathcal{L}_1$ *Control System Architecture*

Since $\mathcal{L}_1$ adaptive controller ensures closeness of the control system to the desired system 14.16, we design the oscillation excitation scheme for the system (14.16). To demonstrate feasibility of such an approach, we use simple nonlinear oscillation excitation of the following form:

$$r_1(t) = \mu_1(t) - k_\mu \mu_2(t), \;\; r_2(t) = \mu_2(t) + k_\mu \mu_1(t)$$

$$\mu_1(t) = \int_0^t k_r \nu_1(\tau)d\tau, \;\; \mu_2(t) = \int_0^t k_r \nu_2(\tau)d\tau$$

$$\dot{\nu}_1(t) = \begin{cases} 0\,, & \begin{array}{l}\text{if } |\nu_1(t)| > \nu_{\max} \text{ and} \\ \text{sign}\Big(\nu_1(t)\big(x_1(t) - b\big)\Big) > 0\,,\end{array} \\ x_{\text{des}_1}(t) - b\,, & \text{otherwise}\,, \end{cases}$$

$$\dot{\nu}_2(t) = \begin{cases} 0\,, & \begin{array}{l}\text{if } |\nu_2(t)| > \nu_{\max} \text{ and} \\ \text{sign}\Big(\nu_2(t)\big(x_2(t) - b\big)\Big) > 0\,,\end{array} \\ x_{\text{des}_2}(t) - b\,, & \text{otherwise}\,. \end{cases}$$

The block diagram of this scheme is shown in Fig. 14.12b. The $\mathcal{L}_1$ adaptive controller is comprised of the state predictor, switching adaptation laws, and the control law. The *state predictor* is given by

$$\dot{\hat{x}}_1(t) = \frac{1}{\tau^*}\Big(\big[T_{12}^{tot}\big]^*\Omega\big(rI_2, K_I, n\big) + \hat{\sigma}_1(t) - \hat{x}_1(t)\Big),$$

$$\dot{\hat{x}}_2(t) = \frac{1}{\tau^*}\Big(\big[T_{21}^{tot}\big]^*\big(1 - \Omega\big(rA_1, K_A, m\big)\big) + \hat{\sigma}_2(t) - \hat{x}_2(t)\big),$$

where $\hat{x}_1(t)$ and $\hat{x}_2(t)$ are the predictions for $[T_{12}A_2](t)$ and $[T_{21}A_1](t)$ respectively; and $\hat{\sigma}_1(t) \in \mathbb{R}$, $\hat{\sigma}_2(t) \in \mathbb{R}$ are the adaptive estimates governed by the following *adaptation laws*:

$$\hat{\sigma}_1(t) = -\Delta_\sigma \operatorname{sgn}\left[\operatorname{dz}_{\epsilon_\sigma}(\tilde{x}_1(t))\right],$$
$$\hat{\sigma}_2(t) = -\Delta_\sigma \operatorname{sgn}\left[\operatorname{dz}_{\epsilon_\sigma}(\tilde{x}_2(t))\right],$$

where $\tilde{x}_1(t) \triangleq \hat{x}_1(t) - [T_{12}A_2](t)$, $\tilde{x}_2(t) \triangleq \hat{x}_2(t) - [T_{21}A_1](t)$; sgn$(\cdot)$ and dz$(\cdot)$ stand for sign and dead–zone functions; $\epsilon_\sigma \in \mathbb{R}^+$ is the dead–zone interval; and $\Delta_\sigma \in \mathbb{R}^+$ is a design constant.

In $\mathcal{L}_1$ adaptive control theory the control signal performs compensation for the system uncertainty within the bandwidth of a lowpass filter. Notice that in our case the plant contains an input nonlinearity. This nonlinearity is invertible within admissible control input ($K_I > 0$ and $K_A > 0$). Therefore, to allow compensation for the system uncertainty, we use a virtual control signals $v_1(t)$ and $v_2(t)$ and define the systems control signals using the *nonlinear inversion compensation*:

$$K_I(t) = \frac{[rI_2](t)}{\left(\frac{1}{v_1(t)} - 1\right)^{\frac{1}{n}}}, \quad K_A(t) = \frac{[rA_1](t)}{\left(\frac{1}{1-v_2(t)} - 1\right)^{\frac{1}{m}}}. \tag{14.17}$$

Notice that upon substituting these control signal into the system equations, we obtain

$$\frac{dx_1}{dt}(t) = \frac{1}{\tau}\big(\nu_1 v_1(t) - \nu_1\big), \quad \frac{dx_2}{dt}(t) = \frac{1}{\tau}\big(\nu_1 v_2(t) - \nu_2\big).$$

The system uncertainty due to variations of parameters of the above equation is compensated with the help of the following *control law*:

$$v_1(s) = k_{g_1} r_1(s) - C(s)\hat{\sigma}_1(s), \quad v_2(s) = k_{g_2} r_2(s) - C(s)\hat{\sigma}_2(s), \tag{14.18}$$

where $k_{g_1} = 1/\nu_1^*$, $k_{g_2} = 1/\nu_2^*$, and $C(s)$ is a stable strictly proper transfer function with unit dc gain $C(0) = 1$.

Stability and Performance Bounds for the $\mathcal{L}_1$ Adaptive Controller

Similar to all $\mathcal{L}_1$ adaptive control architectures from (Hovakimyan and Cao 2010), the analysis can be performed by defining the $\mathcal{L}_1$ reference system, which incorporates the low pass filter and assumes compensation of the system uncertainties only within the available bandwidth of the control channel. Then, the performance bounds can be computed as the distance between the $\mathcal{L}_1$ reference system and the closed–loop adaptive control system for both the system output and the control input. The $\mathcal{L}_1$ reference system is given by

$$\frac{dx_I}{dt} = \frac{1}{\tau^*}\big(\nu_1^* v_1(t) - x_1(t) + \sigma_1(t)\big),$$

$$\frac{dx_2}{dt} = \frac{1}{\tau^*}\big(\nu_2^* v_2(t) - x_2(t) + \sigma_2(t)\big),$$
$$v_1(s) = C(s)(r_1(s) - \sigma_1(s)),$$
$$v_2(s) = C(s)(r_2(s) - \sigma_2(s)),$$

where

$$\sigma_1(t) = \left(\frac{\tau^*}{\tau}[\nu_1 - \nu_1^*\right) v_1(t), \quad \sigma_2(t) = \left(\frac{\tau^*}{\tau}[\nu_2 - \nu_2^*\right) v_2(t).$$

Notice that the $\mathcal{L}_1$ reference system involves the system uncertainty in its definition. Therefore, it can be used only for analysis purposes. This fact also implies that the stability of the $\mathcal{L}_1$ reference system is not guaranteed apriori. Following the same steps of the proof in Sect. 2.4 of (Hovakimyan and Cao 2010) the stability of the $\mathcal{L}_1$ reference system can be ensured *locally* by $\mathcal{L}_1$-norm condition similar to the Sect. 2.4. The derivations and precise equation for the $\mathcal{L}_1$-norm stability condition will be given elsewhere. If the $\mathcal{L}_1$ reference system is stable, then the performance bounds between both the system output and the control signals of the closed–loop adaptive system and the $\mathcal{L}_1$ reference system are given by

$$\|x_1^{\text{rf}} - x_1\|_{\mathcal{L}_\infty} \le \gamma_{x_1}, \ \|x_2^{\text{rf}} - x_2\|_{\mathcal{L}_\infty} \le \gamma_{x_2}, \ \|v_1 - v_1^{\text{rf}}\|_{\mathcal{L}_\infty} \le \gamma_{v_1}, \ \|v_2 - v_2^{\text{rf}}\|_{\mathcal{L}_\infty} \le \gamma_{v_2},$$

where γ_* are computable bounds. These bounds can be obtained by combining the proofs of theorem 1 from (Kharisov and Hovakimyan 2012) and (J. Vanness et al. 2012). The proofs along with precise equations for the bounds γ_* will be given elsewhere. Due to the nature of the input nonlinearity, only local results can be achieved. In other words, the system states must remain positive-valued, which can be achieved by applying $r_1(t)$ and $r_2(2)$ within the admissible set. The precise bounds on the reference commands can be derived from the $\mathcal{L}_1$ reference system along with 14.7 for given conservative knowledge of the uncertainty.

References

Adleman L (1994) Molecular computation of solutions to combinatorial problems. Science 266(11):1021–1024

Anderson JC, Clarke EJ, Arkin AP, Voigt CA (2006) Environmentally controlled invasion of cancer cells by engineered bacteria. J Mol Biol 355(4):619–627

Anderson JC, Voigt CA, Arkin AP (2007) Environmental signal integration by a modular AND gate. Mol Syst Biol 3(133)

Arkin A, Ross J (1994) Computational functions in biochemical reaction networks. Biophys J 67(2):560–578

Benenson Y, Gil B, Ben-Dor U, Adar R, Shapiro E (2004) An autonomous molecular computer for logical control of gene expression. Nature 429(6990):423–429

Epstein IR, Pojman JA (1998) An introduction to nonlinear chemical dynamics: oscillations, waves, patterns, and chaos. Oxford University Press

Érdi P, Tóth J (1989) Mathematical models of chemical reactions: theory and applications of deterministic and stochastic models. Manchester University Press

Horn F, Jackson R (1972) General mass action kinetics. Arch Rational Mech Anal 47:81–116

Hovakimyan N, Cao C (2010) $\mathcal{L}_1$ adaptive control theory. Society for Industrial and Applied Mathematics, Philadelphia

Jiang H, Kharam AP, Riedel MD, Parhi KK (2010) A synthesis flow for digital signal processing with biomolecular reactions. IEEE International Conference on Computer-Aided Design, pp 417–424

Jiang H, Riedel MD, Parhi KK (2011) Synchronous sequential computation with molecular reactions. Design Automation Conference, pp 836–841

Kepper PD, Epstein IR, Kustin K (2008) A systematically designed homogeneous oscillating reaction: the arsenite-iodate-chlorite system. J Am Chem Soc 103(8):2133–2134

Kim J, Winfree E (2011) Synthetic in vitro transcriptional oscillators. Mol Syst Biol 7(465)

Kharisov E, Hovakimyan N (2012) Generalization of $\mathcal{L}_1$ adaptive control architecture for switching estimation laws. In: American Control Conference, Montréal, Canada, June 2012

Parhi KK (1999) VLSI digital signal processing systems. Wiley

Qian L, Soloveichik D, Winfree E (2010) Efficient turing-universal computation with DNA polymers. International Conference on DNA Computing and Molecular Programming

Samoilov M, Arkin A, Ross J (2002) Signal processing by simple chemical systems. J Phys Chem A 106(43):10 205–10 221

Seelig G, Soloveichik D, Zhang DY, Winfree E (2006) Enzyme-free nucleic acid logic circuits. Science 314:1585–1588

Senum P, Riedel MD (2011) Rate-independent biochemical computational modules. Proceedings of the Pacific Symposium on Biocomputing

Shea A, Fett B, Riedel MD, Parhi K (2010) Writing and compiling code into biochemistry. Proceedings of the Pacific Symposium on Biocomputing, pp 456–464

Soloveichik D, Seelig G, Winfree E (2010) DNA as a universal substrate for chemical kinetics. Proc Natl Acad Sci U S A 107(12):5393–5398

Vanness J, Kharisov E, Hovakimyan N (2012) $\mathcal{L}_1$ adaptive control with proportional adaptation law. American Control Conference, Montréal, Canada, June 2012

Venkataramana S, Dirks RM, Ueda CT, Pierce NA (2010) Selective cell death mediated by small conditional RNAs. Proc Natl Acad Sci U S A 107(39):16 777–16 782

Weiss R (2003) Cellular computation and communications using engineering genetic regulatory networks. Ph.D. dissertation, MIT

Weiss R, Homsy GE, Knight TF (1999) Toward in vivo digital circuits. DIMACS Workshop on Evolution as Computation, pp 1–18

Win MN, Smolke CD (2007) A modular and extensible RNA-based gene-regulatory platform for engineering cellular function. Proc Natl Acad Sci U S A 104(36):14283

Win MN, Liang J, Smolke CD (2009) Frameworks for programming biological function through RNA parts and devices. Chem Biol 16:298–310

Yurke B, Turberfield AJ, Mills AP, Simmel FC Jr, Neumann J (2000) A DNA-fuelled molecular machine made of DNA. Nature 406:605–608

Chapter 15
Designing Zinc Finger Proteins for Applications in Synthetic Biology

Shayoni Dutta and Durai Sundar

Abstract Transcription factors capable of regulating the expression repertoire of a cell possessing specific recognition patterns dominating their interaction with its respective DNA, that can be exploited to achieve targeted genome engineering, happens to be the cynosure of most studies encompassing DNA-protein interaction. The mostly widely studied transcription factors are zinc finger proteins that bind to its target DNA via few cardinal residues on its alpha–helix, comprising each finger of the protein. Exploiting the binding specificity and affinity of the interaction between the zinc fingers and the respective DNA can help to generate engineered zinc fingers for therapeutic purposes involving genome targeting. Exploring the structure-function relationships of the existing zinc finger-DNA complexes can aid in predicting the probable zinc fingers that could bind to any target DNA. This chapter describes the interaction of the zinc finger with its respective DNA, its prospective manipulation and application in the field of engineering the genome, various prediction tools dealing with either machine learning or physicochemical parameters for designing customized zinc fingers for any target DNA.

Keywords Genome engineering · Zinc finger protein · Zif-268 · Zinc Finger Nucleases · DNA methylase domains · Cys_2-His_2 · Gag knuckle · Treble clef · Zinc ribbon · Zn_2/Cys_6 · Zinc Finger Targeter (ZiFiT) · ZIFIBI · ZiF-Predict

15.1 Introduction to Zinc Fingers

Presence of various protein-folds that command sequence-specific binding like helix-turn-helix, leucine zipper and zinc finger domain elicits the desire to use them for therapeutic purposes. The coordination of the structure-specific motif of the zinc finger protein with the zinc ion ensures the stability of the fold of these transcription factors. Transcription factors hold the key to regulation of gene expression and the most sought-after transcription factor with a predefined mode of interaction with its

---

S. Dutta · D. Sundar (✉)
Department of Biochemical Engineering and Biotechnology,
Indian Institute of Technology (IIT) Delhi, Hauz Khas, 110016 New Delhi, India
+91-11-26591901
e-mail: sundar@dbeb.iitd.ac.in

V. Singh, P. K. Dhar (eds.), *Systems and Synthetic Biology*,
DOI 10.1007/978-94-017-9514-2_15

target sequence are the zinc finger proteins. These transcription factors are unique in terms of their ability to bind with the target DNA in a sequence-specific manner. Hence understanding the stability of its domains and interaction is of topmost concern to ascertain the design of customized zinc finger proteins for genomic targets of interest. The most common DNA-binding motif in the human genome and most of the multi-cellular animals are the cysteine-histidine (*cys2-his2*) zinc finger.

15.2 Engineering the Genome

15.2.1 What is Genome Engineering?

The principle of genome engineering involves the modification of the nucleotide bases of the genome of any species to (1) uncover the functioning of the respective gene (2) therapeutic aspects like disease treatment (3) production of protein products. Hence it is an extensive and intentional genetic manipulation of a replicating system to serve a much desired purpose. The work on gene targeting by Capecchi, Smithies and colleagues describe using selectable markers along with genes to ensure its transfer to a target specific locus, causing gene disruption by positive-negative selection. Genome engineering acts as the link to modify change or introduce attributes which happen to be the physical manifestation of the genetic content of an individual or species. Since the entire genome sequence of most organisms are available, the need to study cross-species interaction of a gene function followed by an intricate understanding of manipulating the genome with adept precision and efficiency can be enabled only by enhancing the targeted approach of genome engineering. There are three probable strategies to manipulate genes directly (1) insertion: to add an attribute to the genome (2) Correction: replace a defective gene by its functional copy (3) inactivation: stop or block the expression of a gene. Genome engineering is at its inception and tools need to be developed to meet: design DNA templates of choice, construction of designer proteins to manipulate DNA, implementation, testing and debugging. Looking at the current pace of development, the promising applications of the genome targeting tools are alluring if large scale reengineering of genomes are carried out (Carr and Church 2009).

15.2.2 How Zinc Fingers are Instrumental in Mediating Gene Therapy?

The work on zinc finger-DNA interaction concluded the existence of a new protein fold for nucleic acid binding as well as a novel principle of DNA recognition which upon exploitation holds the base for engineering novel zinc fingers. Fingers with different triplet specificity can be engineered by mutating the key amino acid residues hence enabling specificity in DNA recognition by ensuring a large number of combinatorial possibilities. Further, linking these modules or fingers as they function independently can ascertain the recognition of longer DNA stretches. The two main methods currently used to generate engineered zinc finger arrays are modular assembly and bacterial selection methods. Another method called bipartite selection improves the specificity of the domains to bind to a target DNA of interest by fusing individual zinc fingers from two pools of engineered zinc fingers.

15.2.3 Applications of Engineered ZFP

15.2.3.1 Zinc Finger Nucleases

Fusion of engineered zinc finger domain with non-specific nuclease domains can cause double stranded breaks in the target DNA and these molecular scissors are called zinc finger nucleases (ZFN). The ZFNs consist of a Cys2-His2 zinc finger domain fused to the nuclease domain of a type IIs restriction endonuclease *FokI* that cleaves double stranded DNA nonspecifically. It is the zinc finger domain that imparts the specificity to the ZFN for its target DNA. Since double stranded cleavage by *FokI* is only possible upon its dimerization hence the need for the ZFNs to dimerize is a pre-condition for its specific nuclease activity. ZFNs can be used for targeted mutations as well as gene correction. In case of targeted mutation, ZFNs have been employed to target the first coding exon of CCR5 and their introduction in T-cell lines results in its reduced expression in the cells as well as protection from HIV infection. Gene correction which includes the inclusion of the corrected sequence by a donor construct encoding it via homologous recombination into the respective region of the defective gene. ZFNs hence play an important role in gene correction via homologous recombination in case of treating monogenic disorders like SCID, gauchers disease, x-linked disorders etc. ZFPs and ZFNs have therapeutic uses as well, especially where there are successful reports of ZFPs used to target VEGF-A promoters for treating diabetic nephropathy and amyotrophic lateral scelerosis. Developing isogenic cell lines which discern on the basis of presence or absence of drug of interest enhance drug discovery by the usage of ZFPs.

15.2.3.2 DNA Methylase Domains Fused with ZFPs

Another major area of application of ZFP is gene silencing using targeted DNA methylation. This uses the unique property of DNA methylases of methylating a particular base in a specific DNA sequence, thus making the gene (of which this DNA sequence is a part) non-functional (gene silencing). Fusion of the methylase domain of DNA methylases and ZFP (comprising of the designed DNA binding motif), thus, can be used to silence any gene. There are reports wherein such zinc finger-methylase fused proteins proposed to bind to the p53 site from p21WAF1/CIP1 gene have been used to methylate the target oligonucleotides *in vitro* within the gene sequence (Xu and Bestor 1997). Further, methylation of a large number of sequences using designer ZFP has also been reported *in vivo*. Though this area presents huge potential in genetic engineering, the toxicity of methylase domain has always acted to restrict the fast growth of this field.

15.2.3.3 ZFP Transcription Factors

Gene regulation is another major application field of ZFPs. A target gene can be regulated by engineering the proteins responsible for gene regulation e.g. designer transcription factors (an activator domain fused to custom designed ZFP). Such designer transcription factors find widespread use in drug therapy for formalizing genes as the drug target and in human therapeutics. Literature is filled with the explorations of the effects of such artificial transcription factors on promoter regions to regulate endogenous gene expression. Some of these include regulation of human erythropoietin gene (EPO1) (Zhang and Spartt 1997), the erbB-3 protoncogene as well as silencing of the multidrug resistance 1 gene (MDR1) (also known as ABCB1), the erbB-2 protoncogene (Beerli and Dreier 2000), the peroxisome proliferator activated receptor- γ gene (PPARγ) (Ren and Collingwood 2002) and the checkpoint kinase 2 (CHK2). Regulation of specific gene expression in mammalian cell lines using ZFP artificial transcription factors were the first reported successful experiments (Beerli and Barbas 2002). Artificial transcription factors have also been formed by fusing ZFP to the independent repressor domains e.g. Krupel associated box (KRAB), ERF repressor domain (ERD), Mad SID or even parts of TATA box binding protein (TBP) (Li and Yang 2008; Tian and Xing 2009). Active research is also taking place in development of treatment of human peripheral arterial disease using vascular ZFPs. The specific ZFPs activate endothelial growth factor (VEGF) and thus stimulate vascular growth (Klug 2005).

The creation of the above mentioned zinc finger chimeras, be it zinc finger- nucleases or zinc finger-transcription factors, depend on the reliable creation of zinc finger proteins (ZFP) that can specifically recognize a target sequence. Availability of highly reliable design and selection approaches to evolve novel ZFP will help realize the full potential of the above technology in therapeutic applications in the future

15.3 Zinc Finger Proteins

15.3.1 History of Zinc Finger Proteins

Zinc finger domain was discovered as a transcription factor in TF III A, the very first eukaryotic transcription factor to be isolated, during the transcription of 5S RNA gene by RNA PolIII (Pelham and Brown 1980; Klug 2010). This factor which was responsible for initiation of transcription was a 40 kilo-dalton protein purified from oocytes extracts. This protein which was first isolated from the oocyte of *Xenopus laevis*, by deletion mapping experiments its interaction with a 50 bp region called the internal control region was established which protects it from enzymatic attack. It was assumed that TF III A was engaged in the autoregulation of 5S gene transcription by binding to 5S RNA and it cognate DNA. Biochemical studies by Miller established a repeating motif in the protein with zinc as the coordination ion to grasp the target DNA, owing to its typical structure and presence of Zinc atom it was christened as Zinc finger proteins (Miller and McLachlan 1985).

15.3.2 Different Types of Zinc Fingers

Upon the discovery of the DNA-binding motif from *Xenopus laevis,* the term zinc finger was primarily used to define it, but progressively it became associated with structures coordinated by zinc ions. Factors determining the binding affinity of zinc finger domains depend on the amino acid sequence of the fingers and the linkers joining the fingers, number of fingers and the resulting fold which is ideally very stable owing to its coordination with zinc ion lacking quantifiable conformational change even after binding to target. Hence different types of zinc fingers have been documented:

Cys_2-His_2 most common type of zinc finger domains consisting of a simple ββα-fold and binds the target DNA in a sequence specific manner which has been identified as the recognition code.

Gag knuckle The zinc knuckle in continuation with a short loop connects the two short β-strands which defines the fold of this group. Resemblance to the cys_2-his_2 zinc finger can be seen if the considerable region of the helix and β-hairpin are deleted. The drug class called the zinc finger inhibitor targets this gag knuckle in the HIV NC protein.

Treble clef the zinc ion in this motif is coordinated by a ligand- one that of a β-hairpin at the N terminal end and an α-helix at the C terminus, but the presence of loop and a second β-hairpin of varying length and conformation between the N-terminal β-hairpin and the C-terminal α-helix may be a possibility. Sequence and functional dissimilarity enhances the diversity of this group of domains. The nuclear hormone receptors are the best representative of the presence of the treble clef domain.

Zinc ribbon the zinc binding subsites are formed by the two beta hairpins which define this specific domain

Zn_2/Cys_6 six cys residues are coordinated by a binuclear set of zinc ions, this type of domain is frequently found in GAL operon transcription factor.

15.3.3 Crystal Structure of zif-268

The crystal structure of the most common cys_2-his_2 zinc finger is of zif-268 (pdb id 1AAY) (Pavletich and Pabo 1991) which owing to its structural clarity corroborates the existence of two anti-parallel beta sheets and an alpha-helix in the folded individual zinc finger. Presence of anti-parallel beta sheets which include the loop formed by Cys-Cys residues and the alpha-helix which includes the His-His loop at its -COOH terminal imparts the uniqueness to the zinc finger held together by zinc ion (Berg 1990). Structural independence of zinc fingers are considered since they are connected by linkers. In a tandemly repeated array, the zinc finger domains are linked by HC link sequences which are conserved. It is called H-C since the first residue of this conserved sequence is His and the last one is Cys. Further, few of the residues form a type II beta turn. Hereby, for the perfect interaction with the DNA, the array of zinc finger domain connected by the H-C links gives it the characteristic radius and pitch owing to the formation of right handed super helix.

Zif268 contains highly conserved linkers, TGEKP, connecting adjacent zinc fingers. This C-capping motif snap locks when correct DNA sequence is bind to the zinc fingers. Alpha helix makes sequence specific contacts with the DNA bases using amino acids at $-1, +2, +3, +6$. Out of these $-1, +3, +6$ bind to the DNA bases on the target strand while $+2$ binds to the nucleotide on the complementary strand which is also the cross strand interaction (Fairall and Schwabe 1993). Specificity is determined by the side chain base interactions. Affinity is determined by the interaction of the phosphate backbone and between adjacent zinc fingers. Any mutation in the conserved linker can decrease the binding affinity by 20 folds. Adding one more zinc finger by this linker can increase the binding affinity by 1000 folds, making not more than three zinc fingers. These characteristics of the zinc fingers have been utilized in the design of ZFP for any target of interest.

Other, more subtle interactions in the Zif-268 complex were only fully appreciated after the structure was refined to 1.6-°A resolution. It was found that in the complex, the zinc fingers bind in the major groove of B-DNA and wrap partway around the double helix. Although the overall conformation of the DNA in the Zif268 complex is similar to B-DNA, it has several distinctive features that may be important for recognition. The DNA is slightly under wound (with 11.3 bp/turn) and the major groove is somewhat deeper than normal. Each finger has a similar relation to the DNA and makes its primary contacts in a three-base pair sub-site. Residues from the amino-terminal portion of an α helix contact the bases, and most of the contacts are made with the guanine-rich strand of the DNA. This structure provides a framework for understanding how zinc fingers recognise DNA and suggests that this motif may provide a useful basis for the design of novel DNA-binding proteins.

15.3.4 Recognition Code of Zinc Finger Proteins

The crystallized structure of mouse transcript factor Zif268 (1AAY) has three finger domains. This protein's alpha-helix interacts with the DNA major groove where each finger interacts with three successive DNA bases at -1, 3, 6 amino acid residue positions on the helix respectively via hydrogen bond interaction (Fig. 15.1). The 2 amino acid residue position on the alpha-helix interacts with the triplet on the adjacent strand (secondary strand) called cross-strand interaction, hence adding significant

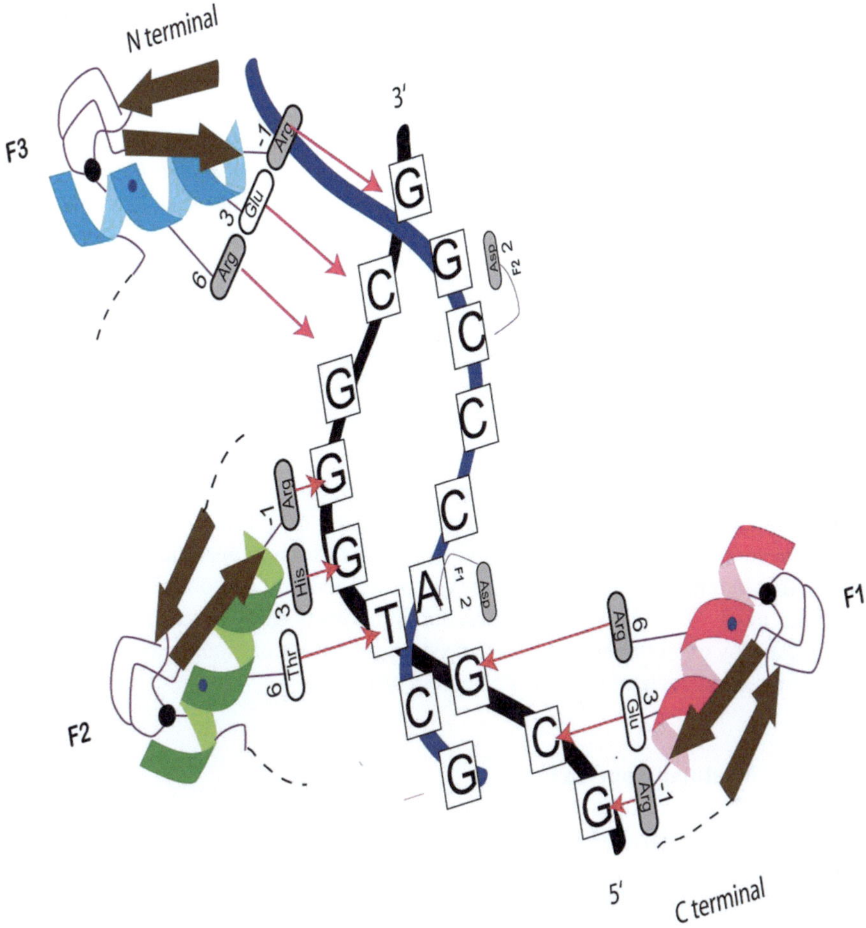

Fig. 15.1 A zinc finger protein bound to its specific target DNA site: A ββα-three zinc finger domain protein (Zif268) wrapping around the target DNA; the zinc finger amino acid side chains at the alpha-helical positions − 1, 3 and 6 makes sequence-specific hydrogen bond interactions with the nucleotides on the primary DNA strand (shown in black); a cross-strand interaction from the amino acid Asp at position 2 (of finger 1 and finger 2) to the complementary DNA strand (blue) is also depicted. The three fingers are labeled as F1, F2 and F3 with the linkages between consecutive fingers shown by extended broken lines. The tetrahedrally coordinated zinc atoms are shown as *small black spheres*

specificity to the interaction and the refined model emphasizes the proteins ability to bind to a 4 bp overlapping sub site. In the above pattern of recognition, the residue at position 6 on the alpha helix co-acts with the 5' base of the primary strand, residue at position 3 with the middle base and residue at position − 1 with the 3' base. This helix which initiates a novel method of DNA recognition is called the recognition helix and the DNA code the recognition code. Hence these 7 conserved amino acid residues ensure the tertiary folding whereas the variable residues are responsible for the specificity of each domain.

Inter-finger side chain-side chain interactions contribute significantly to the DNA-protein interface stability. In case of Zif268, inter-finger interactions between Thr-52 and Arg-74 aids the Arg-74...Asp-76 interaction thereby orienting the Arg-74 residue to interact with the target DNA. The DNA recognition also heavily banks on the intra finger interactions amidst the side chains of individual fingers. The tradeoff between affinity and specificity considering the DNA-protein interaction gives a thoughtful insight into the very application of zinc fingers for regulating gene expression

15.3.5 Modular and Synergistic Modes of ZFP Binding

In various studies of binding of zinc fingers to DNA, the two most prominent models are Modular and Synergistic modes. Although each of the modes account for the same key interacting amino acids in the alpha helix of the fingers, but the difference lies in the definition of the dependency of consecutive fingers to their target codon. Modular mode of binding assumes that binding affinity each finger of the protein is not affected by other fingers, whereas, in synergistic mode of binding the dependency of the fingers on each other is also taken into account.

15.3.5.1 Modular Approach

The modular assembly (MA) method of generating engineered zinc finger proteins (ZFPs) was the first practical method for creating custom DNA-binding proteins which has enabled plethora of exploration of sequence-specific methods and reagents, ushering in the modern era of zinc finger-based applications. Modular approach was used to develop the first zinc finger nuclease to cleave an endogenous site, much like the first artificial transcription factor to enter phase II clinical trials. However, MA is still used widely for many applications. The 3 bp sub-site on the target DNA strand recognised by the cardinal residue positions on the alpha-helix of the respective fingers of the zinc finger protein is called the recognition code. The key residue positions − 1, 3 and 6 on the finger interact with a contiguous set of 3 bp on the target DNA strand. This recognition code inchoated the idea to develop custom made zinc fingers for all possible 64 possible 3 bp DNA sub-sites. Hence once the rules employed in recognition of the respective bases by the residues via hydrogen bonds are completely uncovered, they can be utilised by combination of rational design and combinatorial methods such as phage display to assemble custom designed multiple zinc fingers mutated at their cardinal residues so as to recognise any possible DNA target. The ability to assemble the fingers in any order to assign best possible affinity is completely based on the fact that the fingers bind to their respective 3 bp sub-site as independent modules.

15.3.5.2 Synergistic Approach

The synergistic approach to ascertain the functioning of zinc fingers during interaction with the respective target DNA via their recognition code appears to be highly resourceful and reliable in terms of quantifying the physico-chemical interaction. The ability of zinc fingers to identify overlapping 4 bp sub-sites (Isalan at el. 1997) confirms the sub-site interface specificity to be mediated by possibly different residues positions on adjacent fingers. Moreover, since its only a 4 base pair overlapping subsite it entails dependence on only two different residue positions on the respective zinc fingers. The inter-finger synergism observed for Zif268-like zinc fingers has a parallel in the mode of binding of the DNA-binding domain. The amino acid residue Asp at position 2 of Zif268 finger 3 specifically excludes adenine and cytosine from the 5' position of the middle triplet. Removal of this specific interaction enables the acceptance of these two bases for recognition, although none of the peptides in this study were capable of specifying either nucleotide using an amino acid from position 6 or any other position in the middle finger. Regardless of the outcome of these selections, but particularly if it emerges that residues in position 6 cannot directly specify nucleotides other than guanine, it will be essential to define any possible pairings of synergistic contacts that give rise to sequence specificity. Further, it will be of interest to determine whether any amino acids in position 2 can make non synergistic contacts to the triplet sub-site of a neighboring finger, which by themselves confer sequence specificity to the 5' position.

Moreover, the intermolecular contact between positions − 1 and 2 in a zinc finger shows heightened levels of synergy which can be accounted upon uncovering the absolute recognition code. This expatiates on the possible networks of contacts that occur at the protein-DNA interface in the region of the 4 bp overlapping sub-sites.

15.4 Prediction Tools for Engineering Customized Zinc Finger Proteins

15.4.1 Basic Insight Behind Prediction Models

One of the key goals in zinc finger engineering has been to produce proteins that can specifically recognize a pre-determined DNA sequence. The experimental strategies for engineering zinc finger proteins either through selection or by rational design, although useful, are time consuming, expensive and have techniques not accessible to all laboratories. The regularity in structures of zinc fingers which recognize and bind to nucleotide base triplets, where only the few amino acids at fixed positions interacting with the DNA strand vary, offer the possibility of devising and using a prediction algorithm. As a result of this, ZFPs benefit mostly by computational design. ZFP prediction tools would be valuable for researchers interested in designing specific zinc finger transcription factors and ZFNs for several biological and biomedical applications including targeted gene regulation, enzyme engineering, genome editing, gene therapy etc. The computational design methods can be divided into prediction tools of two types. The first is based on experimental data which can use either the sequence-based or structure-based approach. The second one captures the fundamental science behind the interaction—in this case, chemical binding and specificity. A list of available online web tools for prediction of DNA-binding specificity in ZFP is presented in Table 15.1. Some such bioinformatics models and their approaches for prediction of DNA-binding specificity in zinc fingers are described below:

15.4.2 Machine Learning based Prediction Tools

15.4.2.1 A Structure-Based Approach (2001)

A computational scheme that uses knowledge-based parameters for protein DNA interactions based on data derived from X-Ray structures of such complexes has been described. Application of these parameters to specified binding models, a score that reflects the stereo-chemical complementarity and structural compatibility between a protein sequence and a DNA site can be evaluated. Advantage of this procedure is that it can be used for the prediction of binding sites for newly identified proteins that are clustered to a defined family based on binding data. However, it does not always predict the known site at the first rank because it does not consider other factors that affect binding, such as the sequence context of the binding sites and coupled interactions. Secondly, possible position-dependent effects that are specific to each binding motif are masked (Mandel-Mandel-Gutfreund and Baron 2001).

15.4.2.2 Simple Physical Model (2004)

This model is independent of a prior knowledge of structure or sequence. Protein-design approaches like the combination of simple physical models of macromolecular energetics and rapid algorithms for sampling side-chain conformations provide a powerful, quantitative description of protein-DNA interfaces in their entirety. An all-atom description of both the DNA and protein is used. Model uses a simple physically-based energy function, fixed DNA and protein backbone conformations, and a rotamer-based description of protein side-chain conformation. This model does not include electrostatic and water-mediated interactions dictating affinity and specificity. Also, multiple protein-DNA binding modes have to be considered. This model requires improvements by utilizing backbone sampling and docking techniques. Prediction in structurally homologous complexes is limited by specificity (Havranek and Duarte 2004).

Table 15.1 Web tools for predicting DNA-binding specificity in zinc finger proteins

Tool	Features	Reference
SVM Model http://compbio.cs.princeton.edu/zf	Even in the absence of known binding sites it can quantify degree of DNA-binding preference of particular ZFP–DNA pair	Persikov et al. 2009
ZIFIBI http://bioinfo.hanyang.ac.kr/ZIFIBI/frameset.php	By searching against gene name or the SWISS-PROT database it predicts C_2H_2 ZFP binding sites in cis-regulatory regions of target DNA	Cho et al. 2008
ZiFiT http://zifit.partners.org/ZiFiT	provides modular design of ZFP DNA from chosen standard dataset(s) for target DNA along with options for Context-Dependent Assembly and Oligomerized Pool Engineering are also available	Sander et al. 2007
ZiF-BASE http://web.iitd.ac.in/ sundar/zifbase	Exhaustive database of natural and engineered proteins; which also allows search of DNA sequence for binding sites and known ZFPs	Jayakanthan et al. 2009
ZiF-Predict http://web.iitd.ac.in/ sundar/zifpredict	Allows modular prediction of 2- and 3-finger C_2H_2 ZFPs (and ZFNs) that bind to specificied target DNA sites based on neral networks	Molparia et al. 2010
Zinc Finger Tools http://www.zincfingertools.org	Searches for potential target sites within DNA; predicts ZFP for target; further is capable of predicting target site for given ZFP and finds similarity of particular DNA with target site	Mandell and Barbas 2006

15.4.2.3 Zinc Finger Tools (2006)

Zinc Finger Tools developed by mandell acts as a multiple utility web server that can scan a given DNA sequence for consecutive DNA triplets that can be targeted with zinc finger domains, design a zinc finger protein for a valid DNA sequence and most importantly predict the binding sites in ZFP. The user inputs the amino acid sequence of the ZFP and not the DNA sequence. The algorithm is such that it recognizes only helices and ignores all other sequence to minimize the impact of poor sequence quality or extraneous sequences. This tool can also be used to ensure that the intended targeted site of a designed ZFP is correct (Mandell and Barbas 2006).

15.4.2.4 Zinc Finger Targeter (ZiFiT) (2007)

Zinc Finger Targeter (ZiFiT) was developed by the Zinc Finger Consortium as an effort to provide a simple and easy tool for ZFP and ZFN design. It is a popular web tool that provides an integrated modular design approach by incorporating three different data sets enumerating zinc finger binding-patterns for independent modules developed by Barbas, Sangamo and ToolGen (Segal and Dreier 1999; Dreier and Fuller 2005). The user may enter the query DNA sequence and choose one or more of these sets. The data from the chosen set(s) is then used to identify the best DNA target site in the query and the corresponding zinc finger arrays are returned as a text file. Scores are given alongside predictions as an indication of their chances of success, as measured by a bacterial two-hybrid assay. Recently, options of using CoDA (Context-Dependent Assembly) (Sander and Dahlborg 2011) and OPEN (Oligomerized Pool Engineering) (Maeder and Thibodeau-Beganny 2008) for design have been added. The Consortium advises the use of CoDA due to its simplicity over OPEN and much higher success likelihood than the modular approaches (Sander and Zaback 2007).

15.4.2.5 PWM Prediction: Sensitivity to Docking Geometry (2007)

Transcription factor binding-specificity is described by a consensus sequence or PWM (Position Weight Matrix). Given a TF-DNA complex structure, a scoring function is used to evaluate relative affinities. For scoring, approaches like knowledge-based structural potentials or all-atom modeling of complexes are used. Since PDB for every complex is not available, homology modeling of complexes is employed. Conserved stereo-specific H-bond interactions are informative for template selection. This requires similarity of docking geometry, which is quantified in terms of IAS (Interface Alignment Score). The IAS score and prediction accuracy are related. For modeling side-chains and bases, residue confirmations are iteratively minimized by selecting rotamers, generated by wriggling algorithm that yielded lowest energy of complex (Siggers and Honig 2007).

15.4.2.6 ZIFIBI (2008)

Interaction studies between all possible C_2H_2 zinc finger and its target DNA availed from data available from literature and crystallographic data. Then a 3 Position Weighted Matrix (PWM) was constructed for positions -1, 3 and 6 of the alpha helix and a HMM (Hidden Markov Model) can be used to calculate the most probable state path of three nucleotides sequences. ZIFIBI provides functions to search DNA binding sites and by the gene name, SWISS-PROT ID or SWISS-PROT access number for specific protein and to search target genes. These computations are used to predict C_2H_2 zinc finger transcription factor binding sites in cis-regulatory regions of their target genes. The ZIFIBI database contains proteins with potential binding sites for zinc finger proteins that have not yet been experimentally identified. The average Euclidean distance of ZIFIBI was 0.613929, which is lower than those found in other studies and its predictions were similar to other studies thereby demonstrating its superiority (Cho and Chung 2008).

15.4.2.7 SVM-based Approach (2009)

Support Vector Machine are supervised machine learning classification tool. Using canonical binding model, the C_2H_2 zinc finger protein-DNA interaction interface is modeled by the pairwise amino acid-base interactions. Using a classification framework, known examples of non-binding ZF-DNA pairs are incorporated. Using a linear kernel, information about relative binding affinities of ZF–DNA pairs is incorporated. A polynomial SVM also captures dependencies among the canonical contacts. SVMs search for a weight vector w that best separates binding and non-binding proteins. The advantage of the polynomial kernel over the linear SVM may suggest the limitation of the originally used canonical representation. It feature vectors into a higher dimensional space, thereby making possible implicit inclusion of higher order interactions not listed in the original canonical model. But, use of the polynomial kernel does not allow the incorporation of relative binding information (Persikov and Osada 2009).

15.4.2.8 ZiF-Predict (2010)

ZiF-Predict tool is based on artificial neural network and enables prediction of recognition helices for C_2H_2 zinc fingers binding to specific DNA targets. An exhaustive dataset of seven-residue-long recognition helices of three-finger ZFPs, ZFNs and their corresponding triplets reported in literature were used. In this user-friendly interface, users can input a DNA sequence and select the option to predict two or three zinc fingers for the same. This web tool also incorporates both the modular prediction and the synergistic interactions between the fingers, a feature not available elsewhere. For instance, depending on the position of the finger motifs, binding affinities to the target sequence may differ. The network consisted of an input layer followed by two hidden layers and a single output neuron (Molparia and Goyal 2010).

15.4.3 Prediction Based on Physico-Chemical Approach

Computational tools ease the prediction of such engineered zinc fingers by effectively utilizing information from the available experimental data. A study of literature reveals many approaches for predicting DNA-binding specificity in zinc finger proteins. However, an alternative approach that looks into the physico-chemical properties of these complexes would do away with the difficulties of designing unbiased zinc fingers with the desired affinity and specificity. We have described a physico-chemical approach that exploits the relative strengths of hydrogen bonding between the target DNA and all combinatorially possible zinc fingers to select the most optimum ZFP candidate (Roy and Dutta 2012).

Zif-Predict-IHBE was developed based on the calculation of interfacial hydrogen bond energy to shift to a more realistic and unbiased structure based design approach. The major hindrance to identifying all existing patterns of ZFP-DNA complexes can be overcome by developing an organized and exhaustive database of native as well as engineered ZFPs obtained from other curated databases and literature. Zif-BASE is one such database (Jayakanthan and Muthukumaran 2009).

In continuation to the development of Zif-Predict-IHBE analysis of the Modular mode of binding of the Zinc fingers to target 9 bp long DNA was initiated. The work comprised of building a highly efficient pipeline to construct a database of computationally calculated interfacial hydrogen bond energies for each of possible 64 codons with zinc fingers designs constructed from a reduced set of 392 mutations in a zinc finger. As a result the database developed for each of the top predictions for each possible codon, finally a potential ZFP comprising of three fingers can be calculated based on the modular mode of binding to its specified target. The pipeline built used HADDOCK (de Vries and Dijk 2010) for protein DNA docking and MODELLER to study the reduced set of mutations in each of the docked complexes. The results obtained serve the part of Modular mode of binding of the zinc fingers to DNA in this study. Although the efficiency of the modelling is affected by the individual efficiencies of HADDOCK and MODELLER yet this pipeline is the most efficient to study all possibilities of ZFP to all possible 9 bp targets as the time required to model the complexes is greatly reduced. The tool developed produced a lot of false predictions and hence the pipeline was tested with Synergistic mode of binding also to cross validate how much error if there was due to Modular mode of binding.

There are several approaches to the prediction strategies for designing of zinc finger proteins targeting specific DNA sequences. The basic algorithm for modular mode of binding had an advantage that it was a unbiased structure based study of modular mode of binding. These approaches in the study of binding of two macromolecules involve modelling of the predicted complexes using available computational tool, often limited by the requirement of very large computational time. This could be plainly due to huge number of candidates to be tested, the accuracy of the method or other improvements.

Therefore, in continuation to the previous work done in the lab a strategy was designed to model ZFP-DNA complexes using the synergistic mode of binding and, thereafter, compare the results to that of modular mode with respect to a particular rationale specifically designed considering experimentally validated ZFP designs for a limited set of DNA target sets. This study aims at providing not only any improvements if required, but also with advantages of these two modes over each other. This study also has hurdle of protein-DNA docking owing to the flexibility of the double stranded polynucleotide macromolecule, which makes it the most time consuming step in the designed pipeline.

15.5 Conclusion

The need to develop accurate and effective genome targeting technologies to enable retrograde engineering of causal genetic variants by targeting the individual genetic loci selectively. The presence of many other genome editing technologies with varying claims of precision and efficiency enforces the compulsion to evaluate the best technology complemented by its feasibility. Homing endonucleases encoded by open reading frames rooted within group 1, group2, archael introns and inteins exhibit enormously high DNA binding specificity emanating from DNA targets possessing long stretches, oblivious to reasonable sequence variations in these sites followed by disparate DNA cleavage mechanisms. The HNH and His-cys box homing group 1 endonuclease family are the most common and well studied ones (Takeuchi and Lambert 2011).

Similarly another class of DNA-binding repeat containing proteins which propose effortless engineering to target novel DNA sequences by manipulating host gene expression are the TALENS—Transcription activators-like effectors derived from plant pathogenic bacterial genus *Xanthomonas* (Miller and Tan 2011). These polymorphic repeats identify specific nucleotides with sufficient degeneracy; hence this code capacitates the prediction of genomic binding sites providing the opportunity to customize TAL effectors in DNA targeting. Latest in the league are the type II prokaryotic CRISPR (clustered regularly interspaced palindromic repeats)/Cas adaptive immune system. The CRISPR system can facilitate RNA guided endogenous site specific DNA cleavage by Cas9 nuclease, which are directed by short RNAs (Burgess 2013). Hence to emerge as the best genome editing tool, designer ZFNs (Wood and Lo 2011) need to surpass all the expectations required to study cross species comparison of gene functions majorly inhibited by lack of reverse genetic tools even though complete genome sequences of most organisms are available. The ease, feasibility and the level of specificity in terms of recognition of the target DNA may suffice an impartial comparative study involving the existing genome editing tools we have discussed above.

Acknowledgement The work on zinc finger proteins in the laboratory of DS is supported by grants from Lady Tata Memorial Trust, DuPont and Department of Biotechnology (DBT), Govt. of India, under the IYBA & National Bioscience Award schemes. SD is a recipient of DST INSPIRE Fellowship for her doctoral studies in the laboratory of DS.

References

Beerli RR, Barbas CF 3rd (2002) Engineering polydactyl zinc-finger transcription factors. Nat Biotechnol 20(2):135–141

Beerli RR, Dreier B et al (2000) Positive and negative regulation of endogenous genes by designed transcription factors. Proc Natl Acad Sci U S A 97(4):1495–1500

Berg JM (1990) Zinc finger domains: hypotheses and current knowledge. Annu Rev Biophys Biophys Chem 19:405–421

Burgess DJ (2013) Technology: a CRISPR genome-editing tool. Nat Rev Genet 14(2):80–81
Carr PA, Church GM (2009) Genome engineering. Nat Biotechnol 27(12):1151–1162
Cho SY, Chung M et al (2008) ZIFIBI: prediction of DNA binding sites for zinc finger proteins. Biochem Biophys Res Commun 369(3):845–848
de Vries SJ van Dijk M et al (2010) The HADDOCK web server for data-driven biomolecular docking. Nat Protoc 5(5):883–897
Dreier B, Fuller RP et al (2005) Development of zinc finger domains for recognition of the 5'-CNN-3' family DNA sequences and their use in the construction of artificial transcription factors. J Biol Chem 280(42):35588–35597
Fairall L, Schwabe JW et al (1993) The crystal structure of a two zinc-finger peptide reveals an extension to the rules for zinc-finger/DNA recognition. Nature 366(6454):483–487
Havranek JJ, Duarte CM et al (2004) A simple physical model for the prediction and design of protein-DNA interactions. J Mol Biol 344(1):59–70
Isalan M, Choo Y, Klug A (1997) Synergy between adjacent zinc fingers in sequence-specific DNA recognition. Proc Natl Acad Sci 94(11):5617–5562
Jayakanthan M, Muthukumaran J et al (2009) ZifBASE: a database of zinc finger proteins and associated resources. BMC Genomics 10:421
Klug A (2005) Towards therapeutic applications of engineered zinc finger proteins. FEBS Lett 579(4):892–894
Klug A (2010) The discovery of zinc fingers and their applications in gene regulation and genome manipulation. Annu Rev Biochem 79:213–231
Li Y, Yang D et al (2008) ZNF418, a novel human KRAB/C2H2 zinc finger protein, suppresses MAPK signaling pathway. Mol Cell Biochem 310(1–2):141–151
Maeder ML, Thibodeau-Beganny S et al (2008) Rapid "open-source" engineering of customized zinc-finger nucleases for highly efficient gene modification. Mol Cell 31(2):294–301
Mandel-Gutfreund Y, Baron A et al (2001) A structure-based approach for prediction of protein binding sites in gene upstream regions. Pac Symp Biocomput 6:139–150
Mandell JG, Barbas CF 3rd (2006) Zinc Finger Tools: custom DNA-binding domains for transcription factors and nucleases. Nucleic Acids Res 34(Web Server issue):W516–523
Miller J, McLachlan AD et al (1985) Repetitive zinc-binding domains in the protein transcription factor IIIA from Xenopus oocytes. EMBO J 4(6):1609–1614
Miller JC, Tan SY et al (2011) A TALE nuclease architecture for efficient genome editing. Nat Biotechnol 29(2):143-U149
Molparia B, Goyal K et al (2010) ZiF-Predict: a web tool for predicting DNA-binding specificity in C2H2 zinc finger proteins. Genomics Proteomics Bioinformatics 8(2):122–126
Pavletich NP, Pabo CO (1991) Zinc finger-DNA recognition: crystal structure of a Zif268-DNA complex at 2.1 A. Science 252(5007):809–817
Pelham HR, Brown DD (1980) A specific transcription factor that can bind either the 5S RNA gene or 5S RNA. Proc Natl Acad Sci U S A 77(7):4170–4174
Persikov AV, Osada R et al (2009) Predicting DNA recognition by Cys2His2 zinc finger proteins. Bioinformatics 25(1):22–29
Ren DL, Collingwood TN et al (2002) PPAR gamma knockdown by engineered transcription factors: exogenous PPAR gamma 2 but not PPAR gamma 1 reactivates adipogenesis. Genes Dev 16(1):27–32
Roy S, Dutta S et al (2012) Prediction of DNA-binding specificity in zinc finger proteins. J Biosci 37(3):483–491
Sander JD, Zaback P et al (2007) Zinc Finger Targeter (ZiFiT): an engineered zinc finger/target site design tool. Nucleic Acids Res 35(Web Server issue):W599–605
Sander JD, Dahlborg EJ et al (2011) Selection-free zinc-finger-nuclease engineering by context-dependent assembly (CoDA). Nat Methods 8(1):67–69
Segal DJ, Dreier B et al (1999) Toward controlling gene expression at will: selection and design of zinc finger domains recognizing each of the 5'-GNN-3' DNA target sequences. Proc Natl Acad Sci U S A 96(6):2758–2763

Siggers TW, Honig B (2007) Structure-based prediction of C2H2 zinc-finger binding specificity: sensitivity to docking geometry. Nucleic Acids Res 35(4):1085–1097

Takeuchi R, Lambert AR et al (2011) Tapping natural reservoirs of homing endonucleases for targeted gene modification. Proc Natl Acad Sci U S A 108(32):13077–13082

Tian C, Xing GC et al (2009) KRAB-type zinc-finger protein Apak specifically regulates p53-dependent apoptosis. Nat Cell Biol 11(5):580–U122

Wood AJ, Lo TW et al (2011) Targeted genome editing across species using ZFNs and TALENs. Science 333(6040):307

Xu GL, Bestor TH (1997) Cytosine methylation targetted to pre-determined sequences. Nat Genet 17(4):376–378

Zhang L, Spratt SK et al (2000) Synthetic zinc finger transcription factor action at an endogenous chromosomal site. Activation of the human erythropoietin gene. J Biol Chem 275(43):33850–33860

Chapter 16
Synthetic Biology for the Development of Biodrugs and Designer Crops and the Emerging Governance Issues

Archana Chugh, Pooja Bhatia and Aastha Jain

Pooja Bhatia and Aastha Jain contributed equally.

Abstract Synthetic biology, an amalgamation of different fields including biology, computer science, physics and chemistry is estimated to grow to a value of $ 4.5 billion in 2015. It finds application in diverse areas such as healthcare, energy, agriculture, food additives and industrial chemicals. The role of synthetic biology for production of biodrugs as well as designer crops has been explored in the present study along with regulation of synthetic biology in the stated domains. The initial sections cover the recent developments in the synthetic biology derived biodrugs and designer crops. Emerging socio-economic issues such as biosafety, biosecurity, intellectual property rights form the later part of the study. Authors have also discussed about evolving a consolidated governance system with the aim to stimulate the growth and to minimise the risks involved in the biodrugs as well as designer crops based on synthetic biology.

Keywords Biomedicine · Agriculture · Metabolic engineering · Synthetic organelles · Standardization · Intellectual property rights · Socio-ethics

List of Abbreviations

CAGR	Compound Annual Growth Rate
DNA	Deoxyribonucleic Acid,
FDA	Food and Drug Administration;
IPR	Intellectual Property Rights;
NCDs	Non-communicable Diseases;
SAVE	Synthetic Attenuated Virus engineering;
TRIPS	Trade Related Intellectual Property Rights;
WTO	World Trade Organisation;

A. Chugh (✉) · P. Bhatia · A. Jain
Kusuma School of Biological Sciences,
Indian Institute of Technology Delhi, Delhi, India
e-mail: achugh@bioschool.iitd.ac.in

V. Singh, P. K. Dhar (eds.), *Systems and Synthetic Biology*,
DOI 10.1007/978-94-017-9514-2_16

16.1 Introduction

The current research in biology thrives primarily on genetic engineering, an approach that has enabled researchers to investigate the role of various genes and products thereof, in the proper functioning, development and behaviour of an organism. For several decades, genetic engineering or recombinant DNA technology involved manipulation and study of one gene at a time, however, in the past few years the engineering of the genetic elements has been extended to manipulation of "genetic networks" with the emergence of a new multidisciplinary field of science—Synthetic Biology (Cuccato et al. 2009).

Synthetic biology emerged as a mainstream field in the year 2004 after the first meeting of synthetic biology held at MIT, USA although, the term synthetic biology was coined in 1974 by Polish geneticist Waclaw Szybalski (Szybalski 1974; Tucker and Zilinkas 2006). Synthetic biology is an interdisciplinary science where the principles of genetics, robotics, nanotechnology, systems biology, engineering and computational biology can be applied together to study, manipulate or introduce an entire genetic circuitry into a "chassis" for various applications. It aims at enablement of "creation" of synthetic organisms with user-defined functionality for the benefit of human society (Jain et al. 2012; Saukshmya and Chugh 2010). The construction of a customized, robust as well as reliable genetic circuitry requires standardized parts that is, promoters, regulators, reporters, translational units, terminators, chassis, etc (Marguet et al. 2007). In order to enhance compatibility and access to standard parts, researchers throughout the globe have started a Registry of Standard Parts (or Biobricks) which provides a list of standardized genetic elements that have been constructed so far (Biobricks Foundation 2012). The role of such Registry of Biology Parts in acceptability of synthetic biology by the public has been examined in the later section of this chapter.

The inherent "artificial" nature of synthetic biology holds immense potential in providing solutions in various sectors that are of concern to the society. Various applications of synthetic biology have been reported in the field of healthcare (diagnostics and therapeutics), energy, agriculture, food additives and industrial chemicals (Savage et al. 2008; Schmidt 2010; Purnick and Weiss 2009; Brenner and Arnold 2011). For example in the energy sector, there is an urgent need for replacing non-renewable sources of energy with renewable and efficient sources of energy such as biofuels. Keeping this in view, the initial search on biofuels focused on extraction of oil from plant species such as *Castor* and *Jatropha*, however, since extraction and processing of oils from plant species suffered from several drawbacks, the focus eventually shifted to algae (Okullo et al. 2012). Although, biodiesel production from algae is efficient than plants, however, it has its own disadvantages such as low yield, and lower efficiency of biodiesel as opposed to fossil fuels making the entire process economically expensive (Demirbas and Demirbas 2011). To circumvent such investments, engineering of pathways for production of biofuels (either in the form of ethanol, butanol or lignocellulose) in *Escherichia coli* is being carried out via synthetic biology based approaches (Jang et al. 2012).

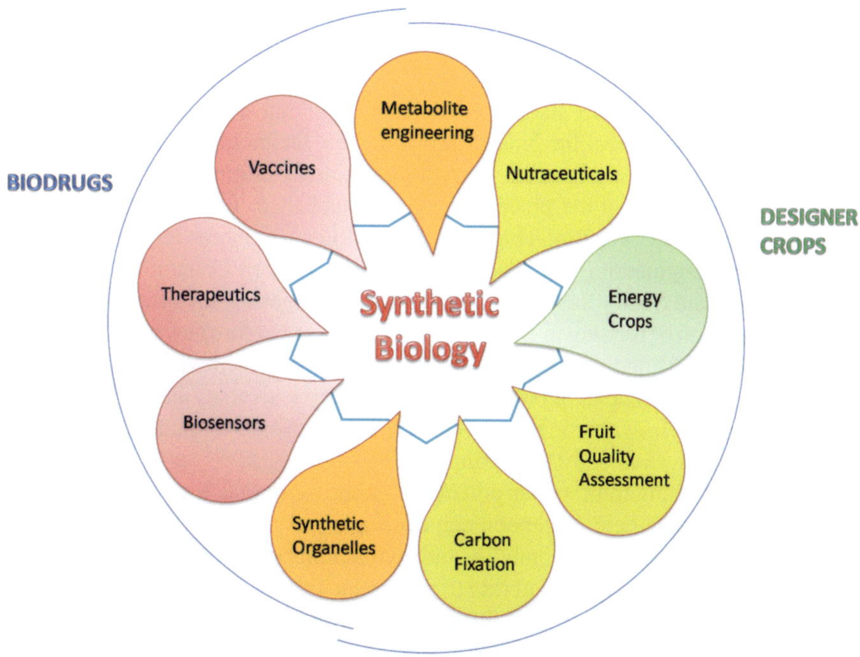

Fig. 16.1 Various applications of synthetic biology in the field of biodrugs and designer crops

Synthetic biology can also play a pivotal role in the development of novel therapeutics, diagnostic methods and prevention strategies (Fig. 16.1). It can pave way for exploitation of naturally occurring medicinally active compounds that are otherwise difficult to extract from their natural source (e.g. plant secondary metabolites), enhance the efficacy of existing antibiotic compounds, explore novel drug targets and commercialise gene therapy (Martin et al. 2003; Lu and Collins 2009). Apart from benefiting the pharmaceutical industry, synthetic biology can be also useful in the agronomical sector where plant based genetic elements can be engineered to enhance the rate of carbon fixation, assess fruit quality and generate synthetic organelles (Bar-Even et al. 2010; Bonacci et al. 2011, 2012; Weber et al. 2009; Fig. 16.1).

It is worth mentioning, that the global synthetic biology market in the year 2011 was worth US$ 1,537.5 million and the value of the market has reached US$ 2,120 million in 2012. The market is expected to reach US$ 16,745 million by 2018 growing at a CAGR of 41.1 % from 2010 to 2018 (Synthetic Biology Market, Global Industry Analysis, Size, Growth, Share and Forecast, 2012–2018, 2012). As indicated earlier, synthetic biology as a transformative technology is gaining recognition and is being explored to find solutions for human needs e.g. NASA, at one of the conferences envisaged using synthetic biology for food and drug production in space during the explorations (Langhoff et al. 2010). Besides academic institutes and government agencies, private sector is also showing increasing interest in the field of synthetic biology (Table 16.1). Some of the companies are, in fact, start-ups, established on synthetic biology based biodrugs and designer crop technologies such as Amyris Inc, Sample6 Technologies. Amyris, USA was founded in 2003, based on the UC Berkley technology for production of Artemisinin, a potent anti-malarial compound (Amyris Inc. 2012). Another start–up Sample6 Technologies, USA is based on a MIT innovative platform useful for biosensing (Sample6 Technologies 2012). Research being investment intensive, an ecosystem to fund start up companies and enterprises in synthetic biology is in the process of being established. In Europe, Technology Strategy Board (TSB), the Biotechnology and Biological Sciences Research Council (BBSRC), the Engineering and Physical Sciences Research Council (EPSRC) and the Economic and Social Research Council (ESRC) together have started a project with $\leq$ 6.5 million of funding to encourage companies to explore new applications for synthetic biology (Funding to explore synthetic biology potential 2012). Similarly SynBio Startup Launchpad, a program of Singularity University intends to nurture start up companies in synthetic biology (Singularity University Announces Inaugural Synthetic Biology Accelerator Program 2012).

Despite possessing tremendous potential to improve human lives, synthetic biology research like the field of genetic engineering poses social as well as economic challenges such as public acceptability and commercialization. As described earlier, synthetic biology has a significant role to play in the future towards the development of biodrugs and designer crops. Ethical questions pertaining to biosafety as well as biosecurity have been raised due to its various applications in these two sectors. Although, public engagement can mitigate the fear of misuse, however, development of an effective governance system remains crucial. Also, although, the allied regulatory frameworks do exist, yet, these are not synthetic biology specific or sufficient to govern the emerging field. Synthetic biology being at a nascent stage, a flexible form of regulation can provide an efficient solution for its growth. Across the globe with increasing collaborations among the industry players and academia, standardization of parts as well as terms can lead to successful joint ventures. Intellectual property rights (IPRs) also play a vital role in collaborations and commercialization of a product. Although, most of the countries are TRIPs compliant, however, as IPRs are territorial, there are differences in the laws at the national level. Therefore, different strategies are being adopted country-wise for the protection of synthetic biology based biodrugs and designer crops. The addressal of the said issues on various aspects of governance can lead to significant investments by major companies and emergence of new players in the field of synthetic biology based commercial products.

The present study describes the recent advances made in the field of synthetic biology based biodrugs and designer crops. Various associated issues that are emerging or have the possibility to emerge in future with the evolution of synthetic biology based applications in the described field have been also highlighted in the study. The intellectual property rights, biosafety and biosecurity related issues that could arise during commercialization of synthetic biology derived products have been discussed. In the later sections, the focus is on the significance of governance of synthetic biology to address these issues at an early stage to balance the growth as well as benefits of the biodrugs and designer crops.

Table 16.1 A non-exhaustive list of companies involved in production of synthetic biology based biodrugs and designer crops

S. No.	Company's Name	Product	System	Website
1.	Amyris, Inc, USA	Artemisinin—anti-malarial Therapeutic	Microbial strain	http://www.amyris.com
2.	Synthetic Genomics Vaccines Inc, USA	Flu Seed Bank	Virus	www.syntheticgenomics.com.
3.	Synthetic Biologics Inc, USA	Monoclonal antibodies	Mice	http://www.syntheticbiologics.com
4.	Codagenix Inc,USA	Vaccines based on SAVE	Virus	http://www.codagenix.com/
5.	Jennerex Biotherapeutics, Inc,USA	Anticancer vaccine	Virus	www.jennerex.com
6.	Amgen inc,USA	Anticancer vaccine	Virus	www.amgen.com
7.	Oncolytics Biotech Inc, Canada	Anticancer vaccine	Virus	www.oncolyticsbiotech.com
8.	Sample6 Technologies, USA	Biosensing	bacteria	http://www.sample6tech.com/technology.html
9.	Evolva holdings SA, Switzerland	Stevia-sweetner	Yeast	http://www.evolva.com
		Vanillin	Yeast	
		Saffron	Yeast	
10.	Sygenta AG, Switzerland	Biofuel	Plants	http://www.syngenta.com
11.	Dupont, France	Biofuel	Plants	http://www2.dupont.com/
		Food		
12.	Monsanto, USA	Biofuel	Plants	http://www.monsanto.com/
		Food		
13.	Bayer CropScience AG, Switzerland	Biofuel	Plants	www.cropscience.bayer.com/
		Food		

16.2 Role of Synthetic Biology in Biodrugs

Diseases, whether communicable or non-communicable, have remained a concern for the human society since centuries. The major burden of diseases occurring globally is due to non-communicable diseases (NCDs) such as cancer, type-2 diabetes, vascular disorders and chronic respiratory diseases (The Global Burden of Disease: 2004 update: WHO report 2008). It has been reported that 63 % of deaths occurred due to NCDs in 2008 and an estimated 52 million people will die annually by 2030 (Global status report on non-communicable disease 2010: WHO report 2011). The other major class of diseases, that is, infectious diseases have caused more than 25 % deaths worldwide with malaria and tuberculosis as leading causes of mortality (Hotez et al. 2004; Tuberculosis: WHO fact sheet 2010; Malaria: WHO fact sheet 2008).

The genre of antibiotics available commercially was speculated to be potent enough to eradicate the most lethal pathogenic organisms, but with a surge of multi-drug resistant organisms and new variants of pathogenic microbes these antibiotics have been rendered inefficient. It has been observed that in the past few decades, emergence of novel, FDA approved antibiotics in the market has suffered a slow-down. Since 2004, only three novel antibiotics have emerged namely Platensimycin, Daptomycin and Linezolid of which, Platensimycin could not reach clinical trials due to its poor pharmacokinetic properties (Norrby et al. 2005; Pearson 2006; Martin and Demain 2011). Due to increase in the multi-drug resistant organisms, and reduced efficacy of the pre-existing antibiotics in the market, there is a need for development of novel antibiotics. Synthetic biology can be of significance in achieving the above mentioned targets that can also circumvent the drug resistant nature of the pathogens and develop novel methods of disease treatment.

As an initiative in this direction, Lu and Collins have described a strategy wherein the activity of pre-existing antibiotics such as quinolones can be enhanced via use of synthetically designed adjuvants. Molecules such as adjuvants, interfere with the survival pathways of the pathogenic organism that are initiated in response to the antibiotic treatment. Lu and Collins have designed a system where a bacteriophage M13mp18 has been engineered with SOS response repressor LexA3. Quinolones act by damaging the DNA of the bacteria in response to which the SOS DNA repair pathway gets activated in the bacteria. It has been shown that phage treated bacteria (*E.coli*) when exposed to the antibiotic has reduced rate of survival *in vitro* and *in vivo* (Lu and Collins 2009). Such high efficiency antibiotics can be developed further, however, their risk assessment at the preclinical and clinical stage is of crucial importance for its commercialization.

Another strategy to inhibit growth of pathogens is by exploiting the phenomenon of quorum sensing. Duan and March explored the signalling mechanism of *Vibrio cholerae* by engineering commensal micro-organisms such as *E.coli* with *V.cholerae* signalling molecules, CAI-1 and CAI-2. An increase in the concentration of CAI-1 and CAI-2 in the environment signals *V.cholerae* to inhibit its growth and stop the release of cholera toxin. *In vivo* analysis showed that the binding of cholera toxin to the intestine reduced to 80 % in mice treated with engineered *E.coli* as compared to the untreated mice (Duan and March 2010). Utilization of commensal microbial population as a chassis for synthetic biology based biomedical research has an advantage of easy administration, and also as these micro-organisms reside naturally in the gut microbiota, they can be given as probiotic formulations. However, it would be essential to study the effect of such engineered organisms on the resident organisms of the body.

Synthetic biology has also been used to control the insect vectors responsible for spreading the diseases as a preventive measure. Two strategies have been developed: a Homing Endonuclease Gene (HEG) based and Release of Insects carrying a Dominant Lethal (RIDL) technology (Windbichler et al. 2011; Fu et al. 2010). These two strategies have been employed on vectors spreading malaria and dengue respectively. HEG or homing endonuclease genes occur once per chromosome of many microbial organisms and they encode an endonuclease that is placed in the center of its recognition site. These endonucleases express specifically in heterozygous condition and can carry out site-specific recombination. The HEGs can be expressed under a male testis-specific promoter and allowed to home in transgenic male mosquitoes. Upon mating with wild type females, the offsprings will be heterozygous and can be induced to become non-viable or resistant to the malarial parasite, thereby, controlling the spread of the infection (Windbichler et al. 2011; Harris et al. 2011). On the other hand, RIDL technology is based on expression of conditional lethal cascade in female *Aedes aegypti*. The genetic circuitry is modified such that the expression of tetracycline-resistant transactivator is specifically restricted to indirect flight muscle under a tissue specific promoter. The gene harbors a sex-specific intron that induces expression of a toxic gene in females specifically in the absence of the antibiotic tetracycline. As a result, the female progeny produced is flightless and hence, not able to spread the parasite responsible for causing dengue to human populations (Fu et al. 2010; Wise de Valdez et al. 2011).

The most commonly practised methodology for preventing diseases from occurring is by vaccination. The field of vaccine development has evolved constantly, with the emergence of first generation vaccines centuries ago, comprising of live, attenuated or killed micro-organisms, to second generation of subunit vaccines and now a third generation of DNA vaccines (Kindt et al. 2000). Each new generation of vaccine is an attempt to encounter the limitations posed by the previous generations, e.g. the setback of reversion of attenuated micro-organisms in first generation vaccines to their virulent form is not prevalent in subunit or DNA vaccines (Robertson 1988; Baxter 2007). However, these new classes of vaccines have their own set of drawbacks that need to be overcome for designing more rational and efficient vaccination strategies. A synthetic biology inspired effort in creating a fourth generation of vaccines is SAVE or Synthetic Attenuated Virus Engineering. It is a technique wherein the viruses are re-encoded such that the wild-type amino acid sequence is retained with simultaneous rearrangement of the codon pairs to obtain attenuated form of the virus. Its efficiency has been tested *in vitro* against influenza and polio viruses (Coleman et al. 2008; Mueller et al. 2010). Codagenix Inc, USA is using the SAVE platform to develop live-attenuated viral vaccines against multiple targets (Codagenix Inc 2012). Although, the technique is useful for generating vaccines for disease prevention, however, there also exists fear of misuse and security threat as the same strategy can be employed for developing virulent forms of a pathogen. Such issues have been dealt with, in detail in the later sections of the chapter.

16.2.1 Synthetic Biology and Non-Communicable Diseases

As mentioned earlier, the global burden of diseases resides with non-communicable diseases (NCDs) such as cancer, diabetes, cardiovascular disorders and respiratory diseases. Currently, synthetic biology based research in biomedicine has major focus on cancer therapeutics. It has been reported that the incidence of occurrence of cancer in developing countries is increasing due to change in the lifestyle and eating habits. It is speculated that the rate of cancer cases diagnosed in developing nations will increase from 56 % of world total in 2008 to 60 % in 2030 whereas the rate of occurrence declined in USA by 1.9 % per year from 2001 to 2005 (Jemal et al. 2010; Seigel et al. 2011). The most practiced therapies for cancer treatment are chemotherapy, radiotherapy, surgical removal and transplantation in decreasing order (www.cancer.gov/cancertopics/treatment/types-of-treatment). Chemotherapy and radiotherapy pose major side-effects as the healthy tissues are also damaged making the patient more susceptible to other ailments. Hence, there is a need to develop novel methods for targeting the tumor specifically and efficiently.

Keeping this insight, development of cancer prevention has become an active area of research in synthetic biology based biodrug development. The following designer biology strategies have been developed for cancer therapy via SynBio-oncolytic designer viruses, tumor-targeting bacteria and cancer specific T-cells (Kirn and Thorne 2009; Forbes 2010; Porter et al. 2011). Oncolytic synthetic viruses are virus particles that contain prodrug convertases and cytokines. They are coated with an envelope to bypass immune system with specificity for cancer cells. Three of these virus types have reached the clinical trial stage. OncoVEX^{GM-CSF} is an attenuated Herpes simplex virus containing granulocyte-macrophage colony stimulating factor (GM-CSF), marketed by Amgen/BioVex, CA (Hu et al. 2006). It is targeted against metastatic melanoma and has reached Phase III of clinical trials. Reolysin®, marketed by Oncolytics Biotech, CA, is the second oncolytic virus to reach Phase III clinical trial stage. It is a reovirus with an engineered genetic circuitry such that it replicates specifically in RAS-activated tumors primarily targeting the head and neck tumors (Hingorani et al. 2011). The third virus—JX−594 marketed by Jennerex Biotherapeutics, USA has reached Phase IIb of clinical trials. It is an engineered virus targeting hepatocellular carcinoma, expressing GM-CSF (Heo et al. 2011).

The second strategy for targeting tumors via synthetic biology is by using bacteria as chassis instead of viruses. Bacteria have been used as a model system for studying synthetic biology based manipulations in genetic circuitry or for introducing a novel genetic network. It has been extensively used as a chassis for engineered pathways targeting tumorous cells. As an example, *E.coli* has been engineered with genes encoding for invasin and lysteriolysin O (proteins involved in invasion of mammalian cells and transporting molecules out of the endosomal vesicles, respectively) alongwith the RNAi machinery for knocking down the expression of CTNNB1, a gene involved in causing oncogenesis. It has been shown that the engineered *E.coli* successfully invaded the colon cancer cells and suppressed the growth of tumors in vitro as well as in vivo (Xiang et al. 2006).

As an alternative to engineering microbial systems, Porter et al. (2011) have engineered T-cell lymphocytes for targeting tumor cells. They have reported that genetically modified T-cells targeted CD-19 through transduction with lentiviral vectors expressing anti-CD19 signalling domains. The delayed development of tumor lysis syndrome and circulation time of three weeks are indicative of the fact that targeting the natural policing system can be a better target for manipulation as compared to microbial populations (Porter et al. 2011).

After cancer, the other major category of NCD affecting human population is type-2 diabetes mellitus. It has been reported that approximately 346 million people have diabetes with an estimated 3.4 million deaths in 2004. WHO has projected that the rate of deaths due to diabetes will double in 2005–2030 with more than 80 % deaths occurring in lower and middle income countries (Diabetes: WHO fact sheet 2012). Currently, there is no treatment available at the clinical level to control diabetes. In most of the chronic cases, the patients suffering from type-2 diabetes mellitus are given insulin injections. Hence, to control diabetes and improve blood glucose homeostasis, a synthetic transcription device based on melanopsin has been designed using synthetic biology approach. Glucagon-like peptide 1 (GLP-1), an anti-hyperglcyemic hormone with promising anti-diabetic activity, was placed under the control of light sensitive melanopsin-dependent promoter which induces a cascade leading to expression of GLP-1 only in the presence of blue light. The expression system in HEK293 cells was subcutaneously implanted in mutant mice and it has been shown that the glucose homeostasis was maintained by increasing the secretion of insulin in hypoglycaemic condition and when the glucose levels reached to normal, the whole cascade shuts down automatically (Ye et al. 2011).

16.3 Synthetic Biology for Designer Crops

Plants are emerging as an important system for synthetic biology based studies. They are not only a source of nutrition but also serve as a reservoir of several biopharmaceutical compounds that can be harnessed for the benefit of human health. Plants are also an important system for protein expression at a large scale. As such, agriculture sustains a number of economies in the world particularly biodiversity rich nations. The increasing demand for food crops due to rise in population can be fulfilled by increasing the yield of crops in a sustainable manner. Since, land acts as a limiting factor in increasing crop plantation, therefore, the crop yield needs to be enhanced by modifying the crops such that the yield per hectare is increased (Qaim and Zilberman 2003). The yield of the food crops can be enhanced by either increasing the rate of carbon fixation or by engineering plants such that they are able to sustain biotic and abiotic stress environments. The present section discusses the potential role of synthetic biology in developing tailor made crops. Synthetic biology can play a vital role in increasing the rate of carbon fixation, thereby, enhancing the agronomical yield of the plant. It can also help in replicating the metabolic pathways leading to production of medicinally active compounds that are otherwise difficult to isolate from plants directly (for example artemisinin) in another micro-organism.

Recently, it has been proposed to increase the rate of carbon fixation via synthetic biology based approach either by directly targeting the Calvin cycle in plants or by targeting bacterial carboxysomes as described below (Bar-Even et al. 2010; Bonacci et al. 2011, 2012). Calvin cycle is the most important step in converting free CO_2 in a consumable form of carbohydrates in C3 plants. Since it determines the agronomical yield of a plant, numerous efforts have been made to increase the efficiency of Calvin cycle. In an atmosphere where there is abundance of raw materials such as sunlight, CO_2 and water, the activity of the enzymes involved in the fixation of carbon can be a limiting factor. Although, efforts have been made at increasing the activity of Rubisco, the primary enzyme of C3 cycle via genetic manipulation, the synthetic biology based approach views the problem from a different dimension. Bar-Even et al. (2010) have suggested a modification of the C3 cycle in a way such that Rubisco is replaced with a more efficient enzyme derived from alternative CO_2 fixing cycles such as C4 cycle. A systematic in silico search of comparison of several enzymes, their activities and pathway specificities, revealed that the MOG pathway or Malonyl-CoA-Oxaloacetate-Glyoxylate pathway was the most specific and would yield highest rate of carbon fixation. It utilizes three carboxylating enzymes: PEP, pyruvate and acetyl-CoA carboxylase with high specificity and activity under saturating CO_2/HCO_3^- conditions to increase rate of carbon fixation (Bar-Even et al. 2010). Incorporation and expression of a designer synthetic carbon fixation cycle in plants would help achieve increase in agronomic yield in a sustainable manner.

In another study, attempts have been made at developing modular carbon-fixing microcompartments in a heterologous host. Carboxysomes from *Halothiobacillus neopolitanus*, with a single operon for fixing CO_2 were heterologously expressed in *E.coli* leading to the production of complexes that are similar to the native host. Carboxysomes have been engineered with pore-forming protein, CsoS1D and it has been observed that CsoS1D expressed and functioned appropriately in engineered *E.coli*. The study illustrates that CO_2 fixation can be also enhanced in bacterial systems and the self-assembling stable structures can be produced in vitro synthetically with the desired modification (Bonacci et al. 2012). The latter studies can be extended to the designing of synthetic organelles and compartmentalized structures (discussed in later section).

16.3.1 Metabolic Engineering of Medicinally Important Compounds

As mentioned above, plants have an enormous potential in the biopharmaceutical industry due to the valuable secondary metabolites produced by them. Plant derived compounds such as taxol, vincristine, vinblastine, curcumin are well known for their anti-tumor and anti-inflammatory properties (Martin et al. 2003; Guo et al. 2006; Aslam et al. 2010). Since these and other biopharmaceutical compounds are usually produced as secondary "products", they are often difficult to harvest directly from the plants at industrial scale as it is a complex and cost-intensive process. Hence, the present research of natural medicinal products aims at bioengineering the metabolic pathways leading to the production of these compounds in microbial systems for their efficient production and ease of extraction. A benchmark example for metabolic engineering of plant secondary metabolites via synthetic biology approach is that of artemisinin overproduction.

Artemisinin is derived from the plant *Artemisia annua* and is known for its anti-malarial activity. Since the molecule is difficult to extract from the plant directly, Keasling and co-workers have engineered the biosynthetic pathway of artemisinin in *E. coli* leading to the production of its precursor artemisinic acid in vitro (Martin et al. 2003). The pathway has been re-engineered in yeast such that artemisinic acid is secreted outside the cells and is easy to harvest. Recently, it has been reported that a cluster of plant and yeast genes in a synthetic expression cassette has been successfully transformed in tobacco plants for commercial production of artemisinin (Graham et al. 2010; Farhi et al. 2011). Similarly, other plant based compounds such as vincristine and taxol can be also produced at an industrial scale following synthetic biology approach. However, before such products enter the market; the regulatory framework for commercialization of such products needs to be in place to enhance user acceptability.

"PhytoMetaSyn" project supported by Genome Canada and Genome Alberta aims at studying the biosynthetic pathways of over seventy-five plant species known for production of high value secondary metabolites. Such biosynthetic pathways will be engineered in yeast cells using synthetic biology based approach. The project aims at creating a public resource of metabolic and genomic information of plants producing high-value products, developing yeast strains producing such products, generation of catalogue of plant enzymes for further applications in synthetic biology and development of novel methods for identification of unknown biosynthetic genes and metabolic pathways. Analysis of the socio-economic impact along with the environmental and legal implications of microbial engineering for metabolite production via synthetic biology will also be addressed in the project (PhytoMetaSyn 2012; Genome Alberta 2012).

A number of secondary metabolites such as carotenoids, terpenoids and isoprenoids are produced in plastids instead of cytoplasm (Croteau et al. 2000). Hence, engineering the pathways leading to enhanced production of such metabolites would require engineering of the organelle *per se*, evolving into an entirely novel concept of synthetic organelles as described below.

16.3.2 Synthetic Organelles

The concept of designing synthetic organelles is very intriguing. Synthetic organelles could be classified as modified organelles, i.e. organelles engineered to perform a novel function or artificially introduced endosymbiotic microbes in eukaryotic systems. Engineered organelles can act as excellent synthetic biology based devices with potential applications as "prosthetic organelles" and biosensors. For instance, vacuoles in plant cells can be engineered to act as subcellular biosensors by sequestering excess free radicals and making cells resistant to highly toxic environments increasing the rate of survival of plants in harsh or toxic environments (Synthetic Biology: Key field of the future 2012).

As mentioned before, carboxysomes, the carbon fixing organelles of bacteria have been heterologously expressed in another bacterial strain (Bonacci et al. 2011, 2012). This can be further extrapolated to the introduction of organelles from one family of organisms to another in order to understand the evolution, behaviour and gaining other fundamental insights into the functioning of the present compartmentalized chambers inside a cell. Silver and co-workers have studied the replication and behaviour of non-pathogenic photosynthetic bacteria *Synechococcus elongatus* in zebrafish embryos. It has been reported that *Synechococcus* bacteria were able to replicate and photosynthesize efficiently even upon transfection in mammalian cells (Agapakis et al. 2011). The bacteria behaved as a synthetic chloroplast inside a mammalian cell hinting at the evolutionary development of eukaryotic organelles such as mitochondria and plastids. Such studies can also form a basis for development of prosthetic organelles in humans. Malfunctioning organelles such as mitochondria lead to manifestation of several disorders such as Leigh's syndrome (Chinnery 2010). The dysfunctioning of mitochondria is also related to heart ailments, diabetes and Alzheimer's disease. Designing a functional mitochondria and transforming it into the cells might help in curing the diseases arising due to a defective copy of the mitochondria. Although, such goals are very far-sighted and futuristic in vision, they could be feasible in practice since organelles such as mitochondria and chloroplast possess their own genetic circuitry and metabolic mechanisms which work independent of the other cellular processes and act as potential sites for introduction of synthetic pathways (Boyle and Silver 2009).

As described in the earlier sections, synthetic biology has potential applications in the field of biodrugs and designer crops. Although, the applications are for the benefit of mankind, they can also pose a risk, if misused. Therefore, regulating the use of synthetic biology based tools as well as methods is critical. A close examination of associated socio-ethical, biosafety, biosecurity and standardization is relevant for economic growth of synthetic biology. These implications have been analysed in the sections ahead.

16.4 Governance of Synthetic Biology based Biodrugs and Designer Crops

Owing to development of synthetic biology in diverse spheres such as biomedicine, biofuels, biomaterials and industrial chemicals, the global synthetic biology market is estimated to reach \$ 4.5 billion over the year 2015 (Synthetic Biology Market—Global Industry Analysis, Size, Growth, Share And Forecast, 2012–2018, 2012).

Many strategies, as discussed in the previous sections, have been used to develop organisms with a specific and defined functionality for a particular purpose under the synthetic biology regime e.g. SAVE strategy has been used by Codagenix Inc, USA for development of live-attenuated viral vaccines against multiple targets (Codagenix Inc. 2012). These engineered organisms designed for desired functionality have been referred to as designer organisms. In order to create designer organisms, the synthetic biology community requires ease of accessibility to different parts and procedures as well as interactions among the community. To achieve this goal, freedom to research is the key. However, it is important to balance the freedom of research to ensure societal benefits and at the same time minimise the risks involved (Fig. 16.2). The manner in which, the therapeutics produced in microorganisms and the genetic constructs used as therapeutics, are interfaced with the patients will determine safety as well as efficacy of synthetic biology *per se* e.g. the mode of delivery of genetic circuits would determine its degradability and ability to elicit immune reactions. Therefore, such biodrugs would need to pass through different phases of clinical trials before being released into the market. At present, in order to commercialize synthetic biology based biodrugs as well as designer crops and to gain public acceptance, it is pertinent that the various aspects related to biosafety, intellectual property, standardisation, regulatory as well as socio-ethical issues are dealt with in accordance. Governance of a technology involves a variety of stakeholders and plays a crucial role in steering the development of the field. The various described aspects of SynBio have to be included to form an effective governance system for synthetic biology, of a nation or for cross border regulation. Although, there are frameworks in place for the regulation of biodrugs and transgenic crops in some nations, however, the applicability of the provisions for synthetic biology derived products needs to be assessed. A critical analysis at this point of time is important, when the research is progressing and the products shall be launched in the market for future use. The present section deals with the above mentioned issues pertaining to the use of synthetic biology in biodrugs and designer crops.

16.4.1 Biosafety

Assessment of risk in terms of biosafety associated with synthetic biology based research in biodrugs production and designer crops is essential at this stage. This subsection first discusses the biosafety aspects related to the biodrugs and then moves to the designer crop biosafety issues in context of synthetic biology. Using synthetic biology, microorganisms can be engineered for use as vaccines or probiotics. They are employed as strategy for eradication of vectors such as mosquitoes. However, due to these approaches, questions regarding interactions of engineered microorganisms with the commensal microorganisms and non-target organisms have been raised. The fear of negative impact of such engineered organisms on the non-target organisms or environment exists, making risk assessment studies more relevant than ever before (Dana et al. 2012). Another area of concern is safety of the workers, the

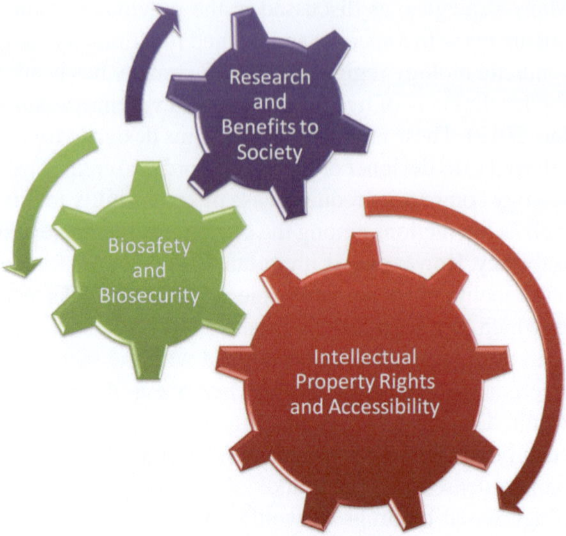

Fig. 16.2 Schematic figure to represent interconnection between freedom to research and acceptability

accidental pricks by needles or inhalation of such manipulated microorganisms pose a health risk (Gutmann et al. 2010). Therefore, a governance system to regulate the use and release of such manipulated organisms is required urgently. Recognizing these concerns, the International Civil Society Working Group on Synthetic Biology submitted a report to the Convention on Biological Diversity's Subsidiary Body on Scientific, Technical and Technological Advice (SBSTTA) on the 'Potential Impacts of Synthetic Biology'. The report emphasized that currently no intergovernmental body focuses on the effect of synthetic biology on land, biodiversity and humans (Potential Impacts of Synthetic Biology 2011). Based on the SBSSTA recommendations, the emerging issues of synthetic biology and its impact on the biodiversity were discussed in one of the events at the COP11 meeting held on October 12, 2012 in India. Emphasis was laid on impact assessment of synthetic biology on conservation and sustainable use of biological diversity and associated social, economic and cultural considerations (Convention on Biological Diversity 2012).

Common concerns have been attributed to the synthetic biology based crops and genetically modified plants. Synthetic biology manipulated plants with better qualities, would require stringent containment to minimise the chances of transfer of genes to other plant varieties and other organisms. Containment would also avoid cross contamination through pollination. Such designer plants may also impact the food chain and the biodiversity (Dana et al. 2012). This issue has also been discussed under the section on regulatory regime for synthetic biology in the present study.

The project "PhytoMetaSyn" focusing on engineering of metabolic pathways of medicinal plants in common yeast using synthetic biology approach, will also deal with the associated socioeconomic, environmental, legal and ethical impact of the research (PhytoMetaSyn 2012). Such initiatives, if replicated, can contribute towards development of risk assessment methodologies, identifying the key constraints and increasing acceptability of synthetic biology based designer crops.

Besides issues of biosafety, there are other important parameters that will play a critical role in determining success of synthetic biology based products. Intellectual property especially patents, are one of the tools that will play significant role in revenue generation for various entities engaged in synthetic biology research. The next section discusses the intellectual property regime for synthetic biology based biodrugs and designer crops.

16.4.2 Intellectual Property: Incentives and Public Benefits

In the emerging biobased economy and increasing demand for bioproducts and enhanced focus on sustainability, synthetic biology is one such field that has the capacity to meet the stated challenges through development of biodrugs and designer crops. Intellectual property would serve as an impetus for the sustainable development of biobased economy. Further, intellectual property has an essential role to play in growth and commercialization of synthetic biology based biodrugs and designer crops. The following section explores the scope of protection for the synthetic biology based biodrugs and designer crops as well as the tools, under intellectual property regime.

16.4.2.1 Patentability of Genetic Circuits and Parts

The number of patent applications filed in the field of synthetic biology is continuously increasing. Methods of developing synthetic DNA strands, compositions, genes or parts of genes to the methods of metabolic engineering are subject matter of protection under the patent regime (Saukshmya and Chugh. 2010). However, the debate on whether isolated DNA sequences are patentable is still continuing. The recent judgement of the Supreme Court of US in the ***Association for Molecular Pathology v. Myriad Genetics*** case reaffirms that the isolated DNA sequences are patentable. Myriad genetics had applied for patent on the *BRCA*1 and *BRCA*2 human genes implicated in breast and ovarian cancer along with method of detecting mutations in these genes. Association for Molecular Pathology filed a declaratory suit claiming that both genes and the method of detection of these genes, do not fall under the purview of patentability criteria. In 2010, the district court had held both genes and the method of detecting as non-patentable, however, this judgement was partly reversed by the US supreme court in 2012 allowing genes to be patentable while maintaining that the method of detection does not involve any transformative step, therefore, rendered non-patentable (Federal circuit decision of August 16, 2012). This judgement would pave way for increased patent filings on gene sequences, in general and genetic circuits in particular for synthetic biology based research. The processes involved in engineering the organisms using genetic circuits will also fall under the purview of such protection. However, there are some exceptions under the national patent laws, which are discussed under the following sub sections.

16.4.2.2 Patents and Synthetic Biology Based Biodrugs

Patentability of an invention is dependent on the three criteria namely novelty, inventive step and industrial application. Synthetic biology based biodrugs can be protected through patents, in case they are able to meet the described criteria. Other than the criteria, the national patent laws have also excluded some inventions from the scope of protection, for example, in case of India, inventions such as the method of treatment or therapeutics based on traditional knowledge are not patentable. Non-patentability of traditional knowledge based therapeutics has been discussed in detail by Jain et al. (2012).

In case of biological inventions, there are two mandatory requirements of patenting to be fulfilled by the applicant. One, the applicant has to disclose the source of the biological material and secondly deposit the biological material in an International Deposition Authority (IDA). Absence of such information can be used as a ground to either oppose or revoke the patent. These requirements can be extended to the synthetic biology based inventions as well. The engineered microorganisms can be deposited in the IDA so that these are available for further research. The disclosure of source will ensure that the origin of the sequence or the biological part is traceable and also serves as an acknowledgment or recognition. The disclosure of source can also play an important role in agricultural innovation, however, the rules for patenting synthetic biology based designer crops, as discussed below may vary from one nation to another.

16.4.2.3 Designer Crops and IP

Although, all the WTO members are TRIPS compliant and they are under obligation to extend protection for products as well as process under the national patent law. However, Article 27(3)(b) of TRIPS Agreement has permitted its member states to opt for not granting patents for "plants and animals, other than microorganisms, and essentially biological processes for the production of plants or animals other than non-biological and microbiological processes." Both India as well as Europe have excluded plants and the biological processes for their production from the scope of protection under the patent law. As per the Section 3(j) of the Patent Act, 1970 (Amendment 2005) reproduced below, plants do not fall under the patentable inventions in India:

> Plants and animals in whole or any part thereof other than micro-organisms but including seeds, varieties and species and essentially biological processes for production or propagation of plants and animals.

As discussed above, the genetic constructs and processes for generating engineered crops would be considered as non-biological process. Therefore, the genetic constructs and the procedure that will result in genetically modified crop may be patentable. However, the designer crop varieties cannot be protected through patents. In order to maintain the balance between incentives and public rights, a weaker form of protection system for the designer crops can be employed. The option of opting for a *sui generis* system for protection of plant varieties was made available to the member states as per the Article 27.3(b) of TRIPs agreement. Majority of the member states including India enacted legislations for protection of plant varieties, to promote development of new plant varieties. Under Section 15 of the Protection of Plant Varieties and Farmers Right Act 2002, a new plant variety can be registered under the Act if it conforms to the criteria of novelty, durability, uniformity and stability. As per the definition of Variety in the Act (Section 2za), a Variety means a plant grouping except micro organism within a single botanical taxon of the lowest known rank, which can be (1) defined by the expression of the characteristics resulting from a given genotype of that plant grouping; (2) distinguished from any other plant grouping by expression of at least one of the said characteristics; and (3) considered as a unit with regard to its suitability for being propagated, which remains unchanged after such propagation, and includes propagating material of such variety, extant variety, transgenic variety, farmers' variety and essentially derived variety. However, it has yet to be ascertained whether transgenic crops will be distinguished from synthetic biology derived crops. The *sui generis* system of protection of plant varieties can play an important role in protection of the varieties developed through synthetic biology provided the variety meets the criteria of novelty, durability, uniformity and stability.

16.4.2.4 Protection of Databases

Copyright is the most common form of intellectual property right considered for protection of software and databases. An open access approach to promote exchange of ideas and promote collaborations is advocated by most scientists. It has been critically debated in various forums whether an open access collaborative mode of research or licensing would be a better option to enhance growth of synthetic biology based biodrugs and designer crops. If all the Biological Parts are patented, it would restrict the freedom to operate; therefore, the first Registry of Parts formed by BioBricks Foundation is under an open common license (Torrance 2010). The same format has been also adapted by the new Registries such as the Joint BioEnergy Institute Inventory of Composable Elements. Accessibility to various tools and databases for production of biodrugs and designer crops is required. Therefore, the open access approach would make the databases more accessible, however, the usage would be dependent on interoperability, therefore, standardization is the key to success. The following section is an overview of the existing and upcoming registries as well as standards in use for synthetic biology.

16.4.3 Standardization of Parts and Registries

Synthetic biology is evolving with focus on enablement of the scientists to tailor-made organisms as per the needs of human beings. To achieve this objective, standardization as well as the ease of customization have been recognized as two important factors. In order to simplify the process and shorten the time of development of designer organism as well as enhance accessibility, a library of parts has been created at MIT, USA. The Registry is an inventory of the parts created and is accessible globally to all the scientists. Availability of parts and tools developed, for use by the researchers is critical for the growth of synthetic biology. An open source Registry that can be used by academia or company has potentially contributed towards synthetic biology based studies. Till now, most of the Registries are maintained by institutes and the companies have capitalised on it (Henkel and Maurer 2007). However, as with the human genome project, a similar situation may arise wherein the Registry is created and maintained by companies which would be accessible by the subscribers only.

The basic steps for using a registry for the development of vaccines or recreating metabolic pathways are shown in (Fig. 16.3). Exchange of information and interoperability of parts is crucial for the development of various designs of genetic circuits in the field of synthetic biology. Also standardisation is required for synthetic biology devices and systems for accurate reproducibility. Standardisation can also help in reducing the investment costs and time associated with the development along with enhanced reliability. Therefore, the existing Registries and the upcoming ones need to have a uniform format for information on the parts. Standard part has been clearly defined by Canton et al. (2008), as a "Standard Biological Part to be a genetically encoded object that performs a biological function and that has been engineered to meet specified design or performance requirements". In order to achieve standardization in synthetic biology, researchers at MIT created 'The Registry of Standard Biological Parts'. Although, it is one of the pioneer Registries, however, it does not guarantee that many such parts can exhibit the same activity in different assemblages (Newman 2012). Another database of engineered biological components, similar to the Registry of Standard Biological Parts is proposed to be created at the Imperial College, UK. An open source Registry to manage information on Biological Parts is the Joint BioEnergy Institute Inventory of Composable Elements (JBEI-ICEs) that records information of parts such as plasmids, microbial host strains as well as DNA that can be employed to produce a designer organism (Ham et al. 2012). Other than the Registry, the format for exchange of information that defines modules, how the parts are interlinked and characterization of behaviour also needs to be standardized. Characterization of long-term performance, behaviour, stability, and fate of synthetic circuits also needs to be mentioned in the datasheet (Cheng and Lu 2012). Standards are gradually being developed for documentation and effective exchange of information among the synthetic biology community. Synthetic Biology Open Language (SBOL) is one such Standard to promote exchange of information related to DNA components employed in synthetic biology. It emphasises on preferred terminology

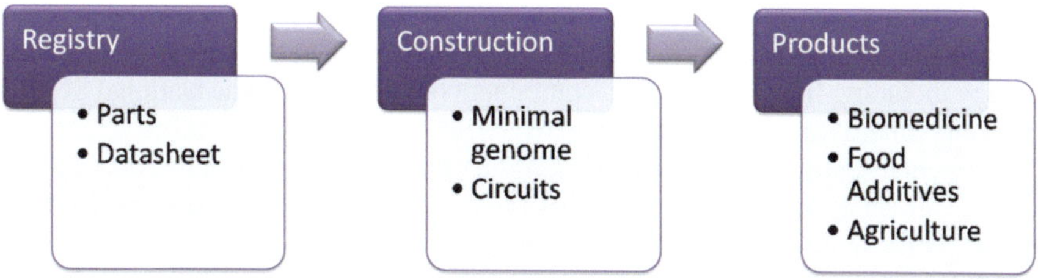

Fig. 16.3 Schematic figure to show the relevance of biological parts registry

for the parts and how the parts are interconnected, so that the same can be reproduced. Another similar Standard is DICOM-SB pertaining to metadata and images related to a Biological Part (Kitney and Freemont 2012). BioBrick has established various standards. Based on these standards, other standards such as BglBrick Standard have been proposed for construction of metabolic pathways in various combinations for improved gene expression (Lee et al. 2011). It is speculated such strategic efforts for standardization of parts will help in accelerating the growth of synthetic biology in the field of biodrugs as well as designer crops by providing the case of interactions and collaboration at the international level. No novel technology or field of research comes without a set of apprehensions among public. Similarly synthetic biology, a multidisciplinary field with a potential to create novel molecules, organisms, has been viewed by the public as "playing God". Therefore, socio-ethical issues for synthetic biology are important area of concern. The section below examines the concerns that public may have with regard to synthetic biology. The measures that can alleviate such concerns among public are also discussed.

16.4.4 Socio-Ethical Issues: Fear of Misuse and Public Engagement

Synthetic Biology can lead to creation of completely new systems or old systems can be reengineered for the human welfare e.g. using the SAVE technique, viruses can be reengineered to form efficient vaccines or systems such as *Mycobacterium* created by Craig Venter and his team in 2010 (Gibson et al. 2010). This enablement of synthetic biology of creating new or reengineered organisms has triggered a debate for its misuse as discussed below.

16.4.4.1 Dual Use of Synthetic Biology

Any new technology that finds application towards benefiting mankind could also be used for its destruction. For example the rocket technology was developed for space explorations, however, the same technology also contributed in development of ballistic missiles or the use of internet for faster pace of communication has also been misused for delivering viruses to corrupt systems. Thus, every emerging technology including life sciences has a component of uncertainty. The National Science Advisory Board on Biosecurity (NSABB), USA defines dual use in life sciences as "biological research with legitimate scientific purpose that may be misused to pose a biologic threat to public health and/or national security" (National Science Advisory Board on Biosecurity 2012). As discussed in the earlier sections, synthetic biology has the potential of engineering viruses to target cancer or engineering *E.coli* for blocking the expression of a gene involved in causing oncogenesis. Although, these strategies are providing new approaches for treatment of diseases, yet, there are chances of misuse of these engineered organisms. The misuse could be intentional with the aim to harm a particular community or nation or unintentional, however, in any of the cases, it still remains a security threat. Significant concerns have been raised on misuse of research tools, procedures, scientific knowledge and genetic material, by disgruntled lab personnel or those without any institutional affiliation (The goldilocks dilemma and polycentric governance risks and regulation in synthetic-biology 2012). Dual use of synthetic biology is an issue that the researchers, policy makers and public have to confront as well as resolve. There is a consensus on importance of measures to mitigate threat that prepares a country to combat bioterrorism arising out of synthetic biology based research (Suk et al. 2011). Monitoring of research, competitions, and especially of 'do it yourself' projects is essential to ensure biosecurity (Edwards and Kelle 2012). If effective strategies to curb easy access to the tools are not developed in the right time, monitoring its use would be even further more challenging. In order to screen the use of sequences of concern, framework such as 'The Screening Framework Guidance for Providers of Synthetic Double Stranded DNA' has been provided by the US Federal Government (The Screening Framework Guidance for Providers of Synthetic Double Stranded DNA 2010). Software such as GenoThreat based on the screening framework by the US Government has been also developed (Adam et al. 2011). In addition International Association Synthetic Biology (IASB) and the International Gene Synthesis Consortium (IGSC) have developed guidelines for the implementation of sequence screening tools as well as mechanisms to monitor the orders of potentially dangerous sequences (Code of conduct for best practices in gene synthesis 2012; Harmonized screening protocol: gene sequence and customer screening to promote biosecurity 2012). However, to what extent the framework is implemented and the guidelines are followed, would require a continuous process of monitoring. A new web based Synthetic Biology Scorecard has been launched by Woodrow Wilson Centre engaged in the synthetic biology project, with the aim to track the developments in governance and risk management framework (Synthetic biology scorecard finds federal agencies responding to presidential bioethics commission report 2012). The score card will also monitor the steps taken towards the implementation of the recommendations made in the Presidential Commission of USA, for the study of bioethical issues related to synthetic biology. More of such initiatives can contribute in developing effective governance of dual use research in synthetic biology.

16.4.4.2 Public Engagement

Difficulty in retrieving the designer organisms or controlling them after their environmental release (intentional or unintentional) is one of the major reasons of concerns among public. Some of the concerns (sometimes fears) associated with synthetic biology can be addressed through various modes of public engagement such as internet, newspapers, scientific reporting and workshops. Ultimately it is the dialogue between the public and the scientists that can bridge the gap of understanding of positive impact of synthetic biology on human health as well as environment. In order to address the concerns of the safety and enhance acceptability of the synthetic biology derived product in medicine and agriculture, public engagement and awareness is important (Anderson et al. 2012). Print media has been instrumental in informing the public about the emerging trends of life science technologies, that is why it is imperative that the media carries out a balanced reporting (Kronberger et al. 2012). Internet is now being increasingly used for public participations such as the European SYNBIOSAFE project used internet to obtain opinion of the public on societal issues of synthetic biology (Schmidt et al. 2008). However, the task does not end at obtaining opinions, but their inclusion in the framework is an important step towards building public trust and promoting a transparent form of governance (Zhang 2012). The following section analyses the amendments that may be required in existing regulatory regime for effective governance of synthetic biology.

16.4.5 Harmonization of Regulatory Regime

As the number of companies (Table 16.1) and collaborations in the field of synthetic biology are increasing, the need for its regulation is also fast emerging. As discussed in the section on biosafety and socio-ethical issues, curbing both accidental as well as intentional release of synthetic biology derived microorganisms calls for a strategic framework and policy decisions for effective governance of synthetic biology. Since the potential application of synthetic biology is in the field of biodrugs and designer crops, the existing regulatory frameworks and laws need to be reviewed to accommodate the issues related to synthetic biology. However, law being territorial in nature, there are inherent differences in the scope of each legislation. Therefore, a harmonized regulatory regime can serve the purpose for synthetic biology based developments that are more on a global platform. As synthetic biology is at a nascent stage and its contribution is crucial in the field of biodrugs and designer crops, an adaptive form of governance that can evolve with the further growth of synthetic biology can be an important alternative path. At the same time, a uniform governance mechanism for monitoring collaborations and trans-border research also needs to be in place (Zhang 2012). In 2010, Presidential Commission for the Study of Bioethical Issues, USA recommended that regulation of synthetic biology can be done on the basis of principle of public benefits, responsible research, freedom to research and public engagement. The Commission emphasized on self regulation however, there was limited focus on risk assessment mechanisms (Letter from Civil Society to President's Commission on Synthetic Biology 2010). In response to the recommendations of the Commission, a group of 111 civil society organizations have agreed to develop a new set of Principles for the Oversight of Synthetic Biology for effective assessment of synthetic biology (The Principles for the Oversight of Synthetic Biology 2012). The principles proposed cover precautionary measure, synthetic biology specific regulations, mechanism for protection of public health, workers' safety and environment. It remains to be ascertained what impact these new set of principles would have on developing governance strategies of synthetic biology, especially in the case of plant based therapeutics and allied products.

16.4.5.1 Bioresource Based Traditional Knowledge and Synthetic Biology

Plant based therapeutic drugs have regained their importance in the modern pharmaceutical era, such as taxol for cancer (Guo et al. 2006), quinine for treatment of malaria (Aslam et al. 2010). Bioresource based traditional knowledge has played an important role in bioprospecting and development of plant derived drugs against diseases. The use of artemisinin as an antimalarial is based on traditional knowledge (Hsu 2006). Artemisinin production has now been engineered into yeast for mass production. However, it remains to be assessed that, though artemisinin is a plant derived product and resourced from traditional knowledge, after its production through yeast, will it still be considered as a traditional medicine based drug. On the other hand, it is noteworthy for most of the plant based traditional medicines, neither pre market approvals are required nor any international legislation governing safety, efficacy and quality exist. In case the product is shown to be based on established traditional knowledge, safety and efficacy are not tested (Bubela et al. 2011). However, in the present case, artemisinin is neither plant derived anymore nor a traditional medicine (generally a complex mixture of various herbal components), the preclinical studies become critical to ensure its safety. As emphasised in the Principles for the Oversight of Synthetic Biology, there is a need to develop synthetic biology specific legislation for its proper regulation. Further, the existing drug approval laws need to be amended to include synthetic biology derived biopharmaceutical products.

16.4.5.2 Synthetic Biology Based Designer Crops

Synthetic biology can be used for developing designer crops with modified flavour, nutritional value or reduced allergenicity. Such crops would be regulated as genetically modified crops or a new set of regulation would be required is an ongoing debateable question (Torrance 2012). As examined above, there are chances of transfer of genes to organisms as in case of genetically modified plants, therefore, there is a similar need of containment of such plants. However, whether the scope of regulation for genetically modified crops would be broadened to include synthetic biology developed crops or specific regulations would be enacted, will become clear with the further development of synthetic biology in the field of agriculture.

16.5 Conclusions

Synthetic biology, a cocktail of various fields of science has profound applications in the field of biodrugs as well as designer crops. The growth of synthetic biology can be ascertained from diverse research projects in different areas. Also the number of spin outs based on university research in the area of synthetic biology is increasing. It is possible that soon some of the synthetic biology based products will reach to market shelves, such as artemisinin will be ready to enter the market by 2015. The success of these products in reaching the market will trigger exploration of different pathways that can be engineered into microorganisms for easy harvesting. Synthetic biology also expands the scope of drug development process by not only providing a strategy for exploiting the natural reservoir of drug molecules but also by offering methods of increasing the efficacy of currently FDA approved drug molecules. Apart from drug development, synthetic biology also provides avenues for engineering “bio-devices” for diagnosing, imaging and parenteral administration of drug molecules. The engineering of bacterial systems for targeting hypoxic cancer tissues is an excellent example of biosensors based on synthetic biology approach.

Initially, the prime focus of applications derived out of synthetic biology was on bioenergy, bioremediation and biomedicine but gradually the agriculture sector is also coming under the purview of synthetic biology based applications. Engineering of biosynthetic pathways for crop improvement such as increasing the yield is one of the primary targets. It would be interesting to know, whether COP 11 takes a proactive role to regulate the release of designer organisms including agricultural crops derived from emerging field of synthetic biology.

The intellectual property assets (patents, copyright or plant variety protection) would determine the freedom to operate that an enterprise may have for commercialization of synthetic biology based biodrugs and designer crops. Safety and standardization of the parts used for the development of products would also form essential component for commercialization of synthetic biology research. The emerging challenges for a synthetic biologist constitute developing robust circuits and pathways with enhanced functionality that can be addressed using principles of engineering. Synthetic biology being multidisciplinary and a field that has implications on health as well as environment, makes it mandatory to increase awareness on safety issues among the non-biologists. Besides enhancing awareness, a synthetic biology specific framework to regulate research and commercialization in this field, will strengthen the growth of synthetic biology based bio-economy.

Acknowledgements Aastha Jain is thankful to the Council for Scientific and Industrial Research (CSIR), Government of India for the award of Senior Research Fellowship.

References

Adam L, Kozar M, Letort G, Mirat O, Srivastava A, Stewart T, Wilson ML, Peccoud J (2011) Strengths and limitations of the federal guidance on synthetic biology. Nat Biotechnol 29:208–210

Agapakis CM, Niederholtmeyer H, Noche RR, Lieberman TD, Megason SG, Way JC et al (2011) Towards a Synthetic Chloroplast. PLoS ONE 6(4):e18877–e18885

Amyris Inc (2012) http://www.amyris.com. Accessed 20 Sept 2012

Anderson J, Strelkowa N, Stan GB, Douglas T, Savulescu J, Barahona M et al (2012) Engineering and ethical perspectives in synthetic biology. EMBO Rep 13:584–590

Aslam J, Khan SH, Siddiqui ZH, Fatima Z, Maqsood M, Bhat MA et al (2010) Catharanthus roseus (L.) G. Don. An important drug: its application and production. Int J Compr Pharm 4:1–16

Bar-Even A, Noor E, Lewis NE, Milo R (2010) Design and analysis of synthetic carbon fixation pathways. Proc Natl Acad Sci U S A 107:8889–8894

Baxter D (2007) Active and passive immunity, vaccine types, excipients and licensing. Occup Med (Lond) 57:552–556

Biobricks Foundation (2012) http://bbf.openwetware.org. Accessed 27 Sept 2012

Brenner K, Arnold FH (2011) Self-organization, layered structure and aggregation enhance persistence of a synthetic biofilm consortium. PLoS ONE 6:e16791–e16798

Bonacci W et al. (2011) Modularity of carbon fixing protein organelle. Proc Natl Acad Sci U S A 109: 478–483

Bonacci W, Teng PK, Afonso B, Niederholtmeyer H, Patricia G, Silver PA et al (2012) Modularity of a carbon fixing protein organelle. Proc Natl Acad Sci 109:478–483

Boyle PM, Silver MA (2009) Harnessing nature's toolbox: regulatory elements for synthetic biology. J R Soc Interface doi: 10.1098/†rsif.2008.0521.focus

Bubela T, Hagen G, Einsiedel E (2011) Synthetic biology confronts publics and policy makers: challenges for communication, regulation and commercialization. Trends Biotechnol 30:132–137

Canton B, Labno A, Endy D (2008) Refinement and standardization of synthetic biological parts and devices. Nat Biotechnol 26:787–793

Cheng A, Lu TK (2012) Synthetic biology: an emerging engineering discipline. Annu Rev Biomed Eng 14:155–178

Chinnery PF (2010) Mitochondrial disorders overview. http://www.ncbi.nlm.nih.gov/books/NBK1224/. Accessed 29 Sept 2012

Codagenix Inc. (2012) http://www.codagenix.com/. Accessed 20 Sept 2012

Coleman JR, Papamichail D, Skiena S, Futcher B, Wimmer E, Mueller S (2008) Virus attenuation by genome-scale changes in codon pair bias. Science 320:1784–1787

Convention on biological diversity. www.cbd.int. Accessed 20 September 2012

Croteau R, Kutchan TM, Lewis NG (2000) Natural products (secondary metabolites). In: Biochemistry and molecular biology of plants. Buchanan B, Gruissem W, Jones R Eds, pp 1250–1319

Cuccato G, Della Gatta G, di Bernardo D (2009) Systems and synthetic biology: tackling genetic networks and complex diseases. Heredity (Edinb) 102:527–532

Dana GV, Kuiken T, Rejeski D, Snow AA (2012) Synthetic Biology: four steps to avoid a synthetic-biology disaster. Nature 483:29

Demirbas A, Demirbas MF (2011) Importance of algal oil as a source of biodesel. Energy Convers Manage 52:163–170

Diabetes: WHO fact sheet (2012) http://www.who.int/mediacentre/factsheets/fs312/en/ Accessed on 28 September 2012

Duan F, March JC (2010) Engineered bacterial communication prevents *Vibrio cholerae* virulence in an infant mouse model. Proc Natl Acad Sci U S A 107:11260–11264

Edwards B, Kelle A (2012) A life scientist, an engineer and a social scientist walk into a lab: challenges of dual-use engagement and education in synthetic biology. Med Confl Surviv 28: 5–18

Farhi M, Marhevka E, Ben-Ari J, Algamas-Dimantov A, Liang Z, Zeevi V et al (2011) Generation of the potent anti-malarial drug artemisinin in tobacco. Nat Biotechnol 29:1072–1074

Federal circuit decision of August 16, 2012. Federal circuit. August 16, 2012. http://www.cafc.uscourts.gov/images/stories/opinions-orders/10-1406.pdf. Accessed 20 Sept 2012

Forbes NS (2010) Engineering the perfect (bacterial) cancer therapy. Nat Rev Cancer 10:785–794.

Fu G, Lees RS, Nimmo D, Aw D, Jin L, Gray P et al. (2010) Female-specific flightless phenotype for mosquito control. Proc Natl Acad Sci U S A 107:4550–4554

Genome Alberta (2012) http://genomealberta.ca/default.aspx. Accessed 20 Oct 2012

Gibson DG, Glass JI, Lartigue C, Noskov VN, Chuang RY, Algire MA et al (2010) Creation of a bacterial cell controlled by a chemically synthesized genome. Science 329:52–56

Global status report on non-communicable diseases 2010, WHO report, 2011. http://www.who.int/nmh/publications/ncd_report_full_en.pdf. Accessed on 29 September 2012

Graham IA, Besser K, Blumer S, Branigan CA, Czechowski T, Elias L et al (2010) The genetic map of Artemisia annua L. identifies loci affecting yield of the antimalarial drug artemisinin. Science 327:328–331

Guo BH, Kail GY, Jin HB, Tang KX (2006) Taxol synthesis. Afr J Biotechnol 5:15–20

Gutmann A, Wagner JW, Yolanda A, Christine Grady RN, Anita LA, Stephen LH, John DA, Raju SK, Barbara FA, Nelson LM, et al (2010) New directions: the ethics of synthetic biology and emerging technologies. US. The presidential commission for the study of bioethical issues 1–192

Ham T S, Dmytriv Z, Plahar H, Chen J, Hillson NJ, Keasling JD (2012) Design, implementation 915 and practice of JBEI-ICE: an open source biological part registry platform and tools. Nucleic Acids Res 40:e141

Harris AF, Nimmo D, McKemey AR, Kelly N, Scaife S, Donnelly CA et al (2011) Field performance of engineered male mosquitoes. Nat Biotechnol 29:1034–1037

Heo J, Breitbach CJ, Moon A, Kim CW, Patt R, Kim MK et al (2011) Sequential therapy with JX-594, a targeted oncolytic poxvirus, followed by sorafenib in hepatocellular carcinoma: preclinical and clinical demonstration of combination efficacy. Mol Ther 19:1170–1179

Henkel J, Maurer SM (2007) The economics of synthetic biology. Mol Syst Biol 3:117

Hingorani P, Zhang W, Lin J, Liu L, Guha C, Kolb EA (2011) Systemic administration of reovirus (Reolysin) inhibits growth of human sarcoma xenografts. Cancer 117:1764–1774

Hotez PJ, Remme JH, Buss P, Alleyne G, Morel C, Breman JG (2004) Combating tropical infectious diseases: report of the disease control priorities in developing countries project. Clin Infect Dis 38:871–878

Hsu E (2006) Reflections on the 'discovery' of the antimalarial *qinghao*. Br J Clin Pharmacol 6:666–670

Hu JC, Coffin RS, Davis CJ, Graham NJ, Groves N, Guest PJ et al (2006) A phase I study of OncoVEXGM-CSF, a second-generation oncolytic Herpes simplex virus expressing granulocyte macrophage colony-stimulating factor. Clin Cancer Res 212:6737–6747

International Association Synthetic Biology Code of conduct for best practices in gene synthesis. http://tinyurl.com/iasbcode/. Accessed 17 Sept 2012

International Civil Society Working Group on Synthetic Biology (2011) Potential impacts of synthetic biology. http://www.cbd.int/doc/emerging-issues/Int-Civil-Soc-WG-Synthetic-Biology-2011-013-en.pdf. Accessed 21 Sept 2012

International Gene Synthesis Consortium Harmonized screening protocol: gene sequence & customer screening to promote biosecurity. http://www.genesynthesisconsortium.org/wp-content/uploads/2012/02/IGSC-Harmonized-Screening-Protocol1.pdf. Accessed 17 Sept 2012

Jang YS, Park JM, Choi S, Choi YJ, Seung do Y, Cho JH et al (2012) Engineering of microorganisms for the production of biofuels and perspectives based on systems metabolic engineering approach. Biotechnol Adv 30:989–1000

Jain A, Bhatia P, Chugh A (2012) Microbial synthetic biology for human therapeutics. Sys Synth Biol 6:9–22

Jemal A, Center MM, DeSantis C, Ward EM (2010) Global patterns of cancer incidence and mortality rates and trends. Cancer Epidemol Biomark Prev 19:1893–1907

Kindt TJ, Goldsby RA, Osborne BA (2000) Vaccines. In: Kuby immunology, 4th edn. W.H. Freeman and Co., New York, pp 455–464

Kirn DH, Thorne SH (2009) Targeted and armed oncolytic poxviruses: a novel multi-mechanistic therapeutic class for cancer. Nat Rev Cancer 9:64–71

Kitney R, Freemont P (2012) Synthetic biology—the state of play. FEBS Lett 586:2029–2036

Kronberger N, Holtz P, Wagner W (2012) Consequences of media information uptake and deliberation: focus groups' symbolic coping with synthetic biology Public Underst Sci 21:174–187

Langhoff S, Cumbers J, Rothschild L, Paavola C, Warden SP (2010) Workshop report on what are the potential roles for synthetic biology in NASA's mission? Available via NASA http://event.arc.nasa.gov/main/home/reports/CP-2011-216430_Synthetic_Bio.v6.pdf. Accessed 17 Sept 2012

Lee TS, Krupa RA, Zhang F, Hajimorad M, Hotlz WJ, Prasad N et al (2011) BglBrick vectors and datasheets: a synthetic biology platform for gene expression. J Biol Eng 5:12

Letter from Civil Society to President's Commission on Synthetic Biology (16 Dec. 2010). http://www.geneticsandsociety.org/article.php?id=5517. Accessed 17 Sept 2012

Lu TK, Collins JJ (2009) Engineered bacteriophage targeting gene networks as adjuvants for antibiotic therapy. Proc Natl Acad Sci U S A 106:4629–4634

Malaria: WHO Fact sheet 2008 http://www.who.int/mediacentre/factsheets/fs094/en/. Accessed 29 Sept 2012

Marguet P, Balagadde F, Tan C, You L (2007) Biology by design: reduction and synthesis of cellular component and behaviour. J R Soc Interface 4:607–624

Martin E, Demain AL (2011) Platensimycin and platencin: promising antibiotics for future application in human medicine. J Antibiot 64:705–710

Martin VJ, Pitera DJ, Withers ST, Newman JD, Keasling JD (2003) Engineering a mevalonate pathway in Escherichia coli for production of terpenoids. Nat Biotechnol 21:796–802

Mueller S, Coleman JR, Papamichail D, Ward CB, Nimnual A, Futcher B et al (2010) Live attenuated influenza virus vaccines by computer-aided rational design. Nat Biotechnol 28:723–726

National Science Advisory Board on Biosecurity (2012). www.biosecurityboard.gov. Accessed 7 Oct 2012

Newman SA (2012) Synthetic biology: life as App store. CNS 23:6–18

Norrby SR, Nord CE, Finch R (2005) Lack of development of new antimicrobial drugs: a potential serious threat to public health. Lancet Infect Dis 5:115–119

Okullo A, Temu AK, Ogwok P, Ntalikwa JW (2012) Physico-chemical properties of biodiesel from Jatropha and Castor oils. Int J Renew Energy Res 2:1–6

Pearson H (2006) Antibiotic faces uncertain future. Nature 441:260–261

PhytoMetaSyn http://www.phytometasyn.ca/index.php?option=com_content&view=frontpage&Itemid=82. Accessed 20 Oct 2012

Porter DL, Levine BL, Kalos M, Bagg A, June CH (2011) Chimeric antigen receptor-modified T cells in chronic lymphoid leukemia. N Engl J Med 365:725–733

Purnick PEM, Weiss R (2009) The second wave of synthetic biology: from module to systems. Nat Rev Mol Cell Biol 10:410–422

Qaim M, Zilberman D (2003) Yield effects of genetically modified crops in developing countries. Science 299:900–902

Robertson JS (1988) New approaches to the development of viral vaccines: scientific and regulatory aspects. J Chem Technol Biotechnol 43:293–300

Sample6 Technologies. http://www.sample6tech.com/technology.html. Accessed 20 Sept 2012

Saukshmya T, Chugh A (2010) Commercializing synthetic biology: socio-ethical concerns and challenges under intellectual property regime. J Commer Biotechnol 16:135–158

Savage DF, Way J, Silver PA (2008) Defossiling fuel: how synthetic biology can transform biofuel production. ACS Chem Biol 3:13–16

Schmidt CW (2010) Synthetic biology: environmental health implications of a new field. Environ Health Perspect 118:A118–A123

Schmidt M, Torgersen H, Ganguli-Mitra A, Kelle A, Deplazes A, Biller-Andorno N (2008) SYNBIOSAFE e-conference: online community discussion on the societal aspects of synthetic biology. Syst Synth Biol 2:7–17

Scottish Enterprise. Funding to explore synthetic biology potential (2012) http://www.scottish-enterprise.com/News/2012/07/Funding-to-explore-synthetic-biology-potential.aspx. Accessed 21 Sept 2012

Seigel R, Ward E, Jemal A (2011) Cancer statistics, 2011: the impact of eliminating socio-economic and racial disparities on premature cancer deaths. Cancer J Clinicians (CA) 61:212–236

Singularity University (2012) Singularity University announces inaugural synthetic biology accelerator program. http://singularityu.org/singularity-university-announces-inaugural-synthetic-biology-accelerator-program/. Accessed 17 September 2012

Suk JE, Zmorzynska A, Hunger I, Biederbick W, Sasse J, Maidhof H et al (2011) Dual-Use research and technological diffusion: reconsidering the bioterrorism threat spectrum. PLoS Pathog 7:e1001253–e1001256

Synthetic Biology: Key Field of the Future (2012) http://ieet.org/index.php/IEET/more/5061. Accessed 29 Nov 2012

Synthetic biology market, global industry analysis, size, growth, share and forecast, 2012–2018. http://www.transparencymarketresearch.com. Accessed on 20 September 2012

Synthetic biology market—global industry analysis, size, growth, share and forecast, 2012–2018. http://www.researchandmarkets.com/research/swfdjt/synthetic_biology. Accessed 17 Sept 2012

Synthetic biology scorecard finds federal agencies responding to presidential bioethics commission report. http://www.synbioproject.org/news/project/6627/. Accessed 17 Sept 2012

Szybalski W (1974/1994) In: Kohn A, Shatkay A (eds) Control of gene expression. Plenum Press, New York, pp 404–405

The global burden of disease (2004) Update, WHO report, 2008. http://www.who.int/healthinfo/global_burden_disease/GBD_report_2004update_full.pdf. Accessed 28 Sept 2012

The Goldilocks Dilemma and Polycentric Governance (2012) Risks and regulation in synthetic biology. http://agilekeys.wordpress.com/2012/05/14/the-goldilocks-dilemma-and-polycentric-governance-risks-and-regulation-in-synthetic-biology/. Accessed 21 Sept 2012

The principles for the oversight of synthetic biology (2012) http://www.biosafety-info.net/file_dir/15148916274f6071c0e12ea.pdf. Accessed 20 Sept 2012

The screening framework guidance for providers of synthetic double stranded DNA (2010) http://www.phe.gov/preparedness/legal/guidance/syndna/Pages/default.aspx Accessed 17 Sept 2012

Torrance AW (2010) Synthesizing law for synthetic biology. Minnesota J Law Sci Technol 11: 629–65

Torrance AW (2012) Planted Obsolescence: Synagriculture and the law. Idaho Law Rev 48:321–352

Tuberculosis: WHO Fact sheet 2010. http://www.who.int/mediacentre/factsheets/fs104/en/. Accessed 29 Sept 2012

Tucker JB, Zilinkas RA (2006) The promise and perils of synthetic biology. New Atl 12:25–45.

Weber W, Luzi S, Karlson M, Fussenger M (2009) A novel hybrid dual-channel catalytic-biological sensor system for assessment of fruit quality. J Biotechnol 139:314–317

Windbichler N, Menichelli M, Papathanos PA, Thyme SB, Li H, Ulge UY et al (2011) A synthetic homing endonuclease-based gene drive system in the human malaria mosquito. Nature 473: 212–215

Wise de Valdez MR Nimmo D Betz J Gong HF James AA Alphey L Black WCt (2011) Genetic elimination of dengue vector mosquitoes. Proc Natl Acad Sci U S A 108:4772–4775

Xiang S, Fruehauf J, Li CJ (2006) Short hairpin RNA-expressing bacteria elicit RNA interference in mammals. Nat Biotechnol 24:697–702

Ye H, Daoud-El Baba M, Peng RW, Fussenger M (2011) A synthetic optogenetic transcription device enhances blood glucose homeostasis in mice. Science 332:1565–1568

Zhang JY (2012) The art of trans-boundary governance: the case of synthetic Biology. Syst Synth Biol 7:107–114

Chapter 17
Advancement of Emerging Tools in Synthetic Biology for the Designing and Characterization of Genetic Circuits

Vijai Singh, Indra Mani and Dharmendra Kumar Chaudhary

Abstract Bioengineering of synthetic metabolic pathways is a valuable tool for production of useful products and basic understanding of the biological complexity. Thus, we require high-throughput cloning and characterization tools and technologies. Standard cloning techniques have limitation for construct size and slow process. Here, we underline recent advancement and development of high-throughput technologies which can help to accelerate the synthetic biology research. It can be useful for the construction of large gene cassettes for production of drugs, bio-fuels, therapeutics and also rewired the natural systems; and created novel gene networks. In this chapter, we confer the different gene cloning methods for assembly of gene networks and their characterization using recent synthetic biology tools. It can be useful for accelerating the synthetic biology research for human welfare.

Keywords Bioengineering · Gene network · Microfluidics · Genetic circuits · Synthetic biology

17.1 Introduction

Gene cloning refers to the process of making multiple copies of DNA molecules. It is commonly used to amplify desire gene encoding the protein or enzyme expression. It could be used to amplify any DNA sequences such as promoters, transcription factors, non-coding sequences and coding genes which can be used in a wide array of biological experiments for large scale of protein production. To amplify any DNA

V. Singh (✉)
Synth-Bio Group, Institute of Systems & Synthetic Biology University of Evry, Genopole Campus 1, Genavenir 6 5 rue Henri Desbruères, 91030 ÉVRY, France
Tel: +33 169475381
e-mail: vijaisingh15@gmail.com

I. Mani · D. K. Chaudhary
National Bureau of Fish Genetic Resources, Canal Ring Road, Dilkusha, Lucknow 226002, India

I. Mani
Department of Biochemistry, Banaras Hindu University, Varanasi 221005, India

V. Singh, P. K. Dhar (eds.), *Systems and Synthetic Biology*,
DOI 10.1007/978-94-017-9514-2_17

sequences in a living organism, this is in under the control of origin of replication (ori) and multiplies the gene copy numbers. There are several types of cloning techniques have been used for construction of recombinant DNA molecule. While in genetic engineering, we generally used conventional cloning techniques that include sticky end, blunt end or TA cloning which could help in the construction of recombinant gene for over production of desire proteins or enzymes. Thus, there is a vital need to engineer microbes or re-engineer/rewire/create biosynthetic pathway using metabolic engineering, protein engineering and synthetic biology for overproduction of drugs, therapeutic and biofuels. Moreover it helps in basic understanding of cellular mechanism.

Metabolic engineering used an improving the cellular activities of host by manipulation of enzymatic, transport and regulatory function. Thus; we require high-throughput cloning methods for assembly of large synthetic construct and exchange of smaller parts such as promoter, ribosomal binding site, operator, transcriptional terminator and fluorescent coding genes. Recently ligase independent cloning has been developed for construction of gene cassette (Marsischky and LaBaer 2004; Sleight et al. 2010; Gibson et al. 2009). These genetic constructs with desirable predictive function that is promoters (Baron 1997; Lutz and Bujard 1997), oscillators (Stricker et al. 2008; Danino 2010), regulatory proteins and RNAs (Rodrigo 2012; Bayer and Smolke 2005; Dueber et al. 2003; Isaacs et al. 2004; Pfleger et al. 2006; Win and Smolke 2007) have been characterized using flow cytometry and microfluidics chip.

In this chapter, we emphasized ligase dependent and independent cloning techniques; and advancement tools for characterization of synthetic genetic networks for better understanding of biological complexity.

17.2 Plasmid

Plasmid is a circular DNA molecule that can replicate independently from chromosome which is present in archaea, bacteria and eukarya. Transfer of plasmid from host to host by direct, mechanical transfer, conjugation or changes in host gene expression allowing the intended uptake of the genetic element. Plasmids carry number of genes that provide resistance to naturally occurring antibiotics or toxic properties expressed under different environmental conditions. It provides the ability to bacteria to fix the atmospheric nitrogen and also degrades the non-degradable, hazardous compounds (Lipps 2008). Plasmid plays key role in genetics and biotechnology research for over expression of useful proteins and enzymes (Russell and Sambrook 2001). There are number of plasmids commercially available that have multiple cloning sites for insertion of desire gene by ligation which transfers into *Escherichia coli*.

Recent advancement in genetic engineering and synthetic biology for constructing new device or redesigning of genetic network using plasmids has been reported. Synthetic biologists are much concern about use of plasmid with known copy number (Table 17.1). Copy number is an important for control of gene expression in cell which

Table 17.1 Origins of replication and copy numbers of various plasmids

Plasmids	Origin of replication	Copy number	Classification
pTZ vectors	pMB1[a]	> 1000	High copy
pUC vectors	pMB1[a]	500–700	High copy
pBluescript vectors	ColE1	300–500	High copy
pGEM vectors	pMB1[a]	300–400	High copy
pBR322 and derivatives	pMB1[a]	15–20	Low copy
pACYC and derivatives	p15A	10–12	Low copy
pSC101 and derivatives	pSC101	~5	Very low copy

[a]The pMB1 origin of replication is closely related to that of ColE1 and falls in the same incompatibility group

has an essential impact on productivity. Some other relevant factors are important for gene expression such as antibiotics resistance, genetic recombination, and stability (Friehs 2004). The origin of replication also determines the plasmid's compatibility: its ability to replicate in conjunction with another plasmid within the same bacterial cell. Plasmids utilize same origin of replication that cannot co-exist in same cell. Co-transform is not possible in same compatibility group plasmids in same cell because it causes genetic recombination (del Solar et al. 1998).

17.3 Host

Escherichia coli are an important strain that can be used in genetic engineering and synthetic biology. Synthetic biologists are more concern about the genotypes of different *E. coli* strains before making experiments. The genotypes of different *E. coli* strains are given (Table 17.2). In these strains, there are several deletion and mutation have been done for developing of more desirable host and minimising the cellular cross talk during gene expression and regulation. There are two methods such as chemical and electroporation that used in transformation of plasmid construct into *E. coli* (Russell and Sambrook 2001).

17.4 Gene Cloning Techniques

17.4.1 Conventional Cloning

Gene cloning is a series of experimental methods in molecular biology that are used to construct recombinant DNA molecules and propagated within host organism (Watson 2007). DNA of interest needs to be isolated to provide a DNA segment of suitable size. The vector is linearised using restriction enzymes and desire gene incubated with DNA ligase. Mainly three cloning techniques are utilized such as sticky end, blunt end and TA cloning. The sticky ends or cohesive ends cloning form

Table 17.2 Genotypes of *Escherichia coli*

Strains	Genotypes
BL21	B F- *dcm ompT hsdS*(r_B^- m_B^-) gal [malB$^+$]$_{K-12}$(λ^S)
BL21 (DE3)	F- ompT gal dcm lon hsdS$_B$(r_B^- m_B^-) λ(DE3 [lacI lacUV5-T7 gene 1 ind1 sam7 nin5]), an *E. coli* B strain with DE3, a λ prophage carrying the T7 RNA polymerase gene and lacI^q
DB3.1	F- gyrA462 endA1 glnV44 Δ(sr1-recA) mcrB mrr hsdS20(r_B^-, m_B^-) ara14 galK2 lacY1 proA2 rpsL20(Sm^r) xyl5 Δleu mtl1
DH5α	F- endA1 glnV44 thi-1 recA1 relA1 gyrA96 deoR nupG Φ80d*lacZ*ΔM15 Δ(*lacZYA-argF*)U169, hsdR17(r_K^- m_K^+), λ–
DH5α Turbo (NEB)	F′ proA+B+ lacI^q Δ lacZ M15/ fhuA2 Δ(lac-proAB) glnV gal R(zgb-210::Tn10)Tet^S endA1 thi-1 Δ(hsdS-mcrB)5
DH10B (Invitrogen)	F- endA1 recA1 galE15 galK16 nupG rpsL ΔlacX74 Φ80lacZΔM15 araD139 Δ(ara,leu)7697 mcrA Δ(mrr-hsdRMS-mcrBC) λ^-
JM107	endA1 glnV44 thi-1 relA1 gyrA96 Δ(lac-proAB) [F′ traD36 proAB$^+$ lacI^q lacZΔM15] hsdR17(R_K^- m_K^+) λ^-
MG1655	F- λ^- ilvG- rfb-50 rph-1
Rosetta(DE3)pLysS	F- ompT hsdS$_B$(R_B^- m_B^-) gal dcm λ(DE3 [lacI lacUV5-T7 gene 1 ind1 sam7 nin5]) pLysSRARE (Cam^R), an *E. coli* B strain with DE3, a λ prophage carrying the T7 RNA polymerase gene and lacI^q
TG1 (Lucigen)	F′ [traD36 proAB$^+$ lacI^q lacZΔM15]supE thi-1 Δ(lac-proAB) Δ(mcrB-hsdSM)5, ($r_K^-m_K^-$)
TOP10 (Invitrogen)	F- mcrA Δ(mrr-hsdRMS-mcrBC) φ80lacZΔM15 ΔlacX74 nupG recA1 araD139 Δ(ara-leu)7697 galE15 galK16 rpsL(Str^R) endA1 λ^-
XL1-Blue (Stratagene)	endA1 gyrA96(nal^R) thi-1 recA1 relA1 lac glnV44 F′[::Tn10 proAB$^+$ lacI^q Δ(lacZ)M15] hsdR17(r_K^- m_K^+)

base pairs of any two complementary cohesive ends can anneal. It is only hydrogen bonds interaction between insert and vector while DNA ligase eventually forms a covalent bond between the sugar-phosphate residues of adjacent nucleotides to join two molecules together. An overhang is a stretch of unpaired bases in the end of a DNA molecule which can be created either 3′ or 5′ overhangs. In a blunt-ended molecule both strands terminate in a base pair. Blunt ends are not always desired in biotechnology since when using a DNA ligase to join two molecules into one, the yield is significantly low. When performing sub-cloning, it also has disadvantage of potentially inserting the insert DNA in the desired orientation. In contrast, blunt ends are always compatible with each other. In case of TA cloning method, when we used normal Taq DNA polymerase for amplification as it creates the 3′ end A nucleotide overhang in PCR product and commercially availability of dT based cloning vector used for TA cloning. Both are ligated together without any use of restriction enzymes that can be ligated using DNA ligase thus; it can transform into *E. coli* cells (Russell and Sambrook 2001). This may be accomplished by means of PCR, restriction fragment analysis and/or DNA sequencing.

17.4.2 Ligase Independent Cloning

Ligase-independent cloning (LIC) is a form of gene cloning that is able to be performed without the use of restriction enzymes, T4 DNA ligase, T4 polynucleotide kinase or alkaline phosphatase. Ligase independent cloning (LIC) is a simple, fast and relatively the cheap method for construction of recombinant gene expression cassette. The 10–15 base single overhangs is created by action of T4 DNA polymerase in plasmid. PCR products with complementary overhangs are created by during amplification and use the T4 DNA polymerase. The annealing of insert and the vector is performed in the absence of ligase by simple mixing of the DNA fragments. It is very efficient method has been developed for efficient cloning. It has been reported for cloning of inter-ALU fragments from hybrid cell-lines and human (*Aslanidis and de Jong* 1990; Haun et al. 1992).

A new cloning technique which allows the assembly of multiple DNA fragments in a single reaction using *in vitro* homologous recombination. The basic mechanism of homologous recombination *in vivo* depends upon a double stranded break, generation of ssDNA by exonucleases. Also homology searches by recombinases, repair of overhangs and gaps by enzymes. To create overhangs for homology recombination, exonucleases used to chew back one strand to reveal ssDNA overhangs. Both vector and insert are joined with T4 DNA polymerase in the absence of dNTPs to generate overhangs, then incubated vector and insert with and without RecA protein and ATP to catalyze homologous recombination (Fig. 17.1). This flexibility allows greater versatility and sensitivity for generation of recombinant expression cassette in synthetic biology research (Li and Elledge 2007).

17.4.3 In-Fusion PCR Cloning

Recently, a variety of methods and expensive kits are available for gene cloning that can be a time-consuming and provoking process. An alternative method has been developed and commercialized by Clontech as In-Fusion PCR cloning kit (http://bioinfo.clontech.com/infusion/). As depicted in Fig. 17.2 the basic mechanism and homology assembly of inserts into vector using In-Fusion PCR cloning method. It helps in rapid, sensitive and cost effective genetic engineering for construction of gene cassette with desirable function. In this technique, no additional treatment of the PCR fragments is required that includes restriction digestion, ligation, phosphorylation, or blunt-end polishing. In brief, primers designing are required for each gene which uses in In-Fusion cloning. The designed primers for insert have 15–20 bases homologous overlapping sequences of linear plasmid vector. Both fragments are amplified using high fidelity DNA polymerase and purified it.

In-Fusion reactions 2:1 insert:vector molar ratio has been used with 100 ng of vector for two-way reactions. The 10 μl volume consisting of insert, vector, and distilled water was transferred into the In-Fusion reaction tube and mixed by pipetting several times. This reaction was transferred into a 0.2 ml PCR tube and incubated in

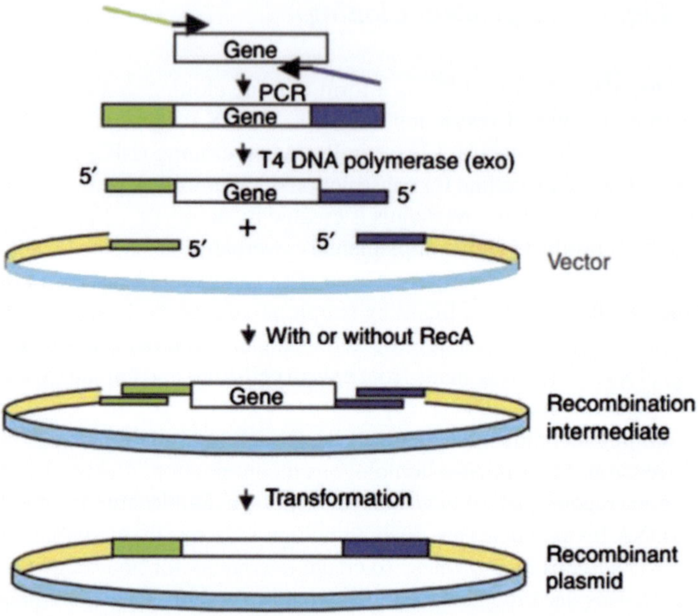

Fig. 17.1 In vitro recombination of MAGIC vectors mediated by RecA for construction of recombinant DNA through in vitro homologous recombination and single-strand annealing. Figure reproduced with permission from Nature Methods (Li and Elledge 2007) ©(2007) Macmillan Publishers Ltd

a thermocycler for 15 min at 37° followed by 15 min at 50° · A volume of 30 μl TE Buffer (pH 8.0) was added to tube and mixed by pipetting several times. A volume of 2.5 μl reaction mixture was transformed in *E. coli* competent cells thus, clones were screened by Blue/white selection (Sleight et al. 2010).

There are numbers of reports available on In-Fusion based cloning for accelerating the construction of gene cassette. It has been used for construction of upto 11 kb gene cassette (Marsischky and LaBaer 2004) A library of promoters has been joined in expression vector that contains 6XHis-tag (Berrow et al. 2007). A highly simplified, reliable, and efficient PCR based cloning technique to insert any DNA fragment or a gene (cDNA) in a vector at any desired position. Vector and insert are separately amplified with only 18 cycles using a high fidelity DNA polymerase. The amplified insert has the ends with 16 bases overlapping with the ends of the amplified vector; and reaction mixture is directly transformed into competent *E. coli* cells to obtain the desired clones (Li et al. 2011). In this report, genetic networks can be assembled from standardized biological parts called as BioBricks which is initiative for accelerating the synthetic biology research. Thus; we need to assemble and exchange the genetic parts that include promoters, ribosomal binding sites, coding sequences and transcriptional terminators. As standard BioBrick assembly normally involves digestion with restriction enzyme and ligation of two BioBricks at a time. However, an alternative assembly strategy allows for two or more PCR amplified BioBricks parts for assembling and re-engineering with any predictive function (Sleight et al. 2010).

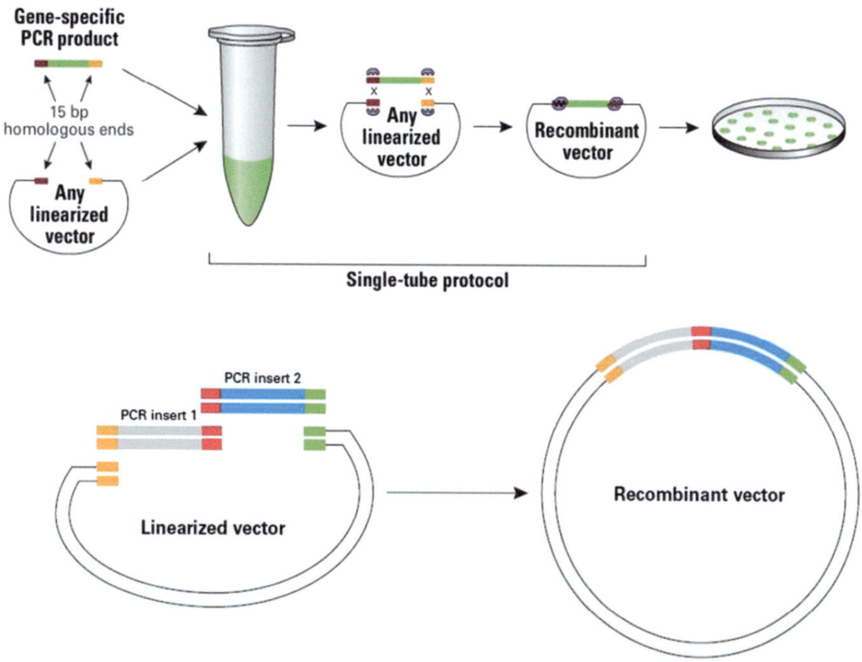

Fig. 17.2 Basic mechanism of In-Fusion cloning technique. **a** fifteen bases homology of insert and vector mixed in a single tube and incubated for Clonte 15 min at 37°C followed by 15 min at 50°C. **b** Multiple insert cloning for construction of large expression cassette. Figure Adapted from Clontech In-Fusion Advantage System

17.4.4 Gibson Assembly for Cloning

Gibson assembly is a DNA assembly method which allows for the joining of multiple DNA fragments in a single step isothermal reaction. It was invented in 2009 by Daniel Gibson at J. Craig Venter Institute (JCVI), USA. The method can simultaneously combine more than ten DNA fragments based on sequence homology. It requires DNA fragments that contain ~ 20–40 base pair overlap with adjacent DNA fragments. These DNA fragments are mixed with a cocktail of three enzymes, along with reaction buffer. It can be used for seamlessly construction of synthetic and natural genes, pathways and entire genomes. The basic process of Gibson assembly is shown in Fig. 17.3. This approach can be used to join DNA molecules that are as large as 583 kb and to clone joined products in *E. coli*. There are three enzymes required that include T5 exonuclease, thermostable DNA polymerase (Phusion polymerase), and Taq DNA ligase. T5 exonuclease chews back DNA from the 5′ end and the resulting single-stranded regions on adjacent DNA fragments annealing while DNA polymerase incorporates nucleotides to fill in any gaps. Finally, Taq ligase covalently joins the DNA of adjacent segments thereby removing any nicks in the DNA. All reagents and enzymes are commercially available and all that is required for DNA

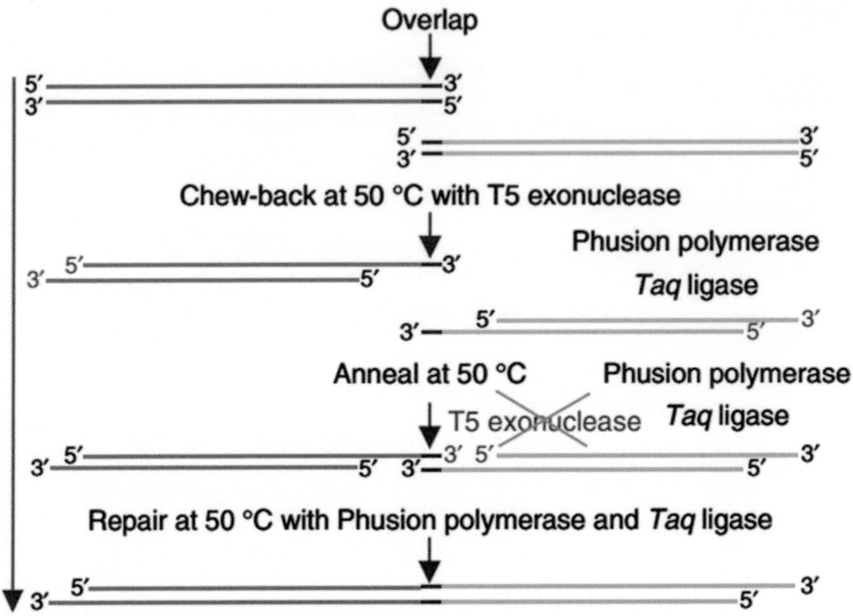

Fig. 17.3 One-step isothermal in vitro recombination. Two adjacent DNA fragments (*magenta* and *green*) sharing terminal sequence overlaps (*black*) were joined into a covalently sealed molecule in a one-step isothermal reaction. T5 exonuclease removed nucleotides from the 5′ ends of doublestranded DNA molecules, complementary single-stranded DNA overhangs annealed, Phusion DNA polymerase filled the gaps and Taq DNA ligase sealed the nicks. T5 exonuclease is heat-labile and is inactivated during the 50°C incubation. Figure reproduced with permission from Nature Methods (Gibson et al. 2009) ©(2009) Macmillan Publishers Ltd

assembly is for the reagent-enzyme mix (which can be stored at − 20° until needed) to be combined with overlapping DNA molecules along with all enzymes. This reaction mixture incubated a 50° for 15–60 min and then transformed into competent *E. coli* cells (Gibson et al. 2009). There are several advantage of this assembly method compared to conventional restriction enzyme/ligation cloning of recombinant DNA. No restriction digestion of the DNA fragments after PCR is necessary. The backbone vector can be digested, or synthesized by PCR. It is far simpler than conventional cloning methods and process also takes less time. Multiple genes fragments can be simultaneously combined in a single reaction.

17.5 Characterization of Genetic Circuits

There is recent advancement of technologies for characterization of genetic circuits in desirable host. It becomes more and more useful for better understanding of cellular mechanism in order to improve our knowledge.

17.5.1 Flow Cytometry and Fluorometry

Flow cytometry is a laser based biophysical technology employed in cell counting, sorting, biomarker detection and protein engineering by suspending cells in a stream of fluid and passing them by an electronic detection apparatus. It is used in the diagnosis of health disorders, blood cancers that can have many other applications in basic and applied research. There is an increasing awareness of technology; flow cytometry has been used by synthetic biologists for measurement of genetic circuits in whole population. Measurement of the activity of fluorescent proteins in living cells uses by flow cytometry which is an easy-to-use and high-throughput technology.

There is recent advancement in the synthetic biology research using the latest tools. There are numbers of report on designing and characterizing the genetic elements, or redesign natural systems with novel predictive function such as promoters (Baron et al. 1997; Lutz and Bujard 1997), regulatory proteins and RNAs (Rodrigo et al. 2012; Bayer and Smolke 2005; Dueber et al. 2003; Isaacs et al. 2004; Pfleger et al. 2006; Win and Smolke 2007). These genetic elements are assembled together and construct the genetic circuits such as synthetic oscillators (Elowitz and Leibler 2000; Stricker et al. 2008), riboregulators (Isaacs et al. 2004) and riboswitch (Winkler et al. 2002; Blount and Breaker 2006). Time-lapse flow cytometry has been performed for studying the fluorescence distribution of a population over time. It can be used accurately the cell population has been synchronized. Stricker et al. (Stricker et al. 2008) has been used time-lapse flow cytometry to measure the period of a synthetic gene oscillator in a synchronized culture of *E. coli*. However, the synchronization of the culture does not last long because the noise inherent in gene expression creates phase diffusion in the oscillators. While each cell continues to oscillate, the relative timing among the members of the population becomes randomized. Subsequently, the dynamics of the population will no longer be measurable from any population average and it becomes essential to measure time-lapse data from single cells (Elowitz et al. 2002; Swain et al. 2002).

17.5.2 Fluorescence Microscopy

Fluorescence microscopy is also used for quantification of gene activity with great accuracy and high resolution to determine the spatial distribution. But the major drawback of fluorescence microscopy is that it cannot quantify fluorescence of thousands of cells simultaneously, which is possible with flow cytometry. It limits the accuracy of calculations that describe the fluorescence distribution of the population. It has a major advantage over flow cytometry in that individual cells can be imaged multiple times for long durations, which allows time-lapse fluorescence measurements. Time-lapse fluorescence microscopy (TLFM) has become more popular methods among synthetic biologists for studying the dynamics of intracellular signalling and gene networks (Locke and Elowitz 2009). There are recent works shown that temporal correlations in gene expression can reveal the structure of the underlying regulatory network because autoregulatory motifs can help to determine the range of gene expression noise (Austin et al. 2006; Simpson et al. 2003). Photobleaching is one of major problem for measurement of genetic circuits at single cell level by TLFM. To avoid this problem, synthetic biologists have adopted the Microfluidics platform for long term measurement of cells.

17.5.3 Microfluidics for Characterization of Genetic Circuits

Microfluidics technology has become popular in synthetic biology for studying the cellular behaviour at single cell level. Microfluidics has been implemented for long-term monitoring of unnatural behavior programmed by the synthetic circuit, which included sustained oscillations in cell density and associated morphological changes. The circuit is induced by IPTG, and subsequently the cell density is detected by a signalling molecule (acyl-homorserine lactone or AHL) which can modulate expression of killer gene (lacZα-ccdB). It controls the rate of cell death. The microchemostat consists of 16 nanolitres fluidic loops with valves. The oscillatory cell growth induced by the synthetic circuit is more stable in the microchemostat than in normal macroscale culture (Balagaddé et al. 2005).

A chip-based bioreactor that uses microfluidic plumbing networks to actively prevent biofilm formation in *E. coli.* It allows semicontinuous, planktonic growth in six independent 16-nanoliter bioreactors with no observable wall growth (Fig. 17.4a). The bacterial cultures monitored *in situ* by optical microscopy to provide automated, real-time, non-invasive measurement of cell density and morphology with single-cell resolution (Fig. 17.4b). The growth loop is composed of 16 individually addressable segments. The microchemostat is operated in one of two alternating states: (i) continuous circulation and (ii) cleaning and dilution (Fig. 17.4c). During cleaning and dilution, the mixing is halted and a segment is isolated from the rest of the reactor with micromechanical valves. A lysis buffer is flushed through the isolated segment for 50 s to expel the cells (Fig. 17.4d). The segment is flushed with sterile growth medium to completely rinse out the lysis buffer. This segment is filled with sterile medium thus; re-united with rest of the growth chamber (Balagaddé et al. 2005).

There are lots of works have been made by synthetic biologists using Microfluidics technology. The uses of microfluidics in biology focused on biochemical assays of small concentrations of DNA and proteins (Khandurina et al. 2000; Sanders and Manz 2000; Lagally et al. 2001; McClain et al. 2003). The complexity of microfluidic chips designed for this purpose has grown quickly (Hong and Quake 2003). The genetic devices have been constructed that consist of an array of chambers which can be controlled individually using valves and switches. It has been used for biochemical experiments (Chiu et al. 2001; Thorsen et al. 2002). For molecular biology experiment, the extraction of mRNA from single cells and the subsequent synthesis of cDNA by PCR while the highly parallel measurement of transcription factor–DNA binding affinities have been earlier reported (Marcus et al. 2006; Maerkl and Quake 2007). Microfluidics platform has been used for long term measurement of genetic

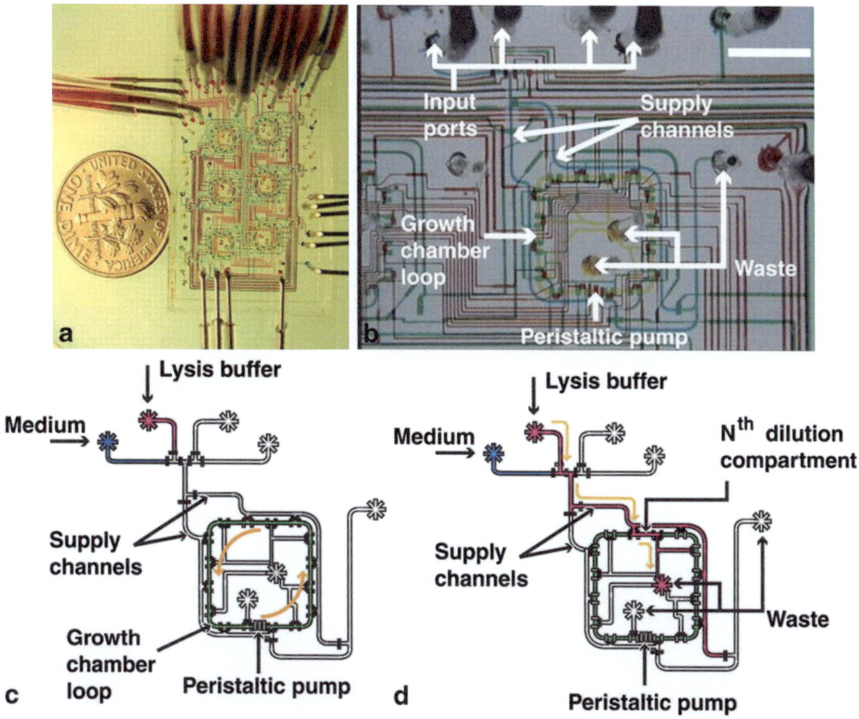

Fig. 17.4 **a** Optical micrograph showing six microchemostats that operate in parallel on a single chip. Various inputs have been loaded with food dyes to visualize channels and sub-elements of the microchemostats. The coin is 18 mm in diameter. **b** Optical micrograph showing a single microchemostat and its main components. Scale bar, 2 mm. **c** A microchemostat in continuous circulation mode. Elements such as the growth loop with individually addressable connected segments, the peristaltic pump, supply channels, and input/output ports are labeled. **d** Isolation of a segment from the rest of the growth chamber during cleaning and dilution mode. A lysis buffer (indicated in *red*) is introduced into the chip through the lysis buffer port. Integrated microvalves direct the buffer through the segment, flushing out cells, including those adhering to chamber walls. The segment is then rinsed with fresh sterile medium and reunited with the rest of the growth chamber. Figure is reproduced with permission from Science (Balagaddé et al. 2005) © (2005) AAAS

oscillators in *E. coli*. Using microfluidic devices tailored for cellular populations at differing length scales, it has been seen for the collective synchronization properties along with spatiotemporal waves occurring at millimetre scales (Stricker et al. 2008; Danino et al. 2010).

17.6 Conclusion and Future Perspective

There are an increasing synthetic gene networks, devices and biosynthetic pathway for basic understanding or predictive cellular behaviour or tunable bio-productions. The common goal of designing and construction of new biological functions and systems is not found in nature. Advancement of recent technologies for gene cloning and characterization allows to rapid construction of gene networks with predictive function. Synthetic biology research approach for the creation of new biological system from different perspectives has been focused on finding how life works or how to use in human welfare. It can provide information, manipulate chemicals, fabricate materials and structures, produce energy and provide food that can also maintain and enhance human health and eco-friendly environment. We can facilitate to find the permanent solution for human diseases such as genetic, neurological, cancer and cardiac using re-reprogramming of stem cells.

Competing Interests There is no competing interest.

Acknowledgements Authors wish to thank A.K. Singh, Satya Prakash and Pritee Singh for providing the suggestions, encouragement and fruitful discussion during preparation of this chapter.

References

Aslanidis C, de Jong PJ (1990) Ligation-independent cloning of PCR products (LIC-PCR). Nucleic Acids Res 18:6069–6074

Austin DW, Allen MS, McCollum JM et al (2006) Gene network shaping of inherent noise spectra. Nature 439:608–611

Balagaddé FK, You L, Hansen CL et al (2005) Long-term monitoring of bacteria undergoing programmed population control in a microchemostat. Science 309:137–140

Baron U, Gossen M, Bujard H (1997) Tetracycline-controlled transcription in eukaryotes: novel transactivators with graded transactivation potential. Nucleic Acids Res 25:2723–2729

Bayer TS, Smolke CD (2005) Programmable ligand controlled riboregulators of eukaryotic gene expression. Nat Biotechnol 23:337–343

Berrow NS, Alderton D, Sainsbury S et al (2007) A versatile ligation-independent cloning method suitable for high-throughput expression screening applications. Nucleic Acids Res 35:e45

Blount KF, Breaker RR (2006) Riboswitches as antibacterial drug targets. Nat Biotechnol 24:1558–1564

Chiu DT, Pezzoli E, Wu H et al (2001) Using three-dimensional microfluidic networks for solving computationally hard problems. Proc Nat Acad Sci U S A 98:2961–2966

Danino T, Mondragón-Palomino O, Tsimring L et al (2010) A synchronized quorum of genetic clocks. Nature 463:326–330

del Solar G Giraldo R Ruiz-Echevarría MJ (1998) Replication and control of circular bacterial plasmids. Microbiol Mol Biol Rev 62:434–464

Dueber JE, Yeh BJ, Chak K et al (2003) Reprogramming control of an allosteric signaling switch through modular recombination. Science 301:1904–1908

Elowitz MB, Leibler S (2000) A synthetic oscillatory network of transcriptional regulators. Nature 403:335–338

Elowitz MB, Levine AJ, Siggia ED et al (2002) Stochastic gene expression in a single cell. Science 297:1183–1186

Friehs K (2004) Plasmid copy number and plasmid stability. Adv Biochem Eng Biotechnol 86:47–82

Gibson DG, Young L, Chuang RY et al (2009) Enzymatic assembly of DNA molecules up to several hundred kilobases. Nat Methods 6:343–345

Haun RS, Serventi IM, Moss J (1992) Rapid, reliable ligation-independent cloning of PCR products using modified plasmid vectors. Biotechniques 13:515–518

Hong JW, Quake SR (2003) Integrated nanoliter systems. Nature Biotechnol 21:1179–1183

Isaacs FJ, Dwyer DJ, Ding C et al (2004) Engineered riboregulators enable posttranscriptional control of gene expression. Nat Biotechnol 22:841–847
Khandurina J, McKnight TE, Jacobson SC (2000) Integrated system for rapid PCR based DNA analysis in microfluidic devices. Anal Chem 72:2995–3000
Lagally ET, Medintz I, Mathies RA (2001) Single molecule DNA amplification and analysis in an integrated microfluidic device. Anal Chem 73:565–570
Li MZ, Elledge SJ (2007) Harnessing homologous recombination *in vitro* to generate recombinant DNA via SLIC. Nat Meth 4:251–256
Li C, Wen A, Shen B et al (2011) FastCloning: a highly simplified, purification-free, sequence- and ligation-independent PCR cloning method. BMC Biotechnol 11:92
Lipps G (ed) (2008) Plasmids: current research and future trends. Caister Academic Press, Norfolk
Locke JC, Elowitz MB (2009) Using movies to analyse gene circuit dynamics in single cells. Nature Rev Microbiol 7:383–392
Lutz R, Bujard H (1997) Independent and tight regulation of transcriptional units in *Escherichia coli* via the LacR/O, the TetR/O and AraC/I1-I2 regulatory elements. Nucleic Acids Res 25:1203–1210
Maerkl SJ, Quake SR (2007) A systems approach to measuring the binding energy landscapes of transcription factors. Science 315:233–237
Marcus JS, Anderson WF, Quake SR (2006) Microfluidic single-cell mRNA isolation and analysis. Anal Chem 78:3084–3089
Marsischky G, LaBaer J (2004) Many paths to many clones: a comparative look at high-throughput cloning methods. Genome Res 14:2020–2028
McClain MA, Culbertson CT, Jacobson SC et al (2003) Microfluidic devices for the high-throughput chemical analysis of cells. Anal Chem 75:5646–5655
Pfleger BF, Pitera DJ, Smolke CD et al (2006) Combinatorial engineering of intergenic regions in operons tunes expression of multiple genes. Nat Biotechnol 24(8):1027–1032
Rodrigo G, Landrain TE, Jaramillo A (2012) De novo automated design of small RNA circuits for engineering synthetic riboregulation in living cells. Proc Natl Acad Sci U S A 109(38):15271–15276
Russell DW, Sambrook J (2001) Molecular cloning: a laboratory manual. Cold Spring Harbor Laboratory, Cold Spring Harbor
Sanders GHW, Manz A (2000) Chip-based microsystems for genomic and proteomic analysis. Trends Analyt Chem 19:364–378
Simpson ML, Cox CD, Sayler GS (2003) Frequency domain analysis of noise in autoregulated gene circuits. Proc Natl Acad Sci U S A 100:4551–4556
Sleight SC, Bartley BA, Lieviant JA et al (2010) In-Fusion BioBrick assembly and re-engineering. Nucleic Acids Res 38:2624–2636
Stricker J, Cookson S, Bennett MR et al (2008) A fast, robust and tunable synthetic gene oscillator. Nature 456:516–519
Swain PS, Elowitz MB, Siggia ED (2002) Intrinsic and extrinsic contributions to stochasticity in gene expression. Proc Natl Acad Sci U S A 99:12795–12800
Thorsen T, Maerkl SJ, Quake SR (2002) Microfluidic large-scale integration. Science 298:580–584
Watson JD (2007) Recombinant DNA: genes and genomes: a short course. WH Freeman, San Francisco
Win MN, Smolke CD (2007) A modular and extensible RNA-based gene-regulatory platform for engineering cellular function. Proc Natl Acad Sci U S A 104:14283–14288
Winkler W, Nahvi A, Breaker RR (2002) Thiamine derivatives bind messenger RNAs directly to regulate bacterial gene expression. Nature 419:952–956

Chapter 18
Metabolic Engineering of Microorganisms for Biosynthesis of Antibiotics

Vijai Singh, Indra Mani and Dharmendra Kumar Chaudhary

Abstract Number of microorganisms produces antibiotics that can inhibit or kill the other microbes. The production of some antibiotics is not sufficient in native host rather difficult to synthesize chemically and to extract in large amounts for commercialization. Metabolic engineering plays an increasingly significant role in the production of antibiotics and its precursors. Thus, we engineer biosynthetic pathways in desire host for the production of sufficient quantity of antibiotics. In this chapter, we illustrated bioengineering of different microbes using synthetic biology and metabolic engineering approaches for production and regulation of antibiotics.

Keywords Biosynthetic pathway · Disease · Antibiotics · Metabolic engineering · Gene regulation · Synthetic biology

18.1 Introduction

There is well-known example of antibiotics discovery was observed in 1929 by Alexander Flemming, when Staphylococcal growth on a petri plate was inhibited by contamination of *Panicillium notatum* culture. *P. notatum* produces antibiotic penicillin that could be used for control of wound infection during World War II. Antibiotics play a key role to inhibit or kill the microbial growth. It was beginning era for antibiotic research. Large numbers of antibiotics have been produced from

V. Singh (✉)
Synth-Bio Group, Institute of Systems & Synthetic Biology, University of Evry,
Genopole Campus 1, Genavenir 6, 5 rue Henri Desbruères,
91030 ÉVRY, France
Tel. +33 169475381
e-mail: vijaisingh15@gmail.com

I. Mani · D. K. Chaudhary
National Bureau of Fish Genetic Resources, Canal Ring Road, P.O. Dilkusha,
Lucknow 226002, India

I. Mani
Department of Biochemistry, Faculty of Science, Banaras Hindu University,
Varanasi 221005, India

V. Singh, P. K. Dhar (eds.), *Systems and Synthetic Biology*,
DOI 10.1007/978-94-017-9514-2_18

microbial origin and also chemical synthesis. The term 'antibiotic' was coined by Selman Waksman in 1942 to describe any substance produced by a microorganism that is antagonistic to the growth of other microorganisms in high dilution (Waksman 1947). Natural host produces very less quantity of antibiotics which is not sufficient to purify and use for commercialization. We have adopted chemical synthesis of antibiotics, but some time lack of chiralilty and functionality. Expensive raw materials are required for chemical synthesis of antibiotics. Thus, metabolic engineering is one of major area of research where we increase and improve the antibiotics production with high potency and function. There are several antibiotics produced from engineered microbes that include a novel amidated polyketide from *Streptomyces coelicolor* (Zhang et al. 2006), Daptomycin from *Streptomyces lividans* (Penn et al. 2006), Clavulanic acid from *Streptomyces clavuligerus* (Li and Townsend 2006) and Erythromycin A from *Saccharopolyspora erythraea* (Chen et al. 2008). There are several strategies for production of antibiotics in natural hosts or engineer microbial hosts.

In this chapter, we emphasized the conventional, recombinant and metabolic engineering approaches for improvement of antibiotics production in desire microorganisms.

18.2 Conventional Methods for Antibiotics Production

Increasing the advancement in medicinal chemistry most of antibacterials compounds are synthetic or semisynthetic modifications of various natural compounds (Von Nussbaum et al. 2006). The β-lactam antibacterials include penicillins, cephalosporins and the carbapenems. Aminoglycosides are isolated from living organisms while other antibacterials such as sulfonamides, quinolones, and the oxazolidinones produced solely through chemical synthesis. Many antibacterials are classified on the basis of chemical/biosynthetic origin as natural, semisynthetic, and synthetic compounds. Another classification is based on biological activity which is divided into two broad groups as per their biological effect on microorganisms: bactericidal agents kill bacteria, and bacteriostatic agents inhibit bacterial growth. Synthetic antibiotic chemotherapy as a science and development of antibacterials began in Germany with Paul Ehrlich in the late 1880s. He proposed the idea that it might be possible to create chemicals that would act as a selective drug that can inhibit or kill bacteria without any side effect on human. Screened the hundreds of dyes against different organisms thus, it has been discovered a medicinally useful synthetic antibacterial Salvarsan (Bosch and Rosich 2008) which is also called arsphenamine.

First sulfonamide is commercially available as antibacterial antibiotic (Prontosil) has been discovered by Gerhard Domagk (1932) at Bayer Laboratories of IG Farben conglomerate in Germany (Bosch and Rosich 2008). Domagk received Nobel Prize (1939) in Medicine for his excellent efforts. Prontosil shows a broad spectrum effect against Gram-positive cocci but not against enterobacteria. The discovery and development of sulfonamide drug opened era of antibacterial antibiotics. In 1939, coinciding with start of World War II, Rene Dubos reported the discovery of the first naturally derived antibiotic gramicidin from *B. brevis*. It was one of the first commercially manufactured antibiotics universally and effectively used for treatment of wounds and ulcers during World War II (Van Epps 2006).

Florey and Chain succeeded in purifying first penicillin, penicillin G procaine in 1942, but it did not become widely available outside Allied military before 1945. Purified penicillin displayed potent antibacterial activity against a wide range of bacteria and had low toxicity in humans. The discovery of a powerful antibiotic was unique and the development of penicillin leads for renewal interest in investigation of antibiotic with similar efficacy and safety (Florey 1945). Discovery and development of penicillin therapeutic drug by Ernst Chain, Howard Florey and Alexander Fleming shared Nobel Prize (1945) in Medicine. Florey credited Dubos with pioneering approach of deliberately and systematically searching for antibacterial compounds, which had led to the discovery of gramicidin and had revived Florey's research in penicillin (Van Epps 2006). These antibiotics discovery motivated researchers for development of new antibiotics and searching the microbes which are able to produce or engineer for over production.

18.3 Development of Resistance Microbes

Antibiotics are used for treatment of human and animal infectious diseases. There is an increasing the rapid evolution of multi-resistant pathogens which requires development of new antibiotics. Among the several thousand of different antibiotics discovered so far, more than two thirds are produced by bacteria. The emergence of resistance of bacteria to antibacterial drugs is a common process due to gene mutation or metabolized by enzyme. Emergence of resistance often reflects the evolutionary processes which takes place during antibacterial drug therapy. Our understanding, how antibiotics induce bacterial cell death is centred on essential bacterial cell function which is inhibited by drug–target interaction. Antibiotics play role in development of drug resistance and also inhibited bacterial growth (Levy 1994). Antibacterial selections from whole bacterial populations for strains having acquired antibacterial-resistance genes have been previously reported (Luria and Delbrück 1943; Witte 2004).

As shown in Fig. 18.1, the basic mechanism of cell death induced by bactericidal antibiotics. Resistance to antibacterials also occurs during horizontal gene transfer which is more likely to happen in locations of frequent antibiotic use. The primary drug–target interactions (aminoglycoside with the ribosome, quinolone with topoisomerase, and β-lactam with penicillin-binding proteins (PBPs)) stimulate the oxidation of NADH through the electron transport chain, which is dependent on the tricarboxylic acid (TCA) cycle. Hyperactivation of the electron transport chain stimulates superoxide (O_2^-) formation. Superoxide damages Fe–S clusters, making ferrous iron available for oxidation by the Fenton reaction which leads to formation of hydroxyl radicals (•OH) that damage DNA, lipids and proteins. This contributes to antibiotic-induced cell death. Quinolones, β-lactams and aminoglycosides also

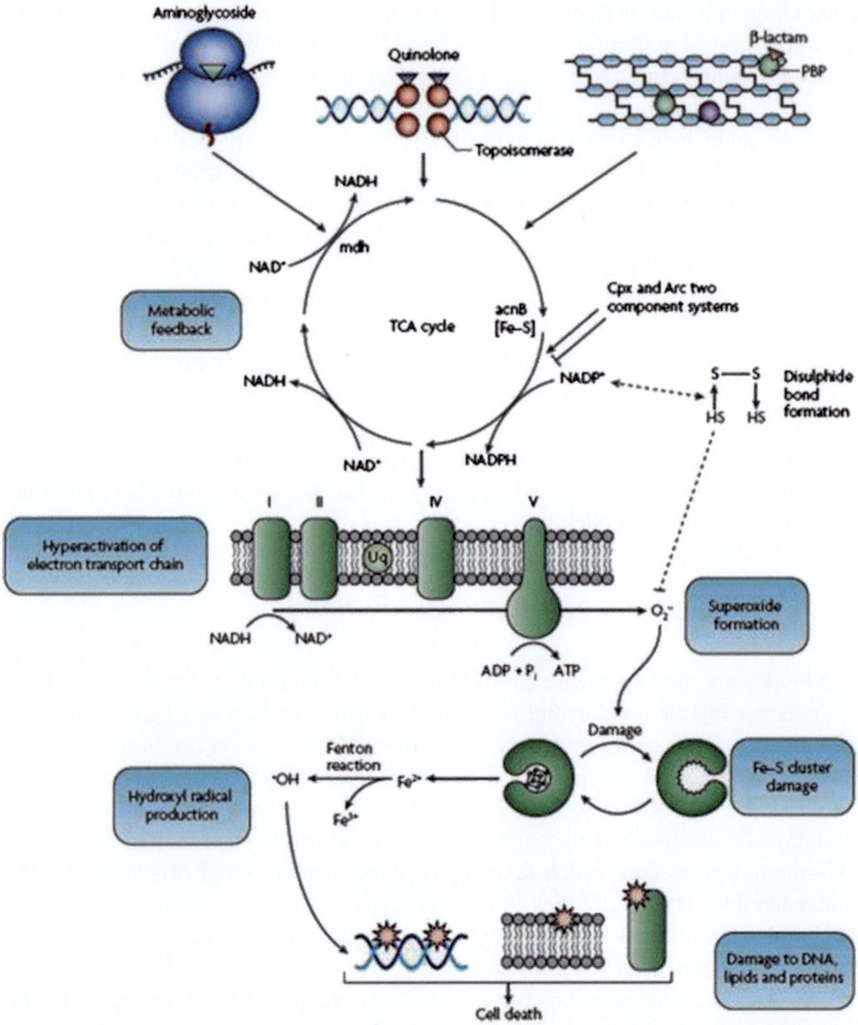

Fig. 18.1 Mechanism of cell death induced by bactericidal antibiotics. Figure reproduced with permission from Nature Review Microbiology (Kohanski et al. 2010) ©(2010) Macmillan Publishers Ltd

trigger hydroxyl radical formation and cell death by envelope (Cpx) and redox-responsive (Arc) two-component systems (Kohanski et al. 2010). Antibiotic mediated cell death is a complex process that only begins with drug–target interaction and the primary effects of respective interactions. The development of new antibiotics and the improvement of current antibacterial drug therapies would benefit from a better understanding of the specific sequences (Kohanski et al. 2010). Penicillin and Erythromycin are used to have high efficacy against many bacterial species and strains. It has become less effective due to increasing resistance. Antibacterial resistance may impose a biological cost thereby reducing fitness of resistant strains, which can limit the spreading of antibacterial-resistant bacteria. Whereas mutations in genes may compensate for this fitness cost and can help in bacterial survival and growth (Andersson 2006).

There are several molecular mechanisms of antibacterial resistance exist. Intrinsic antibacterial resistance may be part of the genetic makeup of bacterial strains (Alekshun and Levy 2007). Antibiotic target may be absent from the bacterial genome. Acquired resistance results from a mutation in the bacterial chromosome or plasmid (Alekshun and Levy 2007). Antibacterial producing bacteria have evolved resistance mechanisms that have been shown to be similar which can be transferred to antibacterial resistant strains (Marshall et al. 1998; Nikaido 2009). The spreading of antibacterial resistance occurs by vertical transmission of mutations during growth and by genetic recombination of DNA through the horizontal genetic exchange. Antibacterial resistance genes can be exchanged between different bacterial strains or species via plasmids (Witte 2004; Baker-Austin et al. 2006). Plasmids carry several different resistance genes which confer resistance to multiple antibacterials. Cross-resistance to several antibacterials may also occur when a resistance mechanism encoded by a single gene conveys resistance to more than one antibacterial compounds (Baker-Austin et al. 2006).

18.4 Metabolic Engineering of Microbes for Antibiotics Production

Advancement of synthetic biology is used in the field of antibiotic production in filamentous fungi and actinomycetes bacteria that include implementation and modification of complex biosynthesis pathway an existing and new production hosts. Antibiotics production is regulated by complex networks and involves intricate multi-step biosynthetic machineries. It reorganizes the metabolic fluxes to redirect cellular metabolic resources towards their biosynthesis. The urgent need for new antibiotics caused by the accelerating emergence of multi-drug resistant pathogens worldwide has led to a strong interest in research community. It has started investigation of various aspects such as metagenomics, combinatorial biochemistry, mathematical and computational modelling, cell engineering, molecular cell biology and biotechnology for the antibiotic production.

Metabolic engineering considers metabolic and cellular system that allows manipulation of genetic networks which distinguishes from simple genetic engineering (Bailey 1991, Stephanopoulos et al. 1998). Development of structurally and functionally diverse antibiotics by metabolic engineering is a great importance to combat against emerging drug-resistant pathogens (Menzella and Reeves 2007; Menzella et al. 2005). General strategies of metabolic engineering develop new drugs and efficient production. Several microorganisms considered as a drug factory for many chemicals and biological molecules. As these drugs are synthesized in only small amounts, it is difficult to obtain them in appropriate amounts. Therefore, metabolic engineering requires for the engineering of microbes for sufficient production of

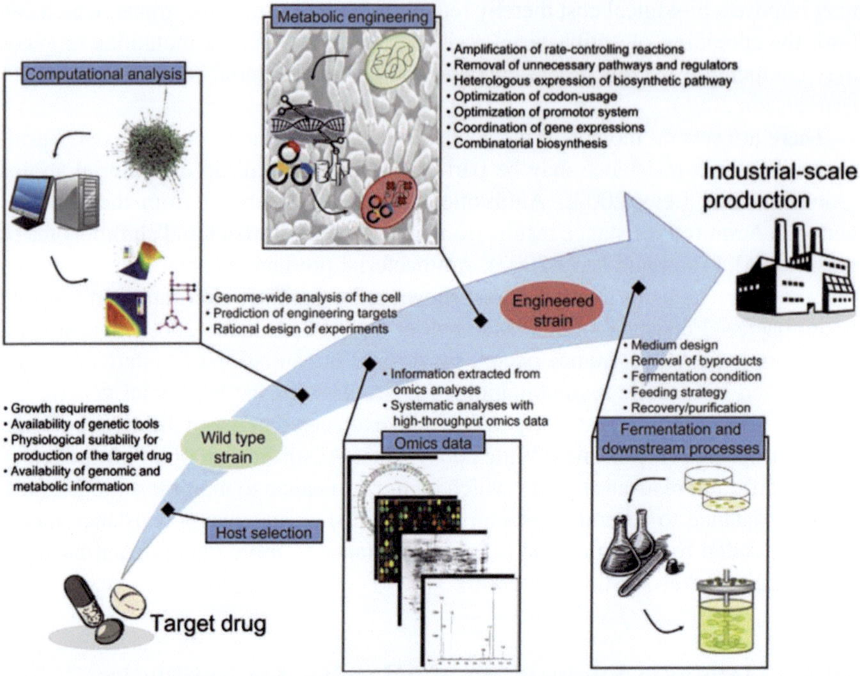

Fig. 18.2 General strategy for the metabolic engineering of microorganisms for antibiotics production. Figure reproduced with permission from Drug Discovery Today (Lee et al. 2009) ©(2009) Elsevier

antibiotics. Recent advances in our understanding on the metabolic pathways for synthesis of these drugs together with the development of various genetic and analytical tools. Also, more systematic and rigorous engineering of microorganisms for enhanced antibiotics production (Fig. 18.2). Antibiotics production through metabolically engineered microorganisms has several advantages over total chemical synthesis or extraction from natural resources.

Chemicals that are used as drugs usually have complex structures including chirality that are rather difficult to synthesize chemically. While the extraction of medically valuable compounds from natural resources is generally inefficient that may cause negative impacts on environment (Chang and Keasling 2006). On the other hand, antibiotics can be produced from engineered microbial fermentation which is relatively from inexpensive substrates in a controlled and consistent manner. Much rapid growth of microbial cells compared with higher organisms is another obvious advantage. Moreover, metabolic engineering of microorganisms can be performed more easily than mammalian and plant cells that can allow easier genetic modification of metabolic pathways for the production of structurally more diverse analogs with potent biological activities like polyketides and non-ribosomal peptides (Nguyen et al. 2006). These advantages are key driving forces for the microbial production of antibiotics and its precursors.

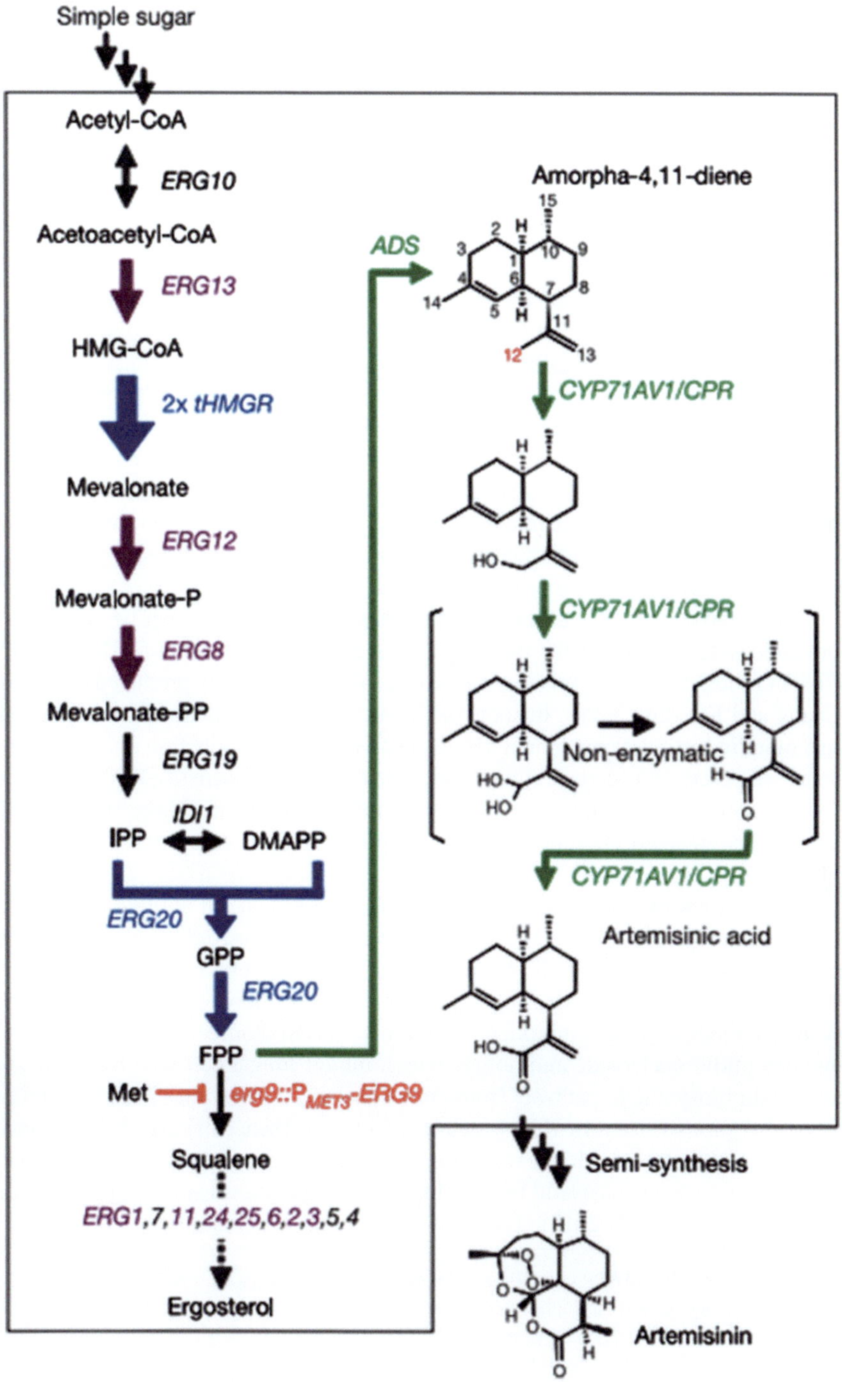

Fig. 18.3 Metabolic engineering of mevalonate pathway in S. cerevisiae that are directly upregulated are shown in blue; those that are indirectly upregulated byupc2-1 expression are in purple; and the red line denotes repression of ERG9 in strain EPY224. The pathway intermediates IPP,

There is a great effort has been made by Keasling and colleague for metabolic engineering of yeast for production of artemisinin. Artemisinin and its derivatives are a group of drugs that possess the most rapid action of all current drugs against *Plasmodium falciparum* which causes malaria (White 1997). Treatments containing an artemisinin derivative (artemisinin-combination therapies, ACTs) are now standard treatment worldwide for the malaria. Metabolic engineering of *S. cerevisiae* to produce high titres of artemisinic acid using an engineered mevalonate pathway, amorphadiene synthase, and a novel cytochrome P450 monooxygenase (*CYP71AV1*) from *A. annua.* It performs a three-step oxidation of amorpha-4, 11-diene to artemisinic acid. Mevalonate pathway in *S. cerevisiae* upregulated by *upc2-11* expression while repression of *ERG9* in strain EPY224. The pathway intermediates IPP, DMAPP and GPP are defined as isopentenyl pyrophosphate, dimethyl allyl pyrophosphate and geranyl pyrophosphate, respectively. Green arrows indicate the biochemical pathway leading from farnesyl pyrophosphate (FPP) to artemisinic acid, which was introduced into *S. cerevisiae* from *A. annua*. The three oxidation steps involve in converting amorphadiene to artemisinic acid through CYP71AV1 and CPR (Fig. 18.3). Artemisinic acid is transported and retained on the outside of the engineered yeast. It means that a simple and inexpensive purification process can be used to obtain the desired product (Ro et al. 2006).

In this study, precursor of antimalarial drug artemisinin that can be chemically converted to final form by the two steps via reduction and oxidation. Thus, highly efficient production of artemisinin is possible using biologically produced precursor (Chang and Keasling 2006). In some cases, drug precursors themselves might also have other industrial or medicinal values, justifying their production on large scale. There are precursors of several antibiotics that include monensin (Vrijbloed et al. 1999), cervimycin C (Herold et al. 2004), and valanimycin (Garg et al. 2008). There are engineered microbes for efficient production of antibiotics was given in Table 18.1.

Advancement of the genetics and biochemistry of bacterial antibiotic synthesis as it becomes possible to make new antibiotics with improved properties via genetic engineering and metabolic engineering of the producing organisms. Antibiotic biosynthesis and regulation in *Streptomyces* bacteria is improved for the production levels for creating new antibiotics. Polyoxins and Nikkomycins are potent antifungal peptidyl nucleoside antibiotics, which inhibit fungal cell wall biosynthesis. Polymyxin biosynthetic pathway from *Paenibacillus polymyxa* has been introduced in *Bacillus subtilis* strain BSK3S which could in the presence of exogenously added L-2, 4-diaminobutyric acid. The recombinant BSK4-rB strain produced high levels of polymyxin which can be useful for development and production of novel polymyxin derivatives (Park et al. 2012). Daptomycin and A21978C antibiotic complex are

Fig. 18.3 (continued) DMAPP and GPP are defined as isopentenyl pyrophosphate, dimethyl allyl pyrophosphate and geranyl pyrophosphate, respectively. Green arrows indicate the biochemical pathway leading from farnesyl pyrophosphate (FPP) to artemisinic acid, which was introduced into S. cerevisiae from A. annua. The three oxidation steps converting amorphadiene to artemisinic acid by CYP71AV1 and CPR. Figure reproduced with permission from Nature (Ro et al. 2006) ©(2006) Macmillan Publishers Ltd

Table 18.1 Antibiotics produced by metabolic engineering of microorganisms

Name of antibiotics	Production host	Metabolic engineering strategy	References
Cephalosporin	*Acremonium chrysogenum*	Inactivation of cefR delays expression of the cefEF gene increases penicillin N secretion and decreases cephalosporin production. Overexpression of the cefR gene decreased up to 60 % penicillin N secretion, saving precursors and resulting in increased cephalosporin C production	Teijeira et al. (2011)
A novel amidated polyketide	*Streptomyces coelicolor*	Heterologous coexpression of amidotransferase OxyD with minimal oxytetracycline polyketide synthase in *S. coelicolor*	Zhang et al. (2006)
Clavulanic acid	*Streptomyces clavuligerus*	Knockout of gap1 and gap2 and addition of arginine in the medium to improve the drug precursors	Li and Townsend (2006)
Daptomycin	*Streptomyces lividans*	Heterologous production of daptomycin in *S. lividans*, inactivation of actinorhodin, and optimization of the medium by adding additional phosphate	Penn et al. (2006)
Oxytetracycline	*Streptomyces rimosus*	Oxidative pentose phosphate pathway (PPP) and nicotinamide adenine dinucleotide phosphate (NADPH) generation, glucose-6-phosphate dehydrogenase (G6PDH), which is encoded by zwf1 and zwf2. Disruption of zwf1 or zwf2 resulted in a higher production of OTC	Tang et al. (2011)
Erythromycin A	*Saccharopolyspora erythraea*	Overexpression of eryK and eryG with copy number ratio of 3:2	Chen et al. (2008)
Fosfomycin	*Streptomyces lividans*	Cloning of fosfomycin biosynthetic cluster from *Streptomyces fradiae* and its heterologous production in S. lividans	Woodyer et al. (2006)

Table 18.1 (continued)

Name of antibiotics	Production host	Metabolic engineering strategy	References
Nargenicin A(1)	*Nocardia* sp. CS682	Heterologous expression of S-adenosylmethionine synthetase (MetK1-sp) in *Nocardia* sp. CS682 enhanced the production of nargenicin A(1) by about 2.8 times	Maharjan et al. (2012)
Lantibiotic subtilin	*Bacillus subtilis*	Expression of subtilin self-protection genes spaIFEG and deletion of a repressor of subtilin gene AbrB	Heinzmann et al. (2006)
Macrolide 6-deoxyerythromycin D	*Escherichia coli*	Production of 6-deoxyerythromycin D in *E. coli* and several generations of activity-based screening assay for further evolution	Lee and Khosla (2007)
Magnoflorine and (S)-scoulerine	*Escherichia coli* and *Saccharomyces cerevisiae*	Production of benzylisoquinoline alkaloids in the co-culture of *E. coli* and *S. cerevisiae* using microbial and plant genes	Minami et al. (2008)
Novobiocin derivatives	*Streptomyces coelicolor*	Engineering of novobiocin biosynthetic gene cluster in *E. coli* using l-Red-medicated recombination and its expression in *S. coelicolor* along with coexpression of the halogenase Clo-hal	Eustaquio et al. (2004)
Tylactone	*Streptomyces venezuelae*	Overexpression of tylosin polyketides synthase in *S. venezuelael* and provision of precursors in the medium	Jung et al. (2006)
Valencene, cubebol, and patchoulol	*Saccharomyces cerevisiae*	Heterologous expression of plant sesquiterpenes biosynthetic genes, downregulation of ERG9 and addition of methionine to the medium	Asadollahi et al. (2008)

lipopeptides produced by *Streptomyces roseosporus* and also in recombinant *Streptomyces lividans* strains (Penn et al. 2006). Fosfomycin is an effective antibiotic against methicillin and vancomycin resistant pathogens. Cloning and characterization of a complete fosfomycin biosynthetic cluster from *Streptomyces fradiae* and their heterologous production of fosfomycin in *S. lividans* has been investigated (Woodyer et al. 2006).

Enhancement of production of subtilin in *Bacillus subtilis* ATCC 6633 has been reported. Insertion of additional copies of subtilin self protection (immunity) genes spaIFEG into the genome that could significantly enhanced the subtilin tolerance level and a repressor of subtilin gene expression which could six-fold enhancement of subtilin production (Heinzmann et al. 2006). There are heterologous production of polymyxin N and nikkixin D antibiotics in *Streptomyces aureochromogenes*. Two of the hybrid antibiotics, polyoxin N and nikkoxin D, were significantly more potent against human or plant fungal pathogens. It may be used for generation of novel peptidyl nucleoside antibiotics in industrial strains (Zhai et al. 2012).

Nargenicin A(1) has been isolated from *Nocardia* sp. CS682 possesses strong antibacterial activity against methicillin-resistant *Staphylococcus aureus*. Heterologous expression of S-adenosylmethionine synthetase (MetK1-sp) in *Nocardia* sp. CS682 enhanced the production of nargenicin A(1) by about 2.8 times due to transcriptional activation of biosynthetic genes. Expression of acetyl-CoA carboxylase genes also improved nargenicin A(1) production by about 3.8 times in *Nocardia* sp. ACC18 compared to that in *Nocardia* sp. CS682 and *Nocardia* sp. NV18 by increasing precursor (Maharjan et al. 2012). There is production of penicillin G antibiotics by *Penicillium chrysogenum* requires the supplementation of the growth medium with the side chain precursor phenylacetate. The growth of *P. chrysogenum* with phenylalanine as the sole nitrogen source resulted in the extracellular production of phenylacetate and penicillin G (Veiga et al. 2012).

The aromatic polyketide antibiotic oxytetracycline (OTC) is produced by *Streptomyces rimosus* as important secondary metabolite. High level production of antibiotics in *Streptomycetes* requires precursors and cofactors which are derived from primary metabolism; therefore it is tricky to engineer the primary metabolism. This has been demonstrated by targeting a key enzyme in the oxidative pentose phosphate pathway (PPP) and nicotinamide adenine dinucleotide phosphate (NADPH) generation, glucose-6-phosphate dehydrogenase (G6PDH), which is encoded by zwf1 and zwf2. While the disruption of zwf1 or zwf2 resulted in a higher production of OTC (Tang et al. 2011). Phloroglucinol derivatives are a major class of secondary metabolites synthesized by some species of *Pseudomonads* which could produce 2, 4-diacetylphloroglucinol (DAPG) that plays an important role in the biological control of many plant pathogens. The combination of conventional biochemistry and molecular biology with new systems biology and synthetic biology tools can provide a better view of phloroglucinol compound biosynthesis and a greater probability of microbial production (Yang and Cao 2012).

Saccharopolyspora erythraea mutB knockout strain (FL2281) contains a block in methylmalonyl-CoA mutase reaction which was found to carry a diethyl methylmalonate-responsive (Dmr) phenotype in an oil-based fermentation medium.

The Dmr phenotype confers the ability to increase erythromycin A production from 250 to 300 %. Although the mutB strain is phenotypically a low level erythromycin producer, diethyl methylmalonate supplementation allowed it to produce up to 30 % more erythromycin than the wild-type strain (Weber et al. 2012).

Tetracyclines are aromatic polyketides biosynthesized by bacterial type II polyketide synthases (PKSs). The gene cluster of oxytetracycline (oxy and otc genes) PKS genes from *Streptomyces rimosus* have sequenced and amidotransferase, OxyD, synthesizes the malonamate starter unit that is a universal building block for tetracycline compounds. *In vivo* reconstitution using strain CH999 revealed that the minimal PKS and OxyD which are necessary and sufficient for biosynthesis of amidated polyketides (Zhang et al. 2006). Clavulanic acid is a potent beta-lactamase inhibitor which uses to combat resistance to penicillin and cephalosporin antibiotics. Clavulanic acid biosynthesis is initiated by condensation of L-arginine and D-glyceraldehyde-3-phosphate (G3P). There are two genes (gap1 and gap2) whose protein products are distinct glyceraldehyde-3-phosphate dehydrogenases (GAPDHs) have been inactivated in *Streptomyces clavuligerus* by targeted gene disruption. A doubled production of clavulanic acid obtained when gap1 was disrupted (Li and Townsend 2006). There are metabolic network of *Streptomyces roseosporus* LC-54-20 proposed for daptomycin production. The analysis of extracellular metabolites throughout the batch fermentation has been evaluated in addition to daptomycin and biomass production. Metabolic flux distributions have been based on stoichiometrical reaction and the extracellular metabolites fluxes. Experimental and theoretical values for both the specific growth rate and daptomycin production rate indicated that *in silico* model proved a powerful tool to analyze the metabolic behaviours based on the analysis under different initial glucose concentrations throughout the fermentation process (Huang et al. 2011). There is used of small-scale culture and bioreactor to compare and improve the heterologous production of the antibiotic erythromycin A across a series of engineered *E. coli* strains. There is a deletion ygfH to increase the biosynthetic pathway carbon flow while strains contain an extra copy of a key deoxysugar glycosyltransferase gene (Zhang et al. 2012).

Synthetic biology is used to develop cell factories for production of chemicals by constructively importing heterologous pathways into industrial microorganisms. There is a retrosynthetic approach to the production of therapeutics with goal of developing an *in situ* drug delivery device in host cells (Carbonell et al. 2011). *Acremonium chrysogenum* cephalosporin biosynthetic genes is encoding a regulatory protein (CefR) containing a nuclear targeting signal and a Fungal_trans domain. Inactivation of cefR delays the expression of cefEF increase the penicillin N secretion and decreases cephalosporin. Overexpression of cefR gene decreased up to 60 % penicillin N secretion, saving precursors and resulting in increased cephalosporin C production (Teijeira et al. 2011).

Naphthomycins (NATs) are 29-membered naphthalenic ansamacrolactam antibiotics with antimicrobial and antineoplastic activities. Their biosynthesis starts from 3-amino-5-hydroxy-benzoic acid (AHBA) by PCR amplification for AHBA synthase and amino-dehydroquinate (aDHQ) synthase, a genomic region containing orthologs of these genes was identified in *Streptomyces* sp. CS. It was confirmed to be involved in naphthomycin biosynthesis by deletion of a large DNA fragment resulting in abolishment of naphthomycin production (Wu et al. 2011). Aminoglycosides are a class of important antibiotic used for various therapeutic applications. *E. coli* affords a widely studied cellular system that could be utilized, not only for understanding but also for attempting to engineer the biosynthetic pathway of secondary metabolites. Production of ribostamycin derivative is engineered host by heterologous expression of recombinants genes encoding the biosynthetic pathway in aminoglycoside-producing strain (Kurumbang et al. 2010).

Doxorubicin (DXR) is an anthracycline-type polyketide, typically produced by *Streptomyces peucetius* ATCC 27952. DXR biosynthesis is tightly regulated, and a very low level of DXR production is maintained in the wild-type strain. DXR is one of the most broadly used and clinically important anticancer drugs; a traditional strain improvement strategy has been practiced via random mutagenesis that enhanced production (Niraula et al. 2010). There is some rule for the selection of host strain and genetic tools required for drug production are essential. Optimization of metabolic pathways and networks is also requires. Host strain is a miniature chemical factory where antibiotics of interest are produced which is one of most important factors to be taken an account. *E. coli* is an attractive host for pharmaceutical production because of its fast cell growth and availability of genetic engineering tools but it lacks enzymes necessary for polyketides assembly (Pfeifer et al. 2001).

Availability of genetic engineering tools such as expression vectors, transformation and chromosomal gene knockout/integration system needs to be considered which can be used in metabolic engineering. It is often necessary to perform overexpression of homologous and/or heterologous genes and knockout of those genes that are accountable for reduced product formation. Natural antibiotics producing microorganisms may not be established gene manipulation systems or are difficult to perform genetic engineering (Weissman and Leadlay 2005). Alternatively, one can select another host microorganism that has less ability to produce a desired product but has much better gene manipulation system available.

The feasibility of cultivating the cells at various scales that incluses flask, small fermentor, pilot-scale fermentor, and industrial-scale fermentor needs to be evaluated depending on the desired scale-up. Mainly ability of host strain to grow on a simple medium using inexpensive carbon substrates needs to be checked because it is highly associated with cost competitiveness in the bioprocesses during production of antibiotics and its precursors. Also, genetic and physiological backgrounds of host microorganisms need to be examined as it is valuable to choose the host that provides suitable intracellular environment, generating sufficient amount of precursors required for regulating genes and successfully biosynthesizing functional antibiotics (Pfleger et al. 2006).

18.5 Conclusion and Future Perspective

Antibiotics resistant microbial infections are becoming more prevalent and are major health issues facing us today. This rises the resistance has limited our repertoire of effective antimicrobials, creating a problematic situation that has been exacerbated by small number of new antibiotics. The complex effects of bactericidal antibiotics discussed in this chapter, we provide a recent progress of metabolic engineering and synthetic biology approach which could enhance the potency of current antibiotics. It will be an important to translate our growing understanding of antibiotic mechanisms into new clinical treatments and approaches thus; we can effectively fight the growing threat from resistant pathogens. Metabolic engineering has enabled sophisticated engineering of various microorganisms for efficient production of various metabolites. Thus, there is an urgent need to engineer microbes or re-engineer/rewire/create biosynthetic pathway using metabolic engineering, protein engineering and synthetic biology for overproduction of novel antibiotics for treatment of emerging diseases. We may provide new insights for the development of novel antibiotics for chemotherapy of diseases.

Competing Interests There is no competing interest.

Acknowledgements Authors wish to thank A.K. Singh, Satya Prakash and Pritee Singh for providing the suggestions, encouragement and fruitful discussion during preparation of this chapter.

References

Andersson DI (2006) The biological cost of mutational antibiotic resistance: any practical conclusions? Curr Opin Microbiol 9:461–465

Alekshun MN, Levy SB (2007) Molecular mechanisms of antibacterial multidrug resistance. Cell 128:1037–1050

Asadollahi MA, Maury J, Møller K et al (2008) Production of plant sesquiterpenes in *Saccharomyces cerevisiae*: effect of ERG9 repression on sesquiterpene biosynthesis. Biotechnol Bioeng 99:666–677

Bailey JE (1991) Toward a science of metabolic engineering. Science 252:1668–1675

Baker-Austin C, Wright MS, Stepanauskas R et al (2006) Co-selection of antibiotic and metal resistance. Trends Microbiol 14:176–182

Bosch F, Rosich L (2008) The contributions of Paul Ehrlich to pharmacology: a tribute on the occasion of the centenary of his Nobel Prize. Pharmacology 82:171–179

Carbonell P, Planson AG, Fichera D et al (2011) A retrosynthetic biology approach to metabolic pathway design for therapeutic production. BMC Syst Biol 5:122

Chang MC, Keasling JD (2006) Production of isoprenoid pharmaceuticals by engineered microbes. Nat Chem Biol 2:674–681

Chen Y, Deng W, Wu J et al (2008) Genetic modulation of the overexpression of tailoring genes eryK and eryG leading to the improvement of erythromycin A purity and production in *Saccharopolyspora erythraea* fermentation. Appl Environ Microbiol 74:1820–1828

Eustaquio AS, Gust B, Li SM et al (2004) Production of 8-halogenated and 8-unsubstituted novobiocin derivatives in genetically engineered *Streptomyces coelicolor* strains. Chem Biol 11:1561–1572

Florey HW (1945) Use of micro-organisms for therapeutic purposes. Br Med J 2:635–642

Garg RP, Xuelei LQ, Lawrence BA et al (2008) Investigations of valanimycin biosynthesis: elucidation of the role of seryl-tRNA. Proc Natl Acad Sci U S A 105:6543–6547

Heinzmann S, Entian KD, Stein T (2006) Engineering *Bacillus subtilis* ATCC 6633 for improved production of the lantibiotic subtilin. Appl Microbiol Biotechnol 69:532–536

Herold K, Xu Z, Gollmick FA et al (2004) Biosynthesis of cervimycin C an aromatic polyketide antibiotic bearing an unusual dimethylmalonyl moiety. Org Biomol Chem 2:2411–2414

Huang D, Jia X, Wen J et al (2011) Metabolic flux analysis and principal nodes identification for daptomycin production improvement by *Streptomyces roseosporus*. Appl Biochem Biotechnol 165:1725–1739

Jung WS, Lee SK, Hong JS et al (2006) Heterologous expression of tylosin polyketide synthase and production of a hybrid bioactive macrolide in *Streptomyces venezuelae*. Appl Microbiol Biotechnol 72:763–769

Kohanski MA, Dwyer DJ, Collins JJ (2010) How antibiotics kill bacteria: from targets to networks. Nat Rev Microbiol 8:423–435

Kurumbang NP, Park JW, Yoon YJ et al (2010) Heterologous production of ribostamycin derivatives in engineered *Escherichia coli*. Res Microbiol 161:526–533

Lee HY, Khosla C (2007) Bioassay-guided evolution of glycosylated macrolide antibiotics in *Escherichia coli*. PLoS Biol 5:e45

Lee SY, Kim HU, Park JH Kim TY et al (2009) Metabolic engineering of microorganisms: general strategies and drug production. Drug Discov Today 14:78–88

Levy SB (1994) Balancing the drug-resistance equation. Trends Microbiol 2:341–342

Li R, Townsend CA (2006) Rational strain improvement for enhanced clavulanic acid production by genetic engineering of the glycolytic pathway in *Streptomyces clavuligerus*. Metab Eng 8:240–252

Luria SE, Delbrück M (1943) Mutations of bacteria from virus sensitivity to virus resistance. Genetics 28:491–511

Maharjan S, Koju D, Lee HC et al (2012) Metabolic engineering of *Nocardia* sp. CS682 for enhanced production of nargenicin A. Appl Biochem Biotechnol 166:805–817

Marshall CG, Lessard IA, Park I et al (1998) Glycopeptide antibiotic resistance genes in glycopeptide-producing organisms. Antimicrob Agents Chemother 42:2215–2220

Menzella HG, Reeves CD (2007) Combinatorial biosynthesis for drug development. Curr Opin Microbiol 10:238–245

Menzella HG, Reid R, Carney JR et al (2005) Combinatorial polyketide biosynthesis by De Novo design and rearrangement of modular polyketide synthase genes. Nat Biotechnol 23:1171–1176

Minami H, Kim JS, Ikezawa N et al (2008) Microbial production of plant benzylisoquinoline alkaloids. Proc Natl Acad Sci U S A 105:7393–7398

Nikaido H (2009) Multidrug resistance in bacteria. Annu Rev Biochem 78:119–146

Niraula NP, Kim SH, Sohng JK et al (2010) Biotechnological doxorubicin production: pathway and regulation engineering of strains for enhanced production. Appl Microbiol Biotechnol 87:1187–1194

Nguyen KT, Ritz D, Gu JQ et al (2006) Combinatorial biosynthesis of novel antibiotics related to daptomycin. Proc Natl Acad Sci U S A 103:17462–17467

Park SY, Choi SK, Kim J et al (2012) Efficient production of polymyxin in the surrogate host *Bacillus subtilis* by introducing a foreign ectB gene and disrupting the abrB gene. Appl Environ Microbiol 78:4194–4199

Penn J, Li X, Whiting A et al (2006) Heterologous production of daptomycin in *Streptomyces lividans*. J Ind Microbiol Biotechnol 33:121–128

Pfeifer BA, Admiraa SJ, Gramajo H et al (2001) Biosynthesis of complex polyketides in a metabolically engineered strain of *E. coli*. Science 291:1790–1792

Pfleger BF, Pitera DJ, Smolke CD et al (2006) Combinatorial engineering of intergenic regions in operons tunes expression of multiple genes. Nat Biotechnol 24:1027–1032

Ro DK, Paradise EM, Ouellet M et al (2006) Production of the antimalarial drug precursor artemisinic acid in engineered yeast. Nature 440:940–943

Stephanopoulos G, Aristidou AA, Nielsen J (1998) Metabolic engineering: principles and methodologies. Academic Press, San Diego

Tang Z, Xiao C, Zhuang Y et al (2011) Improved oxytetracycline production in *Streptomyces rimosus* M4018 bymetabolic engineering of the G6PDH gene in the pentose phosphate pathway. Enzyme Microb Technol 49:17–24

Teijeira F, Ullán RV, Fernández-Aguado M et al (2011) CefR modulates transporters of beta-lactam intermediates preventing the loss of penicillins to the broth and increases cephalosporin production in *Acremonium chrysogenum*. Metab Eng 13:532–543

Van Epps HL (2006) René Dubos: unearthing antibiotics. J Exp Med 203:259

Veiga T, Solis-Escalante D, Romagnoli G et al (2012) Resolving phenylalanine metabolism sheds light on natural synthesis of penicillin G in *Penicillium chrysogenum*. Eukaryot Cell 11:238–249

Von Nussbaum F, Brands M, Hinzen B et al (2006) Medicinal chemistry of antibacterial natural products—exodus or revival? Angew Chem Int Ed Engl 45:5072–5129

Vrijbloed JW, Zerbe-Burkhardt K, Ratnatilleke A et al (1999) Insertional inactivation of methylmalonyl coenzyme A (CoA) mutase and isobutyryl-CoA mutase genes in *Streptomyces cinnamonensis*: influence on polyketide antibiotic biosynthesis. J Bacteriol 181:5600–5605

Waksman SA (1947) What is an antibiotic or an antibiotic substance? Mycologia 39:565–569

Weber JM, Cernota WH, Gonzalez MC et al (2012). An erythromycin process improvement using the diethyl methylmalonate-responsive (Dmr) phenotype of the *Saccharopolyspora erythraea* mutB strain. Appl Microbiol Biotechnol 93:1575–1583

Weissman, KJ, Leadlay PF (2005) Combinatorial biosynthesis of reduced polyketides. Nat Rev Microbiol 3:925–936

White NJ (1997) Assessment of the pharmacodynamic properties of antimalarial drugs in vivo. Antimicrob Agents Chemother 41:1413–1422

Witte W (2004) International dissemination of antibiotic resistant strains of bacterial pathogens. Infect Genet Evol 4:187–191

Woodyer RD, Shao Z, Thomas PM et al (2006) Heterologous production of fosfomycin and identification of the minimal biosynthetic gene cluster. Chem Biol 13:1171–1182

Wu Y, Kang Q, Shen Y et al (2011) Cloning and functional analysis of the naphthomycin biosynthetic gene cluster in *Streptomyces* sp. CS. Mol Biosyst 7:2459–2469

Yang F, Cao Y (2012) Biosynthesis of phloroglucinol compounds in microorganisms-review. Appl Microbiol Biotechnol 93:487–495

Zhai L, Lin S, Qu D et al (2012) Engineering of an industrial polyoxin producer for the rational production of hybrid peptidyl nucleoside antibiotics. Metab Eng 14:388–393

Zhang W, Ames BD, Tsai SC et al (2006) Engineered biosynthesis of a novel amidated polyketide, using the malonamyl-specific initiation module from the oxytetracycline polyketide synthase. Appl Environ Microbiol 72:2573–2580

Zhang H, Skalina K, Jiang M et al (2012) Improved *E. coli* erythromycin A production through the application of metabolic and bioprocess engineering. Biotechnol Prog 28:292–296

Chapter 19
DNA Origami: What, How and Where

Mukta Joshi, Shankar Kundapura, Thirtha Poovaiah and Pawan K. Dhar

Abstract DNA origami is the science of folding DNA molecules to make novel two and three dimensional shapes. The science of folding of DNA into pre-decided shapes was started by Paul Rothemund at California Institute of Technology. Rothemund explored a number of ways by which 7000 base pair viral genome could be folded into three dimensional shapes. He ordered DNA staples from a synthesis company, added them to the hybridization mixture and allowed sing stranded DNA to fold along the path of staples. The result was a smiley face visualized using atomic force microscopy. A number of novel inventions have come up in the recent years, inspired by Rothmund's work. More complex three dimensional shapes like drug-delivery nanocages, alphabets and numbers have been developed by self-folding DNA.

Keywords DNA origami · Self folding · DNA staples

19.1 Introduction

In 2006, Paul Rothmund proposed the concept and invented a method to self-fold DNA to create different structures (Rothemund 2006). He selected DNA due to its specificity and easy foldability. Using his method a long piece of DNA 'SCAFFOLD' is held together in a desired shape by hybridizing it with small DNA strands called as 'STAPLES'. The shape of the origami structure and the staples can be designed using different softwares.

Paul Rothmund introduced what he called as one pot method compiling long DNA strands into desired shapes by using short DNA strands as DNA clips

The entire DNA origami technique was performed in five distinct steps.

1. Building a geometric model of the DNA that resemble a pre-decided shape
2. Raster filling long single strand of DNA

M. Joshi (✉) · S. Kundapura · T. Poovaiah · P. K. Dhar
Centre of Systems and Synthetic Biology, Department of Computational Biology and Bioinformatics, University of Kerala, Trivandrum, India

P. K. Dhar
Department of Life Sciences, School of Natural Sciences, Shiv Nadar University, Dadri, India

V. Singh, P. K. Dhar (eds.), *Systems and Synthetic Biology*,
DOI 10.1007/978-94-017-9514-2_19

3. Designing staple strands that provide the Watson-Crick complements
4. Calculating of scaffold crossover and position changes to minimize the strain
5. Merging adjacent staples to yield fewer and longer staples.

The geometric model of DNA was built using computer programs MATLAB. Using the same program the template DNA was used as scaffold and staples were designed complementary to the scaffold. Because of the complementarily between staple and scaffold, the desired shape was obtained after few trial and error. The size of the staples was found to be critical in determining the stability of the structure. The longer the staple the more was the stability of the structure.

Paul Rothmund used M13mp18 as scaffold DNA. It is a naturally occurring, single stranded, 7249 nt sequence long DNA. Staples were designed based on the sequence of the scaffold strand leading to the folding of DNA into 2D structures like smiley, square, star, triangle etc.

The science of DNA origami has undergone a revolution of sorts. In the last 6 years a number of original methods have evolved, adding features and complexity to the original method. Rcently Castro et al (2011) improved the science of DNA origmi further by designing 3D structure of a robot. The robot was constructed using caDNAno software. Staples were designed using M13mp18 genomic DNA as scaffold. Staple length was programmed to stay between 18 to 50 nucleotide of length to ensure stability. The reaction mixture was set up using scaffold, staples and buffers (materials and methods). After the incubation period the reaction mixture was subjected to agarose gel electrophoresis. The folded DNA was extracted from the gel by selecting the farthest band in the gel as the fully formed structure is expected to be compact in comparison to the partially folded or unfolded as compared to incomplete structures. After the extraction of the structure from the gel it was visualized in TEM (Transmission Electron Microscopy)

Current Software available for designing different shapes of DNA are

1. caDNAno (Shawn Douglas/Wyss Institute for Biologically Inspired Engineering at Harvard University)
2. NUPACK (Prof. Niles A. Pierce/California Institute of Technology)
3. Nanoengineer (Mark Sims)
4. SARSE (centre for Nanotechnology/Aarhus University)

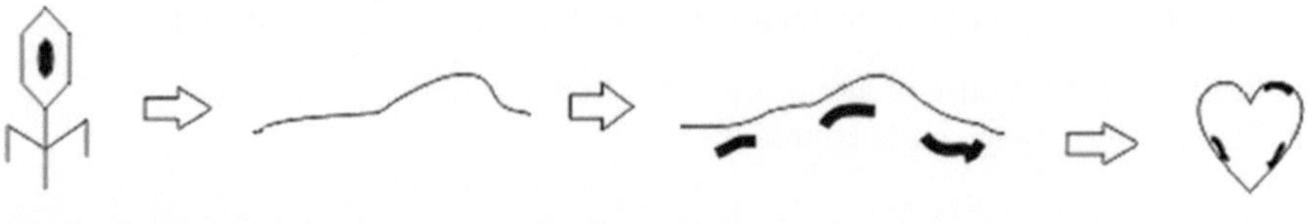

The standard DNA origami protocol usually consists of the following steps

1. Designing a desired shape using a DNA origami software
2. Designing staples depending on the sequence of the scaffold (template)
3. Synthesizing staples and isolating the template

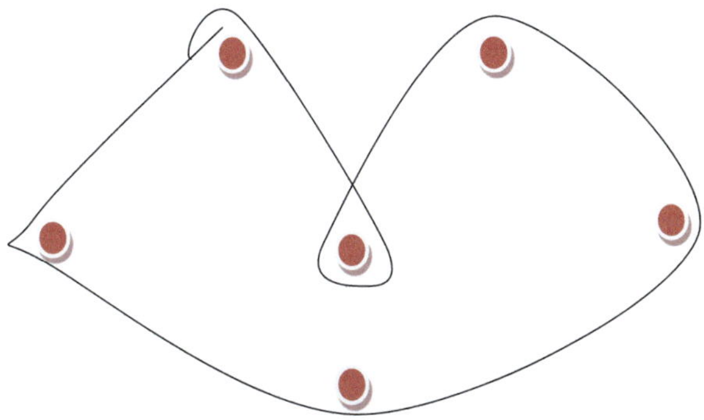

Fig. 19.1 Top view of sticks and string phenomenon

4. Setting up reaction mixture with proper buffer, scaffold, staple concentrations and under appropriate reaction conditions like temperature.
5. Visualization (consists of two steps)
   – Running the hybridized DNA complex through agarose gel electrophoresis. The Band farthest from the gel indicates most compact structure that is fully formed. These bands are isolated and visualized.
   – Either AFM (Atomic force microscopy) or TEM (transmission electron microscopy) can be used to visualize hybridized complex depending the type of structure.

19.2 How to implement DNA Origami?

19.2.1 DNA Origami: Materials and Methods

Let us approach this naively to produce desired shapes of DNA.

What must one need the most? DNA! What kind of DNA would be most suitable: circular DNA, linear DNA, single stranded DNA or double stranded DNA? The answer is a linear single stranded DNA, as flexibility and combinations to make desired shapes would be higher in linear DNA than circular DNA. Why one must choose single stranded DNA would be explained in the following paragraphs. Although, it must be noted that circular DNA is not uncommonly used. It has been used in making 5 point stars and such.

Now, we have the linear DNA, how can one twist, turn, bend this linear DNA into desired shapes?

One of the methods which come to mind is an adaptation of a string and sticks phenomenon where, sticks (red) upright are arranged at desired location and the string (black) is tied around the sticks to give the desired design (Fig. 19.1).

However, this adaptation cannot be used in a biological setup as there aren't any biological 'sticks', one can use.

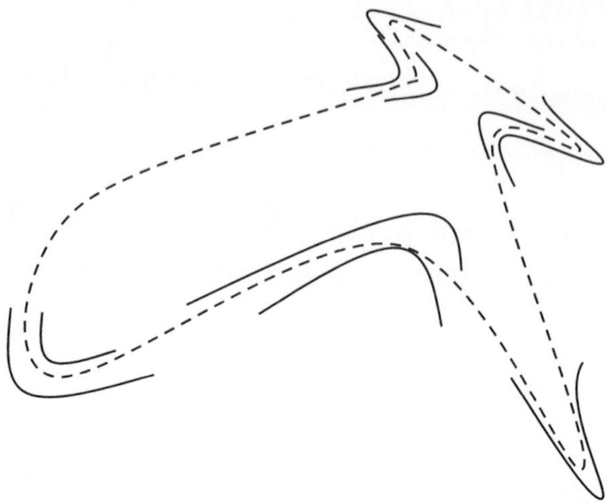

Fig. 19.2 Linear DNA strand (*red dotted line*) being held together in the desired shape with help of staple DNA (*black solid line*)

Another interesting strategy is to hold the turns, bends and twists of the DNA which contributes to the desired shape, firmly. When the bends and turns are held tightly, the DNA would fall into the desired shape (Fig. 19.2).

Now, how do we hold these bends and turns tightly? Since we have considered linear single stranded DNA, we can use short complementary DNA sequences which bind to the DNA sequence of the linear single stranded DNA, also called 'Scaffold DNA', comprising the turns and bends. This shall hold the bends and turns tightly which would force the linear DNA to fall into the desired structure. This short complementary DNA which binds to the Scaffold DNA is called 'Staple DNA', as it staples the linear DNA into desired shape.

Thus we have two important materials required for DNA Origami: scaffold linear DNA and staple DNA. Now throw in some stabilizing agents and mounting agents with appropriate temperature and sufficient time, your desired DNA structure is ready.

Since the core concept is introduced, let's probe into the specifics of the requirements of DNA Origami. The first step involved in DNA Origami is designing of desired shape or pattern, the scaffold DNA needs to adopt. This can be done with the assistance of various softwares that help in designing a desired shape from the scaffold DNA and also give the number, sequences and the position of the staple DNA strands.

One of the most user friendly open source softwares is 'caDNAno' (Douglas et al. 2009a). 'caDNAno' was developed by Shawn Douglas (Wyss Institute for Biologically Inspired Engineering at Harvard University). The caDNAno software has a graphical interface and has no command prompt for operation. This software operates on two lattices, useful for design of shapes, Square lattice and the Honey comb lattice. Using the suitable design lattice one can obtain the desired shape. But the more important task is to obtain the right size and sequence of the staple DNA strands in accordance with the desired shape.

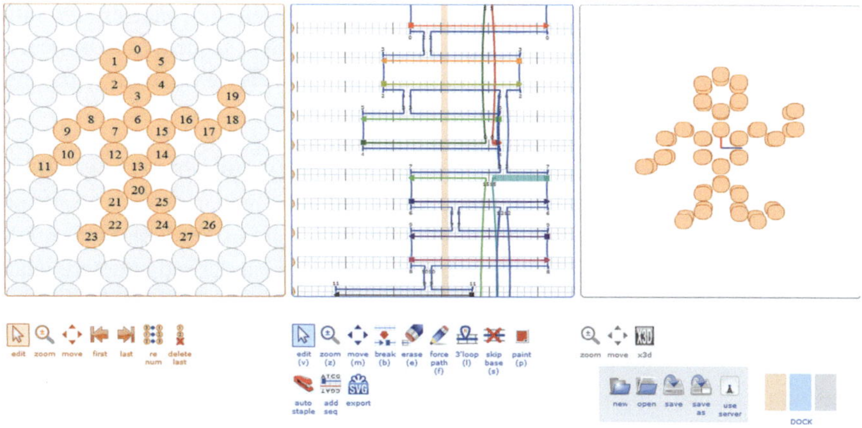

Fig. 19.3 Graphical interface of caDNAno representing the three dock panels. (Courtesy http://cadnano.org)

There are three dock panels provided. The first dock panel is where one can design the desired shape, the second dock panel is where one can modify and manipulate the staple sequences and the third dock panel is where one can visualize the shape being formed in accordance with staples being added (Fig. 19.3).

It must be noted that the scope of designing various structure is based on the skill of the user in the software.

After the sequences and sizes of the staple DNA strands are confirmed and the desired shape approved, the synthesis of staple DNA strands must be initiated and the scaffold DNA must be procured.

Next, the choice of scaffold DNA must be undertaken with considerable thought. The scaffold DNA of choice must have the following properties.

- Linear single stranded DNA
- Known sequence of the DNA strand
- Stable secondary structure

The most common scaffold DNA used in DNA Origami is M13mp18 viral DNA. The M13mp18 viral DNA is a circular 7249 nucleotide DNA, which consisted of a hairpin structure when its secondary structure was examined. This hairpin structure, which was about 73 nucleotides long, caused uncertainty of the staple DNA strands binding to it. Thus this hairpin like structure was deleted to obtain a linear DNA with the help of BsrB1 restriction enzyme. If a circular Scaffold is required, the 73 nucleotide sequence is not considered during the secondary structure formation (Rothemund 2006).

Furthermore, M13mp18 viral DNA is a naturally occurring DNA sequence with a stable secondary structure when compared to other random sequences. It is advised to be stored in -20 oC in 5 mM Tris base and 1 mM EDTA at pH 8 (Rothemund 2006).

The reaction mix or hybridization mix must have a definite concentration ratio of scaffold DNA to staple DNA. When designing the staple sequences, it is ideal to have staple strands of length 18–50. The lower limit is set to 18 nucleotides because stability of staple strands shorter than 18 nucleotides decreases at room temperature and the upper limit is set to 50 nucleotides due to increasing cost of longer sequences (Castro et al. 2011). One can prepare a 1:10 or 1:50 even 1:100 scaffold DNA to staple DNA concentrations (stoichiometric ratio), but an important factor here is, the staple DNA must always be at a higher concentration when compared with the scaffold DNA. This is because of a high probability of obtaining a desired structure when more staple DNA strands are present when compared to scaffold DNA. Another question which might come up is, 'would higher number of Staple DNA strands cause pairing amongst them?' The answer for this is since the staple DNA strands are complementary to the scaffold DNA in terms of Watson and Crick base pairing, the staple DNA strands would bind to their unique location, if the structure designed is symmetrical, there can be a situation where the staple strands would bind to each other. This situation can be avoided if the staples are introduced sequentially. Since more numbers of staple DNA strands are present, there is a higher probability that particular staple DNA would bind successfully to the particular location on the scaffold DNA and the same is true for the other staple DNA strands.

Other than the scaffold DNA and the staples, the reaction mix also contains buffer (TRIS-EDTA), to maintain the pH of the mix. Along with the buffer, magnesium salt (Magnesium chloride or magnesium acetate) is added to neutralize the negative charges of the DNA and enable the single stranded DNA to transform into double stranded helix. Nickel acetate is used to mount the DNA on to the slide during AFM sessions.

Single-stranded M13mp18 DNA is quantitated by UV absorbance at 260 nm. Staple DNA strands are stored at – 20 °C. The desired set of Staple DNA strands are mixed with M13mp18 (it is a good practice to have an 100 fold increase of staple DNA strands when compared with scaffold DNA strand) in a particular volume of 1X Tris-Acetate-EDTA (TAE) buffer with 12.5 mM magnesium acetate (pH 8.3) and annealed from 95 to 20 °C in a PCR machine at a rate of 1 °C/min reduction. The reaction mix now has to be subjected to high temperature which must decreased step wise at the order of 1°C/min from 95 to 20°C in a PCR machine. This is performed, in order for the binding of the staples to the scaffold DNA to obtain the desired shape (Rothemund 2006) (Fig. 19.4).

After the reaction mix is processed, it must be viewed under an Atomic Force Microscope (AFM) for visualization of the desired structure formed. Only folded structures stick to the surface when deposited on nickel acetate, the remaining unbound strands were left behind in the solution. One must note that mica can also be used in fixing the folded structures on the slide and is used extensively for larger origami structures, also the ability to stick for DNA origami on the surface depends on the area available for interaction, and the presence of divalent cations (Mg^{2+} or Ni^{2+}). AFM is suitable for solution based imaging too.

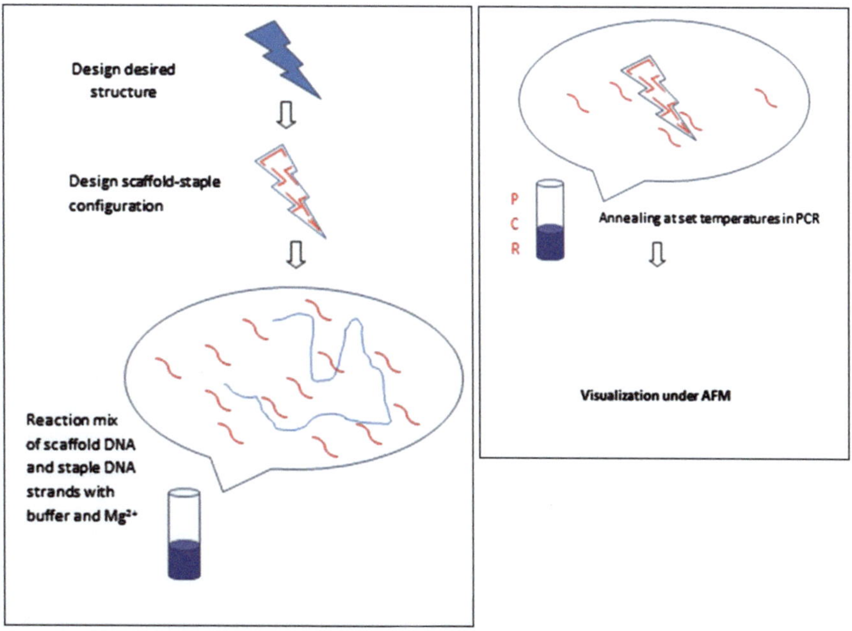

Fig. 19.4 Flow chart of steps involved in DNA Origami. *Blue line* denotes scaffold DNA and *red lines* denote staple DNA strands

Products formed are considered 'well formed' if it has no defects like holes, overlap, indentation and the like, greater than 15 nm in diameter. One can estimate the yield by calculating the proportion of 'well formed' structures among all distinguishable structures (Rothemund 2006).

19.2.2 DNA Origami for 3-D Structures

DNA origami encompasses 3-D structures was perfected by a team led by Carlos Castro (Castro et al. 2011). The steps involved in 3-D DNA origami are fairly similar to that of 2-D DNA origami. Software called CanDo (Castro et al. 2011) can also be used to design 3-D origami structures.

Designing the desired 3-D structure and procuring the length and sequence of the staple DNA strands can be accomplished by method used in 2-D DNA origami.

Here, it is considered good practice to divide the staple DNA strands corresponding to the 3-D structure into different sub-structures; this is in view to make pipetting more organized. The pooling of staple strands must be done in accordance with the sub-structure. Then a master pool must be created using particular concentration and volume.

In the PCR tubes, 20 μl of scaffold DNA, 40 μl of the combined staple pool, 10 μl of a 10 × folding buffer containing 50 mM Tris base, 50 mM NaCl, 10 mM EDTA (pH 8) and 20 μl pure H_2O should be added. The stoichiometric ratio of

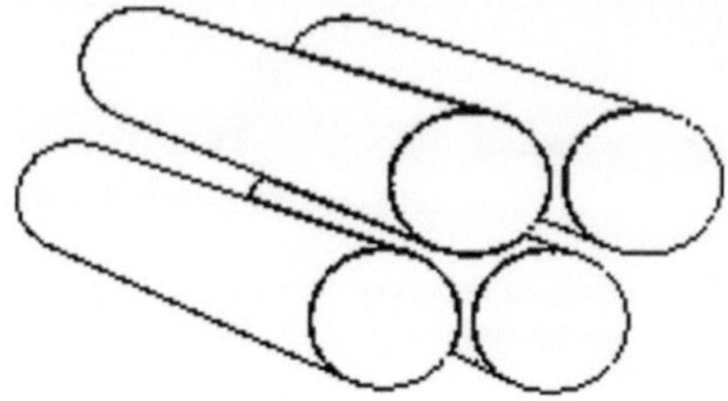

Fig. 19.5 Example of 3-D DNA origami structure

scaffold DNA to staple DNA must be in the order of 1:10 to obtain good resolution of desired structures. It is advised to add, 10 μl of stock solutions containing specific concentrations of $MgCl_2$ dissolved in pure water into the PCR tubes. The reaction mix must be subjected to thermal annealing in a conventional PCR by heating the mixture briefly to 80 °C and cooling it to 60 °C at a rate of 1 °C/5 min, followed by cooling it from 60 to 25 °C at a rate of 1 oC/300 min. This thermal annealing ramp takes about 7 days (Castro et al. 2011).

The next step is to analyze the quality of folding of DNA origami structures and its purification using 2 % agarose gel electrophoresis (Douglas et al. 2009b). The buffer must have magnesium ions to neutralize the negative charges on the scaffold DNA. It has been observed that structures with the least defects travel fastest in the gel. Thus, the most suitable structure for high resolution study would be the ones that are found farthest from the wells of the agarose gel. Therefore, the band containing most suitable DNA origami structure is excised from the gel and dropped into a 1.5 ml tube. The agarose slice is then crunched with a pistil to obtain agarose debris. The agarose debris is subjected to a short spin—the tip of the tube containing the debris is cut off. The inverted tip is put into a freeze 'n' squeeze spin column. It is finally subjected to centrifugation for 10 min in order to obtain the DNA origami structure.

The hybridized complex is then subjected to visualization in negative-stain or cryogenic TEM (Transmission Electron Microscopy). Image processing can help to identify systematic structural flaws. Negative-stain TEM with 2 % uranyl formate as staining agent is a suitable tool for imaging 3-D objects. If flaws are detected, the scaffold-staple configuration has to be rethought and redesigned, following which the entire procedure has to be performed again (Castro et al. 2011; Fig. 19.5).

19.3 What are the Applications DNA Origami?

19.3.1 DNA Origami: Applications

19.3.1.1 Introduction

The answer to this ranges from complex nanoelectronics (Maune et al. 2010) to simple art (Rothemund 2006). In the field of biomedicine, DNA origami has been used to create targeted drug delivery systems, biosensors and diagnostic tools. Most significantly, DNA origami has opened novel and exciting avenues in nanomedicine.

This technique has paved the way for simple, efficient and economic 'bottom up' fabrication of nanostructures. Its unique complimentary nature, geometry and easy availability have made DNA the most popular choice in nanostructure synthesis (Yang et al. 2010). DNA has also shown its potency as a scaffold by binding to other molecules in DNA nanodevices. It can also be used a guide to direct the assembly of nanowires, free standing membranes and crystals (Yang et al. 2010). DNA has successfully amalgamated the fields of molecular biology, material sciences and engineering by its versatility as a building block for nanostructures. The applications of DNA origami are growing at an exponential pace; a few of its notable applications are illustrated below:

19.3.1.2 DNA Rulers

One of the most striking features of DNA origami structures is its inherent stability and precise addressability of its dimensions. Such defined structures find an application as reference standards in calibrating super-resolution microscopy.

Molecules used conventionally to calibrate sophisticated microscopes are loose pieces of DNA or filamentous proteins (actin filaments or microtubules). These structures are inaccurate due to frequent changes in dimensions due to their flexibility. Dr. Tinnefeld and his team at Universität Müchen, Germany, constructed a DNA origami rectangle (100 nm by 70 nm) with staple strands labelled with fluorescent dye molecules. When the DNA folded the two labelled staples fell at opposite ends of the rectangle at precise locations. This ruler finds an application in calibrating super resolution microscopes, which can resolve objects smaller than the diffraction limit of light (roughly 200 nm). Further, origami molecular rulers might serve as quantification standards for super-resolution microscopes and other spectroscopic techniques such as plasmon coupling (Steinhauer et al. 2009).

19.3.1.3 DNA Box

Fabricating nanostructures using DNA origami is not only limited only to 2D models. Well defined 3D structures such as cuboids referred to as "DNA Box" can be efficiently constructed. Going one step ahead, a Danish research group has developed a functional 3D DNA Box with can be opened in the presence of externally supplied 'DNA keys' (Andersen et al. 2009). One strand of DNA holds the lid shut; a separate DNA 'key' springs it open. This technology may lead to applications, in diagnostics and controlled transport of nanocargos.

The DNA box is constructed with fluorescent dyes attached to the two perpendicular surfaces involved in the dual lock key system. The functionality of the box can be measured by the change in the intensity of the fluorescence. When the lid is closed the intensity of fluorescence is at its maximum. As the lid open the intensity falls. Lowest intensity in the fluorescence is seen when both the keys are present. This indicates that a closed box can be programmed to open in response to two external signals (representing a functional AND gate). Similarly the box can be designed to

Fig. 19.6 DNA box

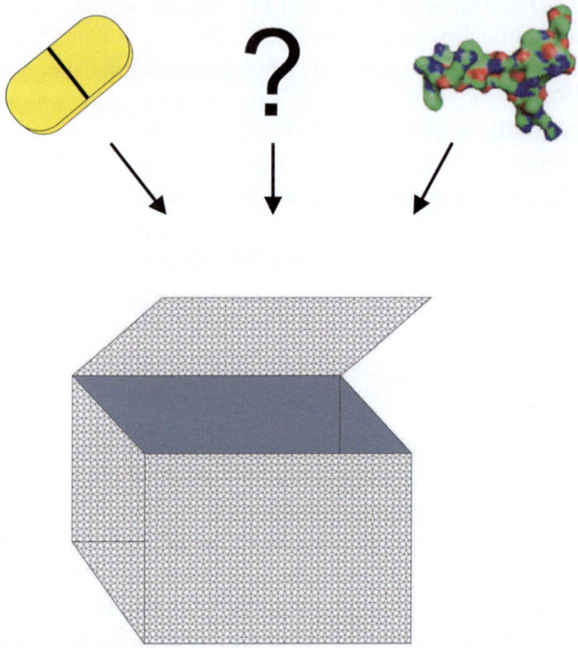

close in the presence of certain signals resulting in a NOT gate. As the DNA box is a cuboid, the box can be programmed to open in response to a single key, resulting in a OR gate. Hence, the lids of the DNA box have the potential to be uniquely programmed to response to complex combinations of oligonucleotide sequences, for instance cellular messenger RNAs or micro RNAs. Another application of this specific response can be implemented in a diagnostic sensor.

Another question that arises with the DNA box is: What does one place a certain object? The object can be a drug, a ribosome, an enzyme or just about anything that fits. The box can be assembled to exhibit tuneable flexible properties. Dynamic changes can be induced by sensing external signals in the environment. Such controlled activity of the box can be used to restrict the movement of cargo in and out of the box. Such 'nano robotic' devices can be used to package enzymes and provide controlled access to the respective substrates (Fig. 19.6).

DNA boxes represent the characteristics of a nano intelligent system, as it has the potential to both sense and act. An example of this function is by the combination of a diagnostic sensor of complex signals with the controlled release of, or access to, a payload.

19.3.1.4 DNA Nanoelectronics

Technology is heading at a rapid pace towards the nano universe. In this view, the main roadblock to the development of nanoelectronic circuitry is the lack of optimized fabrication technology. Current lithographic techniques fail to create structures with dimensions below 22 nm (Kershner et al. 2009). A landmark future invention in the field of nanotechnology would be the coupling of self-assembled molecular nanostructures with conventional microfabrication. Developing such a technology would enable registering individual molecular nanostructures, electronically address them, and integrate them into functional devices.

Conventional DNA origami technique was implemented to design microcircuitry but the results were not promising. The main drawback of using this technology is that DNA origami structure is synthesized in solution and uncontrolled deposition results in random arrangement. This makes it difficult to measure the properties of attached nanodevices or to integrate them with conventionally fabricated microcircuitry. To address this issue, Rothemund and IBM Almaden Research Centre, designed a lithography based strategy to make templates to make templates for self-assembling discreet DNA components (Hung et al. 2009). Examples included the assembly of nanoparticles, carbon nanotubes and nanowires. Electron-beam lithography and dry oxidative etching techniques were used to create DNA origami-shaped binding sites on technologically useful materials, such as SiO_2 and diamond-like carbon. In buffer with100 mM $MgCl_2$, DNA origami structures bound with high selectivity and good orientation: 70–95 % of sites have individual origami aligned. Lithographic templates can also be used to create hierarchical order: the nanostructures they organize can themselves have internal features with dimensions significantly smaller than those of the original template and can serve as scaffolds for the assembly of still smaller components.

As an extension of this technology, the DNA origami can find an application wherever there is a need to place individual molecules in a pattern on a surface. Those molecules need to be coupled with DNA origami structure and is expected to have an enormous impact on single-molecule biophysics.

19.4 Conclusion

Recent studies indicate that DNA origami is a simple and efficient method in fabricating nanostructures. DNA origami technique has found applications in the field of nanomedicne, particularly in diagnostics, nanoelectrotics and single molecule biophysics. Further refinement is needed to (elevate/scale up) this technique to the industrial level. Research should be aimed at reducing cost and time of fabrication (Yang et al. 2010). If these hurdles are effectively overcome, the real time applications of DNA nanostructures may be visible in the society.

References

Andersen ES, Dong M, Nielsen MN, Jahn K, Subramani R, Mamdouh W, Golas MM, Sander B, Stark H, Oliveira CLP, Pedersen JS, Birkedal V, Besenbacher F, Gothelf KV, Kjems J (2009) Self-assembly of a nanoscale DNA box with a controllable lid. Nature 459:73–76

Castro C, Kilchherr F, Kim D, Shiao E (2011) A primer to scaffolded DNA origami. Nat Methods 8(3):221–229

Douglas SM, Marblestone AH, Teerapittayanon S, Vazquez A, Church GM, Shih WM (2009a) Rapid prototyping of 3D DNA-origami shapes with caDNAno. Nucleic Acids Res 37(15):5001–5006

Douglas SM, Dietz H, Liedl T, Hogberg B, Graf F, Shih WM (2009b) Self-assembly of DNA into nanoscale three-dimensional shapes. Nature 459:414–418

Hung AM, Micheel CM, Bozano LD, Osterbur LW, Wallraff GM, Cha JN (2009) Large-area spatially ordered arrays of gold nanoparticles directed by lithographically confined DNA origami. Nat Nanotechnol 5:121–126

Kershner RJ, Bozano LD, Micheel CM, Hung AM, Fornof AR, Cha JN, Rettner CT, Bersani M, Frommer J, Rothemund PWK, Wallraff GM (2009) Placement and orientation of individual DNA shapes on lithographically patterned surfaces. Nat Nanotechnol 4:557–561

Maune HT, Han S-P, Barish RD, Bockrath M, Goddard WA III, Rothemund PWK, Winfree E (2010) Self-assembly of carbon nanotubes into two-dimensional geometries using DNA origami templates. Nat Nanotechnol 5:61–66

Rothemund PWK (2006) Folding DNA to create nanoscale shapes and patterns. Nature 440:297–302

Steinhauer C, Jungmann R, Sobey TL, Simmel FC, Tinnefeld P (2009) DNA origami as a nanoscopic ruler for super- resolution microscopy. Angew Chem Int Ed 48:8870–8873

Yang D, Campolongo MJ, Tran NJN, Ruiz RCH, Kahn JS, Luo D (2010) Novel DNA materials and their Applications. WIREs Nanomed Nanobiotechnol 4:648–669

Chapter 20
Making Synthetic Proteins From Non-coding DNA

Vipin Thomas, Shidhi PR, Deepthi Varughese, Navya Raj and Pawan K. Dhar

Abstract Non-coding DNA describes regions of the genome for which no apparent function has been identified. The term "Junk DNA", was introduced in 1972 by Susumu Ohno for such non-expressed regions of genome. A number of bioinformatics studies have been organized to understand the function of non-coding DNA. However, a clear understanding is lacking. To address this issue, we invented a method to make novel genes from non-coding DNA and study its expression. This chapter describes the general composition of non-coding DNA, describes a novel approach of studying these sequences and provides a first glimpse of some of the interesting results.

Keywords Junk DNA · Proteins · Peptides · Non-coding

The proportion of coding vs. non-coding genomic region varies from species to species. In eukaryotes a large percentage of genome is non-coding as compared to the coding region. For example in humans more than 98 % of the genome size is found to be non-protein coding. However, bulk of it (> 80 %) seems to be RNA coding based on recent reports. Thus, the phrase junk has lost its original meaning even though a comprehensive understanding of its significance is still lacking.

It is in the context of the unknown space that our work of making functional genes from non-coding DNA assumes significance. Before getting into various details let us first look at the composition of non-coding DNA.

Intergenic regions are defined as the genome sequences between the genes. A gene cluster is a set of genes (can be two or more gene) which codes for same or similar protein or RNA products. Intergenic region are non-coding and they do not direct the protein synthesis. However, this region is seems to have role in modulating the expression of adjacent gene. Experimental evidences show that they contain important control sequences such as enhancers and silencers that do not directly

V. Thomas (✉) · S. PR · D. Varughese · N. Raj · P. K. Dhar
Department of Computational Biology and Bioinformatics, University of Kerala, Trivandrum, Kerala, India

P. K. Dhar
Department of Life Sciences, School of Natural Sciences, Shiv Nadar University, Dadri, Uttar Pradesh, India

V. Singh, P. K. Dhar (eds.), *Systems and Synthetic Biology*,
DOI 10.1007/978-94-017-9514-2_20

contribute to the production of protein product. Further, this region has also been found to play critical roles in the process of genetic imprinting, cancers and birth defects.

Introns constitute a major fraction of the noncoding DNA. The occurrence of introns varies between different eukaryotes. The introns are identified by the presence of specific signal sequences called consensus sequences. GT at the start or donor (3′) end and AG at the acceptor (5′) end is the commonly found consensus sequence in eukaryotic genome. The splicing mechanism recognizes this consensus signal for the removal of introns from the DNA.

Pseudogenes are non-functional relatives of genes that were earlier functional but have lost their ability to code for proteins. Some pseudogene do not have introns or promoters (processed pseudogene) but some have retained standard gene features (promoters, CpG islands and splice sites).

miRNAs are short double stranded RNA generated from an endogenous hairpin transcript to produce a mature miRNA. This processed miRNAs bind to the target mRNA thereby modulates its translation. siRNA is a small double-stranded RNA of about 20 base pairs long and generated by cleavage of a dsRNA. The siRNA is processed by an enzyme Dicer and RISC complex. To silence mRNA effectively, siRNAs needs 100 % complementarity with the target mRNA.

Long non-coding RNAs are non-protein coding transcripts longer than 200 nucleotides. While small RNAs exhibit strong conservation across species, the general long ncRNAs lack strong conservation. NcRNAs modulate the function of transcription factors by functioning as co-regulators orby modifying transcription factor activity.

Repeat sequences appear in the form of tandem repeats and Interspersed repetitive DNA. When two or more nucleotides repeat and lie adjacent to each other, they are termed as tandem repeats, which help in determining an individual's inherited traits and also parentage. Tandem repeats can be classified as VNTR (Variable Number of Tandem Repeats), minisatellites and microsatellites (STRs-short tandem repeats) based on the length of the repeats.

20.1 Structural and Functional Characterization of Not- coding DNA

Not-coding regions represent genome regions that encode neither RNA nor proteins. Using bioinformatics methods, not-coding genome sequences have been studied. However, a clear understanding of its origin and function is lacking.

A novel way to address functional significance of not-coding DNA was recently demonstrated (Dhar et al. 2009). A novel approach was invented to artificially synthesize proteins from intergenic templates of *E. coli* using pBADtopo expression vector (Fig. 20.1). The vector provided promoter, start codon, stop codon, ribosome binding site and His tag sequences as a template for expressing sequences of choice. Out of the six novel proteins expressed, one showed growth inhibitory effect which

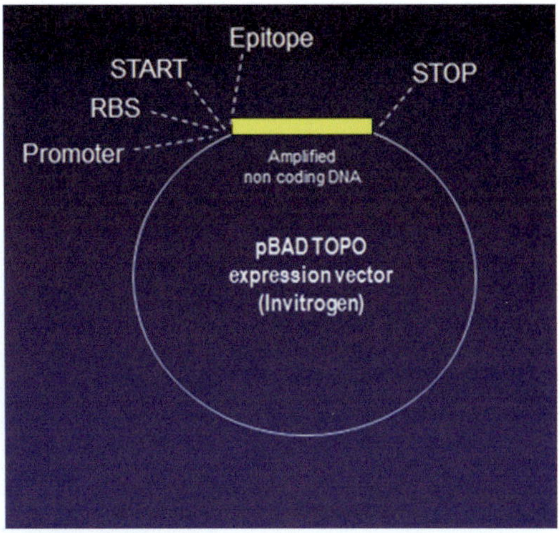

Fig. 20.1 Expression vector

was rescued by switching off the artificial gene expression. Computational structure prediction techniques showed stable tertiary structure conformations for two of these proteins. This work was the first attempt to artificially express regions that have no history of transcription.

After successful demonstration of the method, the question evolved into the following. Out of millions of non-coding DNA sequence combinations how many sequences will actually result in stable and functional proteins? To address this question, every intergenic sequence was computationally profiled in terms of potential protein tertiary structure, function, localization, molecular interaction, protein disorder, similarity with therapeutic peptides and proteins, enzymes, transcription factors, signaling molecules and so on. Currently the knowledgbase houses more than 2000 such sequences from noncoding genomes of *E. coli*, *S. cerevisiae* and *C. elegans*.

Interestingly, some of the peptide hits have shown strong docking against three membrane based Plasmodium falciparum targets i.e., Apical Membrane Antigen-1, Erythrocyte Binding Antigen-175 and Merozoite Surface Antigen 1(19).

Novel drug molecules against malaria A library of potential synthetic peptides was constructed upon translation of 1000 randomly selected not-coding DNA sequences of the yeast genome. The relevant synpeps were screened on the basis of sequence similarity with known ligands that bind to target proteins in their natural setting. This significantly reduced the number of candidate peptides to top nine candidate molecules i.e., three against each target. This number was further reduced to one lead peptide against each target on the basis of structural similarity with the naturally target binding ligands. Subsequently, the three best synpeps were docked with their respective targets to find preferred orientation of binding—important for stable complex formation. The not-coding parts of the yeast genome were identified from Saccharomyces Genome Database via Yeastmine (Table 20.1).

Table 20.1 Comparison of sequences of binding region of natural ligands of targets and prospective syneps

	Similarity (%)	Gap (%)
AMA1-SSBS-1	42.3	11.5
EBA 175-SSBS-2	36.0	0
MSP1 19-SSBS-3	44.4	7.4

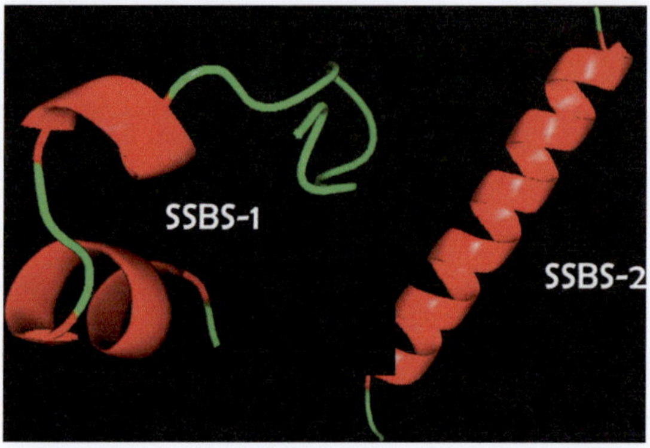

Fig. 20.2 Validated structures of *A* SSBS-1, *B* SSBS-2, *C* SSBS-3

The not-coding sequences were computationally extracted, translated and sequence matched with the regions of natural ligands that bind to the three targets selected. A global sequence similarity of > 30 % and gap less than 12 % were considered for further studies. The selected peptides were submitted to 3-D structure prediction softwares that employ threading and ab-initio modelling methods. The predicted synpep structures were validated for their structural correctness. The validated structures of the selected synpeps and natural ligands were superimposed the Root Mean Square Deviation (RMSD) was calculated. The synpeps whose RMSDs with the natural ligands were less than 1 were chosen with an aim of finding peptides that structurally mimic the binding of natural ligand with the respective target and prevent further Plasmodium infection after entering the blood stream. The structures of the selected not-coding peptides after due validation and structural similarity with the natural ligands of the targets are illustrated in Fig. 20.2. Finally, synpeps were docked against their targets to assess the correctness of fit. Docking jobs were performed using Cluspro and HADDOCK (Fig. 20.3). The docking of AMA-1 with R1 peptide resulted in a docking score of − 1735 as against docking score of − 1616 of AMA-1 and SSBS-1 using Cluspro. The ClusPro results for peptide SSBS-3 and MSP-1(19) was − 8821 and for docking of band3 and MSP-1 (19), it was − 5661. The docking of EBA-175 dimer performed by HADDOCK showed a barely strong interaction at a score of − 19.6 ± − 14.7. However, the docking of EBA-175 and SSBS-2 showed a stronger interaction at − 23.7 ± − 3.7 indicating that EBA-175 might bind with greater affinity to SSBS-2 than its own monomer. Although these findings are encouraging, they are preliminary observations. More computational design of not-coding DNA derived synpeps and their experimental validation would be required for future experimental studies.

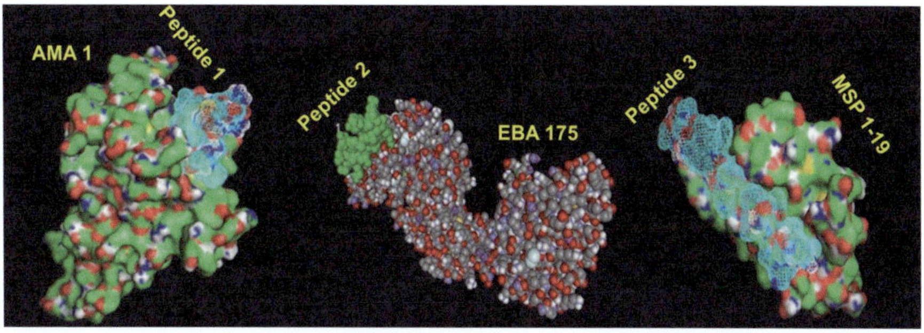

Fig. 20.3 Docking poses of *Plasmodium* surface antigens with their respective *synpeps*

Synthesizing non-natural proteins from the noncoding functional RNAs of prokaryotic and eukaryotic genome and identifying their probable role. The functional properties of the synthesized proteins, the novel application oriented pathway associated with it and the interacting partners could be identified. Expressing the not-coding RNAs using a suitable promoter, followed by the microarray analysis could help to identify the interactions and pathways associated with the expressed proteins.

Reference

Dhar PK, Thwin CS, Tun K, Maurer-Stroh S, Eisenhaber F, Tsumoto Y, Surana U (2009) Synthesizing non-natural parts from natural genomic template. J Biol Engg 3:2

Chapter 21
Engineering Biological Systems: A Brief Overview

Pawan K. Dhar

Abstract Ever since synthetic biology as a formal discipline was launched, several parts, devices and circuits have been designed. The story of engineering-inspired approach to biology probably started from a publication in 2000 that described three gene circuit called repressilator and another one describing toggle switch. While repressilator is given a more detailed chapter in another chapter, this chapter provides a brief overview of key technical innovations in synthetic biology.

Keywords Toggle switch · Never born proteins · Minimal synthetic cell

21.1 Toggle Switch

One of the early ideas of synthetic biology community was a proposal that biological systems could be assembled ground-up, just like lego toys. Though this analogy was inspiring it turned out to be inherently imperfect due to several technical reasons. Nevertheless, the journey of constructing biological systems has begun. An earliest achievement was construction of an artificial toggle switch in *E.coli* that showed property of a bistable system (Gardner et al. 2000).

The switch showed dual stable behaviour and was controlled by a change in temperature or chemical signal, in this case a molecule called IPTG. When the switch was turned ON, the bacterium made green fluorescent protein and glowed green under ultraviolet light. When the switch was turned off, the protein was no longer made and it was dark.

Gardner and colleagues referred to this functional genetic element as a "genetic applet," named in analogy to small programs written in the computer language Java and designed to be self-contained and easily ported between computing platforms

P. K. Dhar (✉)
Department of Life Sciences, School of Natural Sciences,
Shiv Nadar University, Dadri, Uttar Pradesh, India

Department of Computational Biology and Bioinformatics,
Centre for Systems and Synthetic Biology, University of Kerala,
Trivandrum, Kerala, India
e-mail: pawan.dhar@snu.edu.in; pawan@cssb.res.in

V. Singh, P. K. Dhar (eds.), *Systems and Synthetic Biology*,
DOI 10.1007/978-94-017-9514-2_21

and operating systems. A genetic circuit element that behaves like a switch was perceived to be useful in constructing and controlling complex biological systems.

P1	P2	R1	R2	I1	I2	Reporter
1	0	1	0	0	0	0
0	1	0	1	1	0	1
1	0	1	0	0	1	0
1	1	0	0	1	1	1

The toggle switch was designed using a combination of two repressors and two promoters. The idea was simple (as shown in the truth table) but its implementation was non trivial.

The basic design was to construct two mutually repressing promoters. Each promoter was designed to inhibited by the repressor transcribed by the opposing promoter. Since this design used few genes and cis regulatory elements, the robust behaviour of the bistable switch was achieved after several iterations of trial and error. By robust, we mean that state of the toggle switch was not changed randomly over small fluctuations. Thus, the property of bistabilitywas observed achieved over a wide range of parameter values. Although bistability was theoretically possible with a single, autocatalytic promoter, it turns out that such a design would be less robust and little more difficult to tune experimentally. In addition, thechosentoggle design does not require any specialized promoters, such as the P_R/P_{RM} promoter of bacteriophageλ. Bistability is achievable with any set of promoters and repressors as long as they fulfilleda minimum set of conditions.

The bistability of the constructed toggle arose from the mutually inhibitory effect of repressor genes. In the absence of inducers, two stable states were demonstrated—one in which 'promoter 1' transcribed'repressor 2', and another in which 'promoter2' transcribed'repressor1'. Switching was accomplished by transiently introducing an inducer of the existing active repressor. The inducer permitted the opposing repressor to be maximally transcribed until it stably repressed the originally active promoter.

All toggle switches were implemented on *E. coli* plasmids containing ampicillin resistance and containing the pBR322 ColE1 replication origin. The toggle switch genes were arranged as a type IV plasmid, as shown in schematic Fig. 21.1. Although all genes and promoters arecontained on a single plasmid, they could, in principle, be divided into two separate plasmids without altering the functionality of the toggle.

In all the toogle switch variants, the sequence of the three promoters remained unchanged. The rates of synthesis of the repressors or the reporter genes were modified by rearranging the downstream ribosome binding sites.

Fig. 21.1 Schematic design of the toggle switch

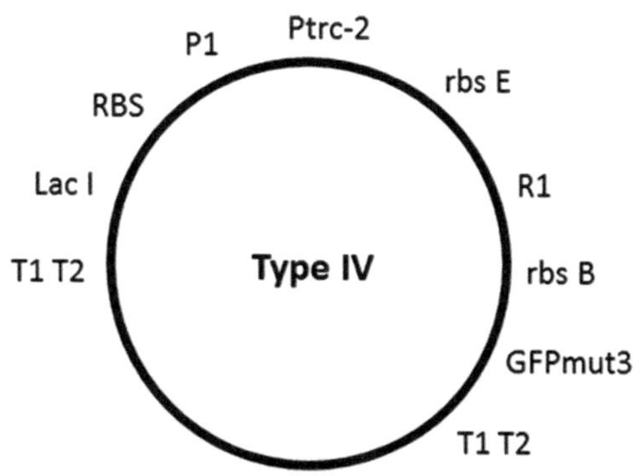

Transcription from Ptrc-2 gene resulted in the repression of P1 gene expression. The opposing state, in which P1 was transcribed and Ptrc-2 repressed was called the 'low' state. The changes in the states of expression were observed by using GFP reporter. To investigate conditions required for bistability, six variants of the toggle switch (four pTAK plasmids and two pIKE plasmids) were constructed by inserting RBS sequences of differing strengths into the RBS1 site.

The computational model of toggle switch predicted switching thresholds and occurrence of monostable behaviour under certain conditions. This was observed experimentally when the transition from bistability to monostabilitywas found to occur in a sharp, discontinuous fashion owing to the existence of a bifurcation. This bifurcation occurred when one of the stable steady states was annihilated by the unstable steady state.

The construction of a genetic toggle switch demonstrated a significant departure from traditional genetic engineering. Here the focus was on creating and communicating new parts, instead of up/down regulating expression of existing parts.

21.2 Never Born Proteins

The number of natural proteins, although large, is significantly smaller than the theoretical number of proteins that can be obtained by combining 20 types of natural amino acids. The difference between theoretically possible proteins and artificially designed proteins, is what has been termed as Never born proteins. The study of the properties of these proteins especially when compared with the natural proteins may provide important information about the relationship between sequence and structure.

In this space, following questions have emerged over last few years:

- What are the criteria based on which existing proteins were chosen for synthesis by nature?
- Do natural proteins have special properties in the parameter space of thermal stability, solubility in water or amino acid composition etc.?

- Can the concept of process of making Never Born Proteins (NBP) be explored towards useful applications?

To answer these questions, we need to figure out (a) fundamental biology behind synthesis of never born proteins, (b) how to distinguish born proteins from never born proteins, and (c) to create a never born protein library for deeper structural and functional analysis. Further, it is important to distinguish the difference between impossible proteins and never born proteins (Figure below).

Based on recent scientific studies, it would be relevant to broadly classify as born, never born and "cannot be born" proteins (currently it is unclear how to address this aspect).

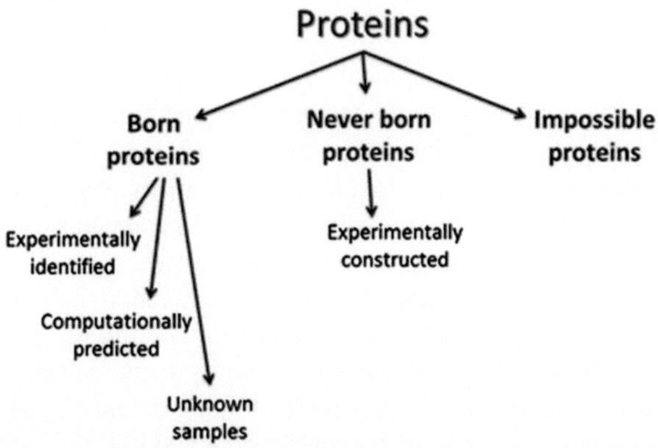

If we do some quick maths, the possible number of proteins considering each polypeptide 50 amino acid residues long equals to 20^{50}. Considering a large pool of unknown proteins the question is: how to determine which proteins can be synthesized non-naturally?

For this purpose, a never-born-protein(NBP) library was made by Prof. Peter Luisi lab. In order to create this library, first the frequency of folding i.e. how many of the never born proteins would actually accomplish a stable tertiary conformation, was determined. The synthesis of NBPs was also accompanied by the synthesis of the corresponding never born mRNAs.

To create a never born protein library, the group uses phage display method that contains approximately 50 amino acid residues long regions. This high throughput screening method helps in studying protein—protein, protein—peptide, and protein—DNA interactions.

A segment of foreign DNA is inserted into either a phagemid or an infectious filamentous phage genome and expressed as a fusion product with a phage coat protein. The technique was first reported by Parmley and Smith in 1988 for E. Coli phage M13 and has been extended to other T4 and λ phages.

On a slightly critical note, it is unclear whether the term Never Born Proteins indicates the real situation, given that the scientific community only knows a tip of the iceberg of this vast protein universe. However, the approach of identifying and practically synthesizing theoretical proteins is very interesting. This novel thought process leads to interesting questions of why some proteins were not naturally created, if at all. What kind of decisions went into selecting some proteins for evolution? Did nature create and retire a subset of never born proteins? What is the boundary condition of artificially creating novel proteins. It is hoped that in the coming years, more work will be published from origin and application point-of-view.

21.3 Minimal Synthetic Cell

A number of genomes from organisms of various complexities have been completely sequenced. However, given such a huge number and variety of organisms, it appears that sequencing projects will run forever. In parallel to sequencing efforts, researchers have been asking a simple question—what is a bare minimum gene set to start life? The minimal cell has been defined as the "smallest possible group of genes that are sufficient to sustain cellular life in the presence of all essential nutrients and in the absence of environmental stress". Though the issue is yet unsettled, a new trend has recently emerged. Can we design genome de novo?

During the last decade, persistent efforts have been made in the direction of designing a minimal synthetic cell by Craig Venter's group. The group used *Mycoplasma genitalium*—the smallest genome containing microbe, as a template. *M. genitalium* contains 482 protein encoding genesin one circular chromosome of 5.82 Kb genome.

In May 2010, Craig Venter made a press release of successfully creating the first self-replicating bacterial synthetic cell by replacing the entire genome of *M. capricolum* with 1.08 mega base pair synthetic genome of *Mycoplasma mycoides*. Even though the overall experiement took nearly a decade to finish (end-to-end) and extremely cost-prohibitive for most of the labs, it gave rise to immense excitement in the community.

To ensure that the synthesized genome was verifiable and tractable with well-defined genetic signatures, Venter team added names of their key scientists as watermark sequences as non-translatable DNA sequences in the chemically synthesized genome. Once sequences were chemically synthesized in 10 kb DNA cassettes, these DNA cassettes were assembled into 100 kb blocks in yeast and blocks were stitched together by restriction enzymes in yeast leading to construction of complete ~ 1 Mb microbial *M. mycoides* genome. The chemically synthesized and assembled mycoides genome was transferred to *M. capricolum* cell. Of several transformants, they found a population of cells (grown on X-gal medium) that showed expected blue colorphenotypic and cell division properties.

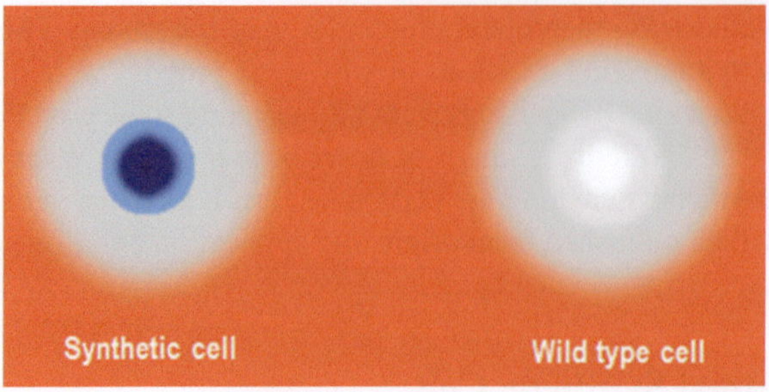

Broadly speaking there are three ways to create a minimal synthetic cell (a) transfer one genome to another cell, (b) chemically synthesize the genome in blocks and stich the blocks together to create a complete version and (c) reduce existing genome to the point where it looks significantly different from its parent.

Irrespective of the method used, one needs to work out minimal regulatory and metabolic content for a cell to live and divide. To minimize the overheads it would be essential to grow the cell in an environment that is provides continuous supply of nutrients. If the cell stays in a physically and chemically secure environment, the role of DNA repair machinery may significantly decrease over time. Thus genes that have been exclusively delegated the work of repairing DNA may not be required over time. In any case, core metabolic, signalling and regulatory processes combined with nutrient transport and cell division capability would be required to construct a minimal cell.

Overall, when one looks back it appears that a quest to reduce the life to its participating components lead not only to an enhanced understanding of its molecular chassis but also a forward engineering approach of synthetic biology. The switch from observation to creation has been more challenging as unlike engineering disciplines, biology is based on temporal, spatial, contextual complexities—all embedded in one system. Even though enormous progress has been made in high throughput sequencing and annotation, a large number of components are still missing or inaccurately understood. In future, one would expect to see a lot more work in the area of annotation to fill in this missing link.

Acknowledgements I would like to thank Ms. Nafisa Bulsara, Mr. Shivananda Naikar and Ms. Snehal Kamble for their useful inputs.

References

Fraser CM et al (1995) The minimal gene complement of *Mycoplasma genitalium*. Science 270:397–403

Gardner TS, Cantor CC, Collins JJ (2000) Construction of a genetic toggle switch in *Escherichia coli*. Nature 403:339–342

Gibson DG et al (2010) Chemical synthesis of the mouse mitochondrial genome. Nat Methods 7(11):901–913

Gibson DG et al (2010) Creation of a bacterial cell controlled by a chemically synthesized genome. Science 329:52–56

Karas BJ et al (2013) Direct transfer of whole genomes from bacteria to yeast. Nat Methods 10:410–412

Luisi PL (2006) The emergence of life—from chemical origins to synthetic biology. Cambridge University Press, Cambridge

Index

V. Singh, P. K. Dhar (eds.), *Systems and Synthetic Biology*,
DOI 10.1007/978-94-017-9514-2

MIX
Papier aus verantwortungsvollen Quellen
Paper from responsible sources
FSC® C105338

If you have any concerns about our products,
you can contact us on
ProductSafety@springernature.com

In case Publisher is established outside the EU,
the EU authorized representative is:
Springer Nature Customer Service Center GmbH
Europaplatz 3, 69115 Heidelberg, Germany

Printed by Libri Plureos GmbH
in Hamburg, Germany